Ying Liu, Aixin Sun, Han Tong Loh, Wen Feng Lu and Ee-Peng Lim (Eds.)

Advances of Computational Intelligence in Industrial Systems

Studies in Computational Intelligence, Volume 116

Editor-in-chief
Prof. Janusz Kacprzyk
Systems Research Institute
Polish Academy of Sciences
ul. Newelska 6
01-447 Warsaw
Poland
E-mail: kacprzyk@ibspan.waw.pl

Further volumes of this series can be found on our homepage: springer.com

Vol. 95. Radu Dogaru
Systematic Design for Emergence in Cellular Nonlinear Networks, 2008
ISBN 978-3-540-76800-5

Vol. 96. Aboul-Ella Hassanien, Ajith Abraham and Janusz Kacprzyk (Eds.)
Computational Intelligence in Multimedia Processing: Recent Advances, 2008
ISBN 978-3-540-76826-5

Vol. 97. Gloria Phillips-Wren, Nikhil Ichalkaranje and Lakhmi C. Jain (Eds.)
Intelligent Decision Making: An AI-Based Approach, 2008
ISBN 978-3-540-76829-9

Vol. 98. Ashish Ghosh, Satchidananda Dehuri and Susmita Ghosh (Eds.)
Multi-Objective Evolutionary Algorithms for Knowledge Discovery from Databases, 2008
ISBN 978-3-540-77466-2

Vol. 99. George Meghabghab and Abraham Kandel
Search Engines, Link Analysis, and User's Web Behavior, 2008
ISBN 978-3-540-77468-6

Vol. 100. Anthony Brabazon and Michael O'Neill (Eds.)
Natural Computing in Computational Finance, 2008
ISBN 978-3-540-77476-1

Vol. 101. Michael Granitzer, Mathias Lux and Marc Spaniol (Eds.)
Multimedia Semantics - The Role of Metadata, 2008
ISBN 978-3-540-77472-3

Vol. 102. Carlos Cotta, Simeon Reich, Robert Schaefer and Antoni Ligeza (Eds.)
Knowledge-Driven Computing, 2008
ISBN 978-3-540-77474-7

Vol. 103. Devendra K. Chaturvedi
Soft Computing Techniques and its Applications in Electrical Engineering, 2008
ISBN 978-3-540-77480-8

Vol. 104. Maria Virvou and Lakhmi C. Jain (Eds.)
Intelligent Interactive Systems in Knowledge-Based Environment, 2008
ISBN 978-3-540-77470-9

Vol. 105. Wolfgang Guenthner
Enhancing Cognitive Assistance Systems with Inertial Measurement Units, 2008
ISBN 978-3-540-76996-5

Vol. 106. Jacqueline Jarvis, Dennis Jarvis, Ralph Rönnquist and Lakhmi C. Jain (Eds.)
Holonic Execution: A BDI Approach, 2008
ISBN 978-3-540-77478-5

Vol. 107. Margarita Sordo, Sachin Vaidya and Lakhmi C. Jain (Eds.)
Advanced Computational Intelligence Paradigms in Healthcare - 3, 2008
ISBN 978-3-540-77661-1

Vol. 108. Vito Trianni
Evolutionary Swarm Robotics, 2008
ISBN 978-3-540-77611-6

Vol. 109. Panagiotis Chountas, Ilias Petrounias and Janusz Kacprzyk (Eds.)
Intelligent Techniques and Tools for Novel System Architectures, 2008
ISBN 978-3-540-77621-5

Vol. 110. Makoto Yokoo, Takayuki Ito, Minjie Zhang, Juhnyoung Lee and Tokuro Matsuo (Eds.)
Electronic Commerce, 2008
ISBN 978-3-540-77808-0

Vol. 111. David Elmakias (Ed.)
New Computational Methods in Power System Reliability, 2008
ISBN 978-3-540-77810-3

Vol. 112. Edgar N. Sanchez, Alma Y. Alanís and Alexander G. Loukianov
Discrete-Time High Order Neural Control: Trained with Kalman Filtering, 2008
ISBN 978-3-540-78288-9

Vol. 113. Gemma Bel-Enguix, M. Dolores Jimenez-Lopez and Carlos Martín-Vide (Eds.)
New Developments in Formal Languages and Applications, 2008
ISBN 978-3-540-78290-2

Vol. 114. Christian Blum, Maria José Blesa Aguilera, Andrea Roli and Michael Sampels (Eds.)
Hybrid Metaheuristics, 2008
ISBN 978-3-540-78294-0

Vol. 115. John Fulcher and Lakhmi C. Jain (Eds.)
Computational Intelligence: A Compendium, 2008
ISBN 978-3-540-78292-6

Vol. 116. Ying Liu, Aixin Sun, Han Tong Loh, Wen Feng Lu and Ee-Peng Lim (Eds.)
Advances of Computational Intelligence in Industrial Systems, 2008
ISBN 978-3-540-78296-4

Ying Liu
Aixin Sun
Han Tong Loh
Wen Feng Lu
Ee-Peng Lim
(Eds.)

Advances of Computational Intelligence in Industrial Systems

With 164 Figures and 37 Tables

Ying Liu
Department of Industrial and Systems Engineering
The Hong Kong Polytechnic University
Hung Hom, Kowloon
Hong Kong, China
mfyliu@polyu.edu.hk

Han Tong Loh
Department of Mechanical Engineering
National University of Singapore
9 Engineering Drive 1
Singapore 117576
mpelht@nus.edu.sg

Ee-Peng Lim
School of Computer Engineering
Nanyang Technological University
Nanyang Avenue
Singapore 639798
aseplim@ntu.edu.sg

Aixin Sun
School of Computer Engineering
Nanyang Technological University
Nanyang Avenue
Singapore 639798
axsun@ntu.edu.sg

Wen Feng Lu
Department of Mechanical Engineering
National University of Singapore
9 Engineering Drive 1
Singapore 117576
mpelwf@nus.edu.sg

ISBN 978-3-642-09698-3 e-ISBN 978-3-540-78297-1

Studies in Computational Intelligence ISSN 1860-949X

Cover design: Deblik, Berlin, Germany

Printed on acid-free paper

9 8 7 6 5 4 3 2 1

springer.com

Preface

Computational Intelligence (CI) has emerged as a rapid growing field over the last decade. CI techniques, e.g., artificial neural networks, genetic algorithms, fuzzy theory, evolution computing, machine learning, and many others, are powerful tools for intelligent information processing, decision making and knowledge management in organizations ranging from manufacturing firms to medical and healthcare institutions. Some of the recent examples of applying CI techniques include case-based reasoning in medical diagnosis and clinical decision support, customer relationship management through data mining, fault detection and quality improvement using artificial neural networks, design rationale assessment through semantic modeling of design documents, and so on.

The editors of this book have been working on data and text mining, machine learning, information retrieval, digital libraries and related research fields. Some of them have conducted CI research to address problems in industrial application scenarios, e.g., product design and manufacturing process control, for a few years. They are fully aware of the demand for advanced tools to support daily routine tasks like process monitoring as well as to support decision making in strategic planning. Therefore, it is timely to have a book that presents the latest advances in CI topics with a strong focus on industrial applications. This book aims to report and demonstrate how industrial organizations can gain a competitive advantage by applying different CI techniques. The industrial organizations here refer to a wide range of enterprises, bodies from design studios and manufacturing firms to hospitals and non-profit organizations.

The book includes 17 chapters selected from more than 30 submissions based on a thorough and strict peer review process. The 17 chapters accepted were contributed by international researchers from 14 countries and regions. They present a snapshot of the cutting-edge research in CI and the state-of-the-art industrial CI applications. These 17 chapters are broadly grouped into two main sections, namely, Section I: Theory and Foundation, and Section II: Industrial Applications.

In the Theory and Foundation section, Das et al. give a comprehensive overview of particle swarm optimization (PSO) and differential evolution (DE) algorithms. This chapter describes several schemes to control the convergence behaviors of PSO and DE by carefully selecting and tuning their parameters. Furthermore, the authors have explored the possibility of a hybrid approach by combining PSO and DE with other soft computing algorithms which leads to a powerful global search algorithm. Its application in various engineering domains has been highlighted.

The chapter by Maamar et al. covers Web services and their integration with context and policies, a topic not discussed much in CI. In their framework, a three-level approach to compose Web services has been proposed, i.e., component, composite and resource level. In their work, context is defined as the information about the environment within each level, i.e., context of Web service, context of composite Web service and context of resource, and jumping from one level to another requires policy activation, either engagement policies or deployment policies.

Qiu focuses on privacy protection when delegating data mining tasks in industrial systems, one of the prevailing topics in the area. Qiu introduces a Bloom filter-based approach for privacy preserving in association rule mining. His findings show that the privacy security level and analysis precision are positively correlated using the Bloom filter. To increase the privacy security level, more data storage is needed. The additional storage required is in a linear relation with respect to the privacy level or the analysis precision. Further research on other data mining tasks and testing in a distributed database environment are suggested.

Tsivtsivadze et al. present a comprehensive overview of various kernels for text analysis, including bag-of-words kernel, string kernel, gappy string kernel, convolution kernel, and graph kernel. What intrigues us is its unique feature of incorporating prior domain knowledge into kernels for text analysis. An evaluation using different features, e.g., grammatical bigram, word and part-of-speech tag, link type, link length, and random walk features from graph feature representation, shows the importance and contribution of grammatical bigram and link bigram and their merits in embedding prior domain knowledge in feature representation.

Wang et al. report their work on the discovery of frequent patterns from long sequences which may contain many events. Different from the existing approaches in sequential pattern mining that simply count the occurrence of patterns in disjoint sequences, Wang et al.'s work adopts the concept of maximum cardinality to describe the number of pattern occurrences in an individual sequence. The advantages of their approach include not requiring sliding window sizes and efficient implementation with sound precision.

Lin's work on gauging image and video quality in industrial applications highlights an important area which receives little attention in CI. Based on a unified framework and its computational modules, three perceptual visual quality metrics and an embedded just-noticeable difference estimator have been proposed. More applications of perceptual models in industrial scenarios are expected to further stress the importance of Lin's study.

Finally, after a nice review on model construction, Stein presents an interesting methodology regarding model construction for knowledge intensive engineering tasks. Based on the observation that domain experts and knowledge engineers often develop a new and appropriate model by modifying some existing ones, Stein proposes a means of horizontal modeling construction. A few real-world case studies are provided to demonstrate its practical value.

In Sect. II, we emphasize the state-of-the-art application of CI techniques in various industrial scenarios. We start with a few chapters focusing on medical and healthcare areas. Almeida et al. adopt case based reasoning and genetic algorithm in the implementation of a virtual medical office for medical diagnosis and knowledge transfer and management purpose. In their case study, promising results are achieved.

Chan et al. address the important issue of using data mining approach in clinical decision support. They highlight the complicated reality that a clinical decision is often drawn by synthesizing knowledge from multiple disciplines while information needed is presented in a number of different formats, e.g., magnetic resonance images, somatosensory evoked potentials and radiographs. Hence, Chan et al. propose the DICOM model to form a synergy platform which aims to extract and present relevant features from multiple disciplinary knowledge sources for medical knowledge discovery and clinical decision support purposes. An example is given subsequently.

Caldwell et al. tackle the problem of promoter prediction using neural network for promoter prediction (NNPP) algorithm. They have raised a point that DNA sequence information, which is released through DNA sequence quantitative measurements instead of DNA sequence pattern information, can significantly improve the precision of computational promoter prediction. Their extended algorithms based on three coding region lengths, e.g., TLS–TSC length, TSS–TLS length, and TSC–TTS length, have shown their merits. Future research directions have been outlined.

Das et al. investigate the anomaly detection problem using two successful negative selection algorithms, i.e., the self-organizing RNS algorithm and the V-detector algorithm. For the latter, both a single-stage and a multi-stage proliferating V-detector algorithm are studied to create such vectors.

Braga et al. report their study using a number of pattern classification algorithms, e.g., multi-layer perceptron, RBF neural network, support vector machines, and decision tree, to automatically monitor the dressing procedure in the grinding process of an industrial CNC machine. They also examine the robustness and effectiveness of different algorithms when the controlled noise is purposely introduced in simulation.

Clifton et al. provide a nice survey of the existing methods for detecting novelty or something abnormal in the industrial environment. Using an SVM core, novelty classifiers are set up in a way that their output is calibrated into probabilities, resulting in a probabilistic novelty threshold being specified automatically. This is in contrast to the existing distance-based methods where heuristics are needed and the models are manually constructed. Their validation and experiments are conducted based on real-world jet-engine data.

Khosravi et al. explore another important problem – predicting the origin of power sags in distributed substations using a multiway principal component analysis (MPCA), and an extension of the classic principal component analysis (PCA). In their study, MPCA classifiers are able to achieve better performance in differentiating high-voltage and medium-voltage sags compared to other possibilities, e.g., multi-layer perceptron, RBF neural network, and decision tree.

Cheong and Tan report their study of using a multi-objective multi-colony ant algorithm (MOMCAA) in solving the berth allocation problem, an important topic in logistics and supply chain management. In their work, berth allocation problem has been taken as an optimization of complete shipping schedules with minimum service time and delay for the departure of ships, subject to a number of temporal and spatial constraints, and hence a multi-objective and multi-modal combinational optimization problem. The proposed MOMCAA framework possesses several features, such as using ant groups in searching for each candidate group, concurrent optimization, flexibility and expandability of incorporating other meta-heuristics, e.g., genetic algorithm and simulated annealing, etc.

Yu's chapter focuses on the application of neural network on dynamic system identification and adaptive control. Two examples are given, i.e., rotorcraft acoustic data modeling and DC voltage controller.

Hammiche et al. describe their work of query rewriting for semantic multimedia data retrieval, a domain knowledge based approach to retrieve multimedia data formatted according to MPEG-7 standard. The basic idea is to add a semantic layer on top of the MPEG-7 metadata layer. To fulfill the retrieval task, users' queries are pre-processed, rewritten and translated into the XQuery queries using the mapping rules and a set of XQuery functions defined over the MPEG-7 descriptions.

We are very grateful to all our reviewers: Danilo Avola, Peter Bannister, Andre Ponce de Leon F. de Carvalho, Kit Yan Chan, Wing Chi Chan, Tan Kay Chen, David Clifton, Sanjoy Das, Chi Keong Goh, Jason J. Jung, Dariusz Krol, Wenyuan Li, Weisi Lin, Yan-Xia Lin, Ying Liu, Kok-Leong Ong, Ling Qiu, Benno Stein, Evgeni Tsivtsivadze, Changzhou Wang, Kevin Wang, Jianshu Weng, Xiao-Hua (Helen) Yu, Yun-Shuai Yu, and Zhiwen Yu. Their backgrounds include data and text mining, information retrieval, digital library, evolution algorithm and agents, image and video processing, machine learning, neural network and genetic algorithms, applied statistics, Web services and ontology engineering, control engineering, optimization and operation research, and knowledge management. Reviewers' diversified expertise has greatly strengthened our confidence in reviewing the state-of-the-art of CI applications in various industrial systems. Finally, we express our special thanks to Springer, in particular, Dr. Thomas Ditzinger – the series editor, and Heather King, for giving us the opportunity to gather the latest research findings and for publishing this book. Throughout the year, the editors have been working very closely with the two Springer editors on this book. Their quick endorsement on the book proposal and assistance are deeply appreciated by the book's editors.

China
Singapore
2007

Ying Liu
Aixin Sun
Han Tong Loh
Wen Feng Lu
Ee-Peng Lim

Contents

Particle Swarm Optimization and Differential Evolution Algorithms: Technical Analysis, Applications and Hybridization Perspectives

Swagatam Das[1], Ajith Abraham[2], and Amit Konar[1]

[1] Department of Electronics and Telecommunication Engineering, Jadavpur University, Kolkata 700032, India, swagatamdas19@yahoo.co.in, konaramit@yahoo.co.in
[2] Center of Excellence for Quantifiable Quality of Service, Norwegian University of Science and Technology, Norway, ajith.abraham@ieee.org

Summary. Since the beginning of the nineteenth century, a significant evolution in optimization theory has been noticed. Classical linear programming and traditional non-linear optimization techniques such as Lagrange's Multiplier, Bellman's principle and Pontyagrin's principle were prevalent until this century. Unfortunately, these derivative based optimization techniques can no longer be used to determine the optima on rough non-linear surfaces. One solution to this problem has already been put forward by the evolutionary algorithms research community. Genetic algorithm (GA), enunciated by Holland, is one such popular algorithm. This chapter provides two recent algorithms for evolutionary optimization – well known as particle swarm optimization (PSO) and differential evolution (DE). The algorithms are inspired by biological and sociological motivations and can take care of optimality on rough, discontinuous and multimodal surfaces. The chapter explores several schemes for controlling the convergence behaviors of PSO and DE by a judicious selection of their parameters. Special emphasis is given on the hybridizations of PSO and DE algorithms with other soft computing tools. The article finally discusses the mutual synergy of PSO with DE leading to a more powerful global search algorithm and its practical applications.

1 Introduction

The aim of optimization is to determine the best-suited solution to a problem under a given set of constraints. Several researchers over the decades have come up with different solutions to linear and non-linear optimization problems. Mathematically an optimization problem involves a fitness function describing the problem, under a set of constraints representing the solution space for the problem. Unfortunately, most of the traditional optimization techniques are centered around evaluating the first derivatives to locate the optima on a given constrained surface. Because of the difficulties in evaluating

S. Das et al.: *Particle Swarm Optimization and Differential Evolution Algorithms: Technical Analysis, Applications and Hybridization Perspectives*, Studies in Computational Intelligence (SCI) **116**, 1–38 (2008)
www.springerlink.com

the first derivatives, to locate the optima for many rough and discontinuous optimization surfaces, in recent times, several derivative free optimization algorithms have emerged. The optimization problem, now-a-days, is represented as an intelligent search problem, where one or more agents are employed to determine the optima on a search landscape, representing the constrained surface for the optimization problem [1].

In the later quarter of the twentieth century, Holland [2], pioneered a new concept on evolutionary search algorithms, and came up with a solution to the so far open-ended problem to non-linear optimization problems. Inspired by the natural adaptations of the biological species, Holland echoed the Darwinian Theory through his most popular and well known algorithm, currently known as genetic algorithms (GA) [2]. Holland and his coworkers including Goldberg and Dejong, popularized the theory of GA and demonstrated how biological crossovers and mutations of chromosomes can be realized in the algorithm to improve the quality of the solutions over successive iterations [3]. In mid 1990s Eberhart and Kennedy enunciated an alternative solution to the complex non-linear optimization problem by emulating the collective behavior of bird flocks, particles, the boids method of Craig Reynolds and socio-cognition [4] and called their brainchild the particle swarm optimization (PSO) [4–8]. Around the same time, Price and Storn took a serious attempt to replace the classical crossover and mutation operators in GA by alternative operators, and consequently came up with a suitable differential operator to handle the problem. They proposed a new algorithm based on this operator, and called it differential evolution (DE) [9].

Both algorithms do not require any gradient information of the function to be optimized uses only primitive mathematical operators and are conceptually very simple. They can be implemented in any computer language very easily and requires minimal parameter tuning. Algorithm performance does not deteriorate severely with the growth of the search space dimensions as well. These issues perhaps have a great role in the popularity of the algorithms within the domain of machine intelligence and cybernetics.

2 Classical PSO

Kennedy and Eberhart introduced the concept of function-optimization by means of a particle swarm [4]. Suppose the global optimum of an n-dimensional function is to be located. The function may be mathematically represented as:

$$f(x_1, x_2, x_3, \ldots, x_n) = f(\vec{X})$$

where $\vec{x}$ is the search-variable vector, which actually represents the set of independent variables of the given function. The task is to find out such a $\vec{x}$, that the function value $f(\vec{x})$ is either a minimum or a maximum denoted by f* in the search range. If the components of $\vec{x}$ assume real values then the task is to locate a particular point in the n-dimensional hyperspace which is a continuum of such points.

Example 1. Consider the simplest two-dimensional sphere function given by

$$f(x_1, x_2) = f(\vec{X}) = x_1^2 + x_2^2,$$

if x_1 and x_2 can assume real values only then by inspection it is pretty clear that the global minima of this function is at $x_1 = 0,\ x_2 = 0$, i.e., at the origin (0, 0) of the search space and the minimum value is f(0, 0) = f* = 0. No other point can be found in the $x_1 - x_2$ plane at which value of the function is lower than f* = 0. Now the case of finding the optima is not so easy for some functions (an example is given below):

$$f(x_1, x_2) = x_1 \sin(4\pi x_2) - x_2 \sin(4\pi x_1 + \pi) + 1$$

This function has multiple peaks and valleys and a rough fitness landscape. A surface plot of the function is shown in Fig. 1. To locate the global optima quickly on such a rough surface calls for parallel search techniques. Here many agents start from different initial locations and go on exploring the search space until some (if not all) of the agents reach the global optimal position. The agents may communicate among themselves and share the fitness function values found by them.

PSO is a multi-agent parallel search technique. Particles are conceptual entities, which fly through the multi-dimensional search space. At any particular instant, each particle has a position and a velocity. The position vector of a particle with respect to the origin of the search space represents a trial solution of the search problem. At the beginning, a population of particles is initialized with random positions marked by vectors $\vec{x}_i$ and random velocities $\vec{v}_i$. The population of such particles is called a "swarm" S. A neighborhood

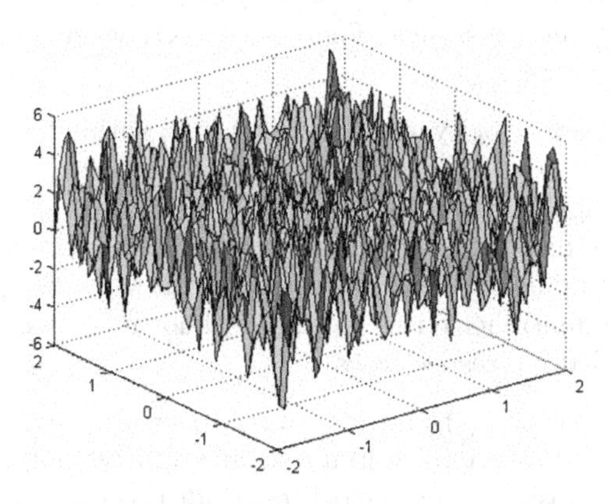

Fig. 1. Surface plot of the above-mentioned function

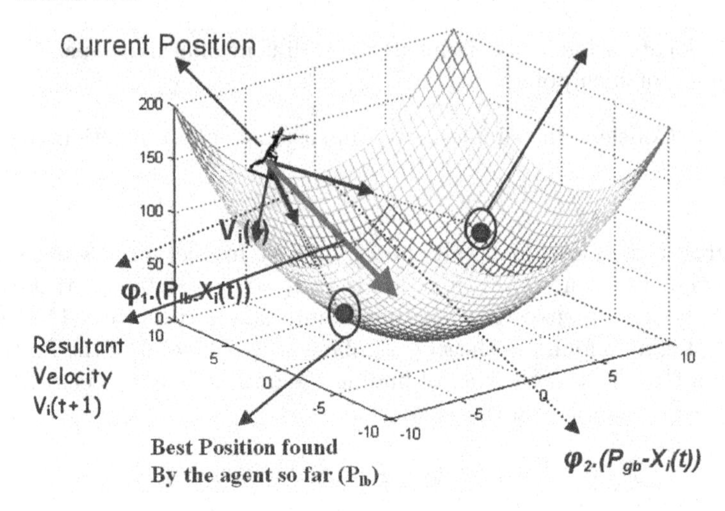

Fig. 2. Illustrating the dynamics of a particle in PSO

relation N is defined in the swarm. N determines for any two particles P_i and P_j whether they are neighbors or not. Thus for any particle P, a neighborhood can be assigned as $N(P)$, containing all the neighbors of that particle. Different neighborhood topologies and their effect on the swarm performance will be discussed later. However, a popular version of PSO uses $N = S$ for each particle. In this case, any particle has all the remaining particles in the swarm in its neighborhood. PSO dynamics is illustrated in Fig. 2.

Each particle P has two state variables viz., its current position $\vec{x}(t)$ and its current velocity $\vec{v}(t)$. It is also equipped with a small memory comprising its previous best position (one yielding the highest value of the fitness function found so far) $\vec{p}(t)$, i.e., personal best experience and the best $\vec{p}(t)$ of all $P \in N(P)$: $\vec{g}(t)$, i.e., the best position found so far in the neighborhood of the particle. When we set $N(P) = S$, $\vec{g}(t)$ is referred to as the globally best particle in the entire swarm. The PSO (PSO) scheme has the following algorithmic parameters:

(a) $V_{\max}$ or maximum velocity which restricts $\vec{V}_i(t)$ within the interval $[-V_{\max}, V_{\max}]$
(b) An inertial weight factor ω
(c) Two uniformly distributed random numbers φ_1 and φ_2 that respectively determine the influence of $\vec{p}(t)$ and $\vec{g}(t)$ on the velocity update formula.
(d) Two constant multiplier terms C_1 and C_2 known as "self-confidence" and "swarm confidence", respectively.

Initially the settings for $\vec{p}(t)$ and $\vec{g}(t)$ are $\vec{p}(0) = \vec{g}(0) = \vec{x}(0)$ for all particles. Once the particles are all initialized, an iterative optimization process begins, where the positions and velocities of all the particles are altered by the following recursive equations. The equations are presented for the dth dimension of the position and velocity of the ith particle.

$$\left.\begin{aligned} V_{\mathrm{id}}(t+1) &= \omega \cdot v_{\mathrm{id}}(t) + C_1 \cdot \varphi_1 \cdot (P_{\mathrm{id}}(t) - x_{\mathrm{id}}(t)) + C_2 \cdot \varphi_2 \cdot (g_{\mathrm{id}}(t) - x_{\mathrm{id}}(t)) \\ x_{\mathrm{id}}(t+1) &= x_{\mathrm{id}}(t) + v_{\mathrm{id}}(t+1). \end{aligned}\right\} \quad (1)$$

The first term in the velocity updating formula represents the inertial velocity of the particle. "ω" is called the inertia factor. Venter and Sobeiski [10] termed C_1 as "self-confidence" and C_2 as "swarm confidence". These terminologies provide an insight from a sociological standpoint. Since the coefficient C_1 has a contribution towards the self-exploration (or experience) of a particle, we regard it as the particle's self-confidence. On the other hand, the coefficient C_2 has a contribution towards motion of the particles in global direction, which takes into account the motion of all the particles in the preceding program iterations, naturally its definition as "swarm confidence" is apparent. φ_1 and φ_2 stand for a uniformly distributed random number in the interval [0, 1]. After having calculated the velocities and position for the next time step $t + 1$, the first iteration of the algorithm is completed. Typically, this process is iterated for a certain number of time steps, or until some acceptable solution has been found by the algorithm or until an upper limit of CPU usage has been reached. The algorithm can be summarized in the following pseudo code:

The PSO Algorithm

Input: Randomly initialized position and velocity of the particles: $\vec{X}_i(0)$ and $\vec{V}_i(0)$

Output: Position of the approximate global optima $\vec{X}^*$

Begin
While terminating condition is not reached **do**
Begin
for i = 1 to number of particles
Evaluate the fitness:= $f(\vec{X}_i)$;
Update $\vec{p}_i$ and $\vec{g}_i$;
Adapt velocity of the particle using equations (1);
Update the position of the particle;
increase i;
end while
end

The swarm-dynamics has been presented below using a humanoid agent in place of a particle on the spherical fitness-landscape.

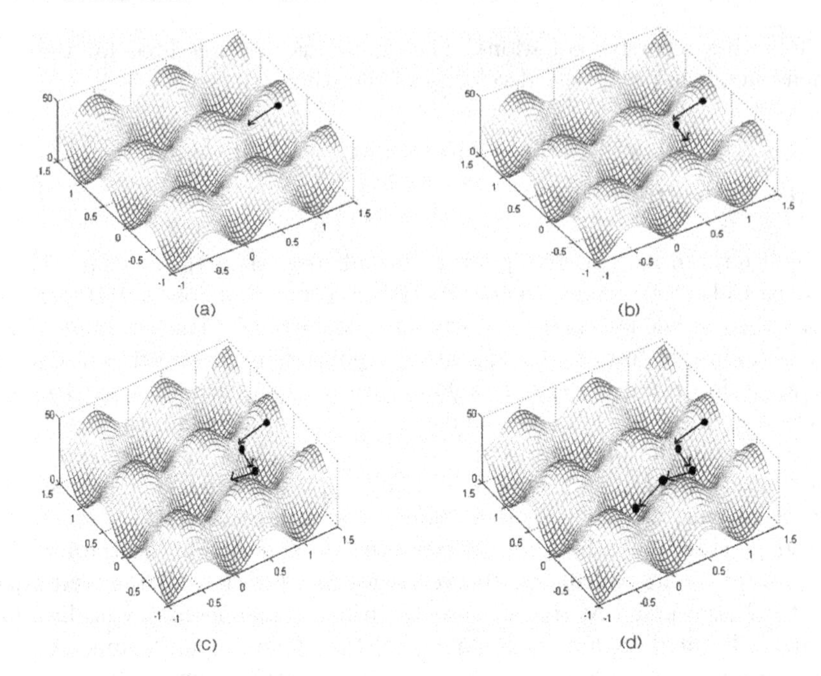

Fig. 3. Trajectory of the best particle for the 2-D Rastrigin function. (**a**) After 40 iterations (**b**) after 80 iterations (**c**) after 150 iterations (**d**) after 200 iterations

Example 2. Consider the two-dimensional function given by

$$f(\vec{x}) = 20 + x_1^2 + x_2^2 - 10(\cos 2\pi x_1 + \cos 2\pi x_2).$$

This function is known as the Rastrigin function [11] and has a global minimum value 0 at $x_1 = 0$ and $x_2 = 0$. A PSO is run to optimize the function. We used 30 particles and randomly initialized their positions and velocities in the interval $[-10, 10]$. We used the following parameter set-up: $C_1 = C_2 = 2.00$, $\omega = 0.729$, and the maximum particle velocity $V_{\max} = 10$. Figure 3 depicts the trajectory of the globally best particle towards the global minima of the function over different iterations.

3 Selection of Parameters for PSO

The main parameters of the canonical PSO model are ω, C_1, C_2, $V_{\max}$ and the swarm size S. The settings of these parameters determine how it optimizes the search-space. For instance, one can apply a general setting that gives reasonable results on most problems, but seldom is very optimal. Since the same parameter settings not at all guarantee success in different problems, we must have knowledge of the effects of the different settings, such that we can pick a suitable setting from problem to problem.

3.1 The Inertia Weight ω

The inertia weight ω controls the momentum of the particle: If $\omega << 1$, only little momentum is preserved from the previous time-step; thus quick changes of direction are possible with this setting. The concept of velocity is completely lost if $\omega = 0$, and the particle then moves in each step without knowledge of the past velocity. On the other hand, if ω is high (>1) we observe the same effect as when C_1 and C_2 are low: Particles can hardly change their direction and turn around, which of course implies a larger area of exploration as well as a reluctance against convergence towards optimum. Setting $\omega > 1$ must be done with care, since velocities are further biased for an exponential growth (see Fig. 2). This setting is rarely seen in PSO implementation, and always together with V_{max}. In short, high settings near 1 facilitate global search, and lower settings in the range [0.2, 0.5] facilitate rapid local search.

Eberhart and Shi have studied ω in several papers and found that "when V_{max} is not small (≥3), an inertia-weight of 0.8 is a good choice" [12]. Although this statement is solely based on a single test function, the Schaffer f_6 function, this setting actually is a good choice in many cases. The authors have also applied an annealing scheme for the ω-setting of the PSO, where ω decreases linearly from $\omega = 0.9$ to $\omega = 0.4$ over the whole run [13]. They compared their annealing scheme results to results with $\omega = 1$ obtained by Angeline [14], and concluded a significant performance improvement on the four tested functions. The decreasing ω-strategy is a near-optimal setting for many problems, since it allows the swarm to explore the search-space in the beginning of the run, and still manages to shift towards a local search when fine-tuning is needed. This was called PSO-TVIW method (PSO with Time varying inertia weight) [15].

Finally, Eberhart and Shi devised an adaptive fuzzy PSO, where a fuzzy controller was used to control ω over time [16]. This approach is very interesting, since it potentially lets the PSO self-adapt ω to the problem and thus optimizes and eliminates a parameter of the algorithm. This saves time during the experimentation, since fine-tuning of ω is not necessary anymore. At each time-step, the controller takes the "Normalized Current Best Performance Evaluation" (NCBPE) and the current setting of ω as inputs, and it outputs a probabilistic change in ω.

3.2 The Maximum Velocity $V_{\max}$

The maximum velocity $V_{\max}$ determines the maximum change one particle can undergo in its positional coordinates during an iteration. Usually we set the full search range of the particle's position as the $V_{\max}$. For example, in case, a particle has position vector $\vec{x} = (x_1, x_2, x_3)$ and if $-10 \le x_i \le 10$ for $i = 1$, 2 and 3, then we set $V_{\max} = 20$. Originally, $V_{\max}$ was introduced to avoid explosion and divergence. However, with the use of constriction factor χ (to be discussed shortly) or ω in the velocity update formula, $V_{\max}$ to some degree has become unnecessary; at least convergence can be assured without it [17]. Thus, some researchers simply do not use V_{max}. In spite of this fact, the maximum velocity limitation can still improve the search for optima in many cases.

3.3 The Constriction Factor χ

In 2002, Clerc and Kennedy proposed an adaptive PSO model [17] that uses a new parameter 'χ' called the constriction factor. The model also excluded the inertia weight ω and the maximum velocity parameter V_{max}. The velocity update scheme proposed by Clerc can be expressed for the dth dimension of ith particle as:

$$\left.\begin{aligned} V_{id}(t+1) &= \chi[V_{id}(t) + C_1 \cdot \varphi_1 \cdot (P_{id}(t) - X_{id}(t)) + C_2 \cdot \varphi_2 \cdot (g_{id}(t) - X_{id}(t))] \\ X_{id}(t+1) &= X_{id}(t) + V_{id}(t+1), \end{aligned}\right\} \tag{2}$$

where,

$$\chi = \frac{2}{\left|4 - \varphi - \sqrt{\varphi^2 - 4\varphi}\right|} \quad \text{With} \quad \varphi = C_1 + C_2$$

Constriction coefficient results in the quick convergence of the particles over time. That is the amplitude of a particle's oscillations decreases as it focuses on the local and neighborhood previous best points. Though the particle converges to a point over time, the constriction coefficient also prevents collapse if the right social conditions are in place. The particle will oscillate around the weighted mean of p_{id} and p_{gd}, if the previous best position and the neighborhood best position are near each other the particle will perform a local search. If the previous best position and the neighborhood best position are far apart from each other, the particle will perform a more exploratory search (global search). During the search, the neighborhood best position and previous best position will change and the particle will shift from local search back to global search. The constriction coefficient method therefore balances the need for local and global search depending on what social conditions are in place.

3.4 The Swarm Size

It is quite a common practice in the PSO literature to limit the number of particles to limit the number of particles to the range 20–60 [12, 18]. Van den Bergh and Engelbrecht [19] have shown that though there is a slight improvement of the optimal value with increasing swarm size, a larger swarm increases the number of function evaluations to converge to an error limit. Eberhart and Shi [18] illustrated that the population size has hardly any effect on the performance of the PSO method.

3.5 The Acceleration Coefficients C_1 and C_2

A usual choice for the acceleration coefficients C_1 and C_2 is $C_1 = C_2 = 1.494$ [18]. However, other settings were also used in different papers. Usually C_1 equals to C_2 and ranges from [0, 4]. Ratnaweera et al. have recently investigated the effect of varying these coefficients with time in [20]. Authors adapted C_1 and C_2 with time in the following way:

$$\begin{aligned} C_1 &= (C_{1f} - C_{1i})\frac{iter}{MAXITER} + C_{1i} \\ C_2 &= (C_{2f} - C_{2i})\frac{iter}{MAXITER} + C_{2i} \end{aligned}, \tag{3}$$

where C_{1i}, C_{1f}, C_{2i}, and C_{2f} are constants, iter is the current iteration number and MAXITER is the number of maximum allowable iterations. The objective of this modification was to boost the global search over the entire search space during the early part of the optimization and to encourage the particles to converge to global optima at the end of the search. The authors referred this as the PSO-TVAC (PSO with time varying acceleration coefficients) method. Actually C_1 was decreased from 2.5 to 0.5 whereas C_2 was increased from 0.5 to 2.5.

4 The Neighborhood Topologies in PSO

The commonly used PSOs are either global version or local version of PSO [5]. In the global version of PSO, each particle flies through the search space with a velocity that is dynamically adjusted according to the particle's personal best performance achieved so far and the best performance achieved so far by all the particle. While in the local version of PSO, each particle's velocity is adjusted according to its personal best and the best performance achieved as far within its neighborhood. The neighborhood of each particle is generally defined as topologically nearest particles to the particle at each side. The global version of PSO also can be considered as a local version of PSO with each particle's neighborhood to be the whole population. It has been suggested that the global version of PSO converges fast, but with potential to converge to the local minimum, while the local version of PSO might have more chances to find better solutions slowly [5]. Since then, a lot of researchers have worked on improving its performance by designing or implementing different types of neighborhood structures in PSO. Kennedy [21] claimed that PSO with small neighborhoods might perform better on complex problems while PSO with large neighborhood would perform better for simple problems.

The k-best topology, proposed by Kennedy connects every particle to its k nearest particles in the topological space. With $k = 2$, this becomes the circle topology (and with k = swarmsize-1 it becomes a gbest topology). The wheel topology, in which the only connections are from one central particle to

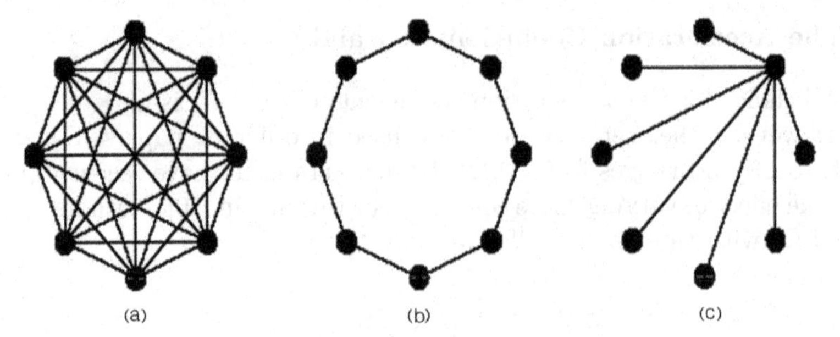

Fig. 4. (**a**) The fully connected org_best topology (**b**) The k-best nearest neighbor topology with $k = 2$ and (**c**) the wheel topology

the others (see Fig. 4). In addition, one could imagine a huge number of other topologies.

5 The Binary PSO

A binary optimization problem is a (normal) optimization problem, where the search space S is simply the set of strings of 0s and 1s of a given fixed length, n. A binary optimization algorithm solves binary optimization problems, and thus enables us to solve many discrete problems. Hence, faced with a problem-domain that we cannot fit into some sub-space of the real-valued n-dimensional space, which is required by the PSO, odds are that we can use a binary PSO instead. All we must provide is a mapping from this given problem-domain to the set of bit strings.

The binary PSO model was presented by Kennedy and Eberhart, and is based on a very simple modification of the real-valued PSO [22]. As with the original PSO, a fitness function f must be defined. In this case, it maps from the n-dimensional binary space B^n (i.e., bit strings of length n) to the real numbers: $f : \mathrm{B}^n \rightarrow \Re^n$. In the binary PSO the positions of the particles naturally must belong to B^n in order to be evaluated by f. Surprisingly, the velocities till belong to $V = [-V_{\max}, V_{\min}] \subset \Re$. This is obtained by changing the position update formula (1) and leaving the velocity update formula unchanged. Now, the ith coordinate of each particle's position is a bit, which state is given by

$$\begin{aligned} X_i(t+1) &= 1 \quad \text{if} \quad \rho < s(\vec{V_i}(t)), \\ &= 0 \quad \text{otherwise,} \end{aligned} \tag{4}$$

where ρ is a random and uniformly selected number in [0, 1], which is re-sampled for each assignment of $\vec{X}_i(t)$. S is a sigmoid function that maps from the real numbers into [0, 1]. It has the properties that $S(0) = \frac{1}{2}$ and $S(x) \rightarrow 0$ as $x \rightarrow \infty$. It is mathematically formulated as

$$S(x) = \frac{1}{1 + e^{-x}}$$

The position $\vec{X}_i(t)$ has changed from being a point in real-valued space to being a bit-string, and the velocity $\vec{V}_i(t)$ has now become a probability for $\vec{X}_i(t)$ to be 1or 0. If $\vec{V}_i(t) = 0$ the probability for the outcome $\vec{X}_i(t+1) = 1$ will be 50%. On the other hand, if $\vec{V}_i(t) > 0$ the probability for $\vec{X}_i(t+1) = 1$ will be above 50%, and if $\vec{V}_i(t) < 0$ a probability below 50%. It is not quite clear, at least intuitively, how and why this changed approach will work. In particular, to the best of our knowledge, there has not been published any papers concerning with the changed meaning of the ω parameter in the binary PSO. The discussion on the meaning of ω above must therefore be regarded as a novel discovery and an independent research result in its own right. In conclusion, since a binary version of the PSO is a key point to practical and commercial use of the PSO in discrete problems solving, this topic definitely needs a lot more attention in the coming years.

6 Hybridization of PSO with Other Evolutionary Techniques

A popular research trend is to merge or combine the PSO with the other techniques, especially the other evolutionary computation techniques. Evolutionary operators like selection, crossover and mutation have been applied into the PSO. By applying selection operation in PSO, the particles with the best performance are copied into the next generation; therefore, PSO can always keep the best-performed particles [14]. By applying crossover operation, information can be swapped between two individuals to have the ability to "fly" to the new search area as that in evolutionary programming and GAs [23]. Among the three evolutionary operators, the mutation operators are the most commonly applied evolutionary operators in PSO. The purpose of applying mutation to PSO is to increase the diversity of the population and the ability to have the PSO to escape the local minima [24–27]. One approach is to mutate parameters such as χ, C_1 and C_2, the position of the neighborhood best [26], as well as the inertia weight [25]. Another approach is to prevent particles from moving too close to each other so that the diversity could be maintained and therefore escape from being trapped into local minima. In [25], the particles are relocated when they are too close to each other. In [24, 27], collision-avoiding mechanisms are designed to prevent particle from colliding with each other and therefore increase the diversity of the population. In addition to incorporating evolutionary operations into PSO, different approaches to combine PSO with the other evolutionary algorithms have been reported. Robinson et al. [28] obtained better results by applying PSO first followed by applying GA in their profiled corrugated horn antenna optimization problem. In [29], either PSO algorithm or GA or hill climbing search algorithm can be applied to a different sub-population of individuals which each individual is dynamically assigned to according to some pre-designed rules. In [30], ant colony optimization is combined with PSO. A list of best positions found so far is recorded and the neighborhood best is randomly selected from the list instead of the current neighborhood best.

In addition, non-evolutionary techniques have been incorporated into PSO. In [31], a cooperative particle swarm optimizer (CPSO) is implemented. The CPSO employs cooperative behavior to significantly improve the performance of the original PSO algorithm through using multiple swarms to optimize different components of the solution vector cooperatively. The search space is partitioned by splitting the solutions vectors into smaller vector. For example, a swarm with n-dimensional vector is partitioned into n swarms of one-dimensional vectors with each swarm attempting to optimize a single component of the solution vector. A credit assignment mechanism needs to be designed to evaluate each particle in each swarm. In [23], the population of particles is divided into subpopulations which would breed within their own sub-population or with a member of another with some probability so that the diversity of the population can be increased. In [32], deflection and stretching techniques as well as a repulsion technique.

7 The Differential Evolution (DE)

In 1995, Price and Storn proposed a new floating point encoded evolutionary algorithm for global optimization and named it DE [9] owing to a special kind of differential operator, which they invoked to create new offspring from parent chromosomes instead of classical crossover or mutation. Easy methods of implementation and negligible parameter tuning made the algorithm quite popular very soon. In the following section, we will outline the classical DE and its different versions in sufficient details.

7.1 Classical DE – How Does it Work?

Like any other evolutionary algorithm, DE also starts with a population of *NP* D-dimensional search variable vectors. We will represent subsequent generations in DE by discrete time steps like $t = 0, 1, 2, \ldots, t, t + 1$, etc. Since the vectors are likely to be changed over different generations we may adopt the following notation for representing the ith vector of the population at the current generation (i.e., at time $t = t$) as

$$\vec{X}_i(t) = [x_{i,1}(t), x_{i,2}(t), x_{i,3}(t) \ldots \ldots x_{i,D}(t)].$$

These vectors are referred in literature as "genomes" or "chromosomes". DE is a very simple evolutionary algorithm.

For each search-variable, there may be a certain range within which value of the parameter should lie for better search results. At the very beginning of a DE run or at $t = 0$, problem parameters or independent variables are initialized somewhere in their feasible numerical range. Therefore, if the jth parameter of the given problem has its lower and upper bound as x_j^L and x_j^U, respectively, then we may initialize the jth component of the ith population members as

$$x_{i,j}(0) = x_j^L + rand\ (0,1) \cdot (x_j^U - x_j^L),$$

where rand (0,1) is a uniformly distributed random number lying between 0 and 1.

Now in each generation (or one iteration of the algorithm) to change each population member $\vec{X}_i(t)$ (say), a Donor vector $\vec{V}_i(t)$ is created. It is the method of creating this donor vector, which demarcates between the various DE schemes. However, here we discuss one such specific mutation strategy known as DE/rand/1. In this scheme, to create $\vec{V}_i(t)$ for each ith member, three other parameter vectors (say the r_1, r_2, and r_3th vectors) are chosen in a random fashion from the current population. Next, a scalar number F scales the difference of any two of the three vectors and the scaled difference is added to the third one whence we obtain the donor vector $\vec{V}_i(t)$. We can express the process for the jth component of each vector as

$$v_{i,j}(t+1) = x_{r1,j}(t) + F \cdot (x_{r2,j}(t) - x_{r3,j}(t)) \ldots\ldots \quad (5)$$

The process is illustrated in Fig. 5. Closed curves in Fig. 5 denote constant cost contours, i.e., for a given cost function f, a contour corresponds to $f(\vec{X}) =$ constant. Here the constant cost contours are drawn for the Ackley Function.

Next, to increase the potential diversity of the population a crossover scheme comes to play. DE can use two kinds of cross over schemes namely "Exponential" and "Binomial". The donor vector exchanges its "body parts", i.e., components with the target vector $\vec{X}_i(t)$ under this scheme. In "Exponential" crossover, we first choose an integer n randomly among the numbers [0, D–1]. This integer acts as starting point in the target vector, from where the crossover or exchange of components with the donor vector starts. We also choose another integer L from the interval [1, D]. L denotes the number of components; the donor vector actually contributes to the target. After a choice of n and L the trial vector:

$$\vec{U}_i(t) = [u_{i,1}(t), u_{i,2}(t) \ldots u_{i,D}(t)] \quad (6)$$

is formed with

$$\begin{aligned} u_{i,j}(t) &= v_{i,j}(t) \quad \text{for} \quad j = < n >_D, < n+1 >_D, \ldots < n-L+1 >_D \\ &= x_{i,j}(t), \end{aligned} \quad (7)$$

where the angular brackets $<>_D$ denote a modulo function with modulus D. The integer L is drawn from [1, D] according to the following pseudo code.

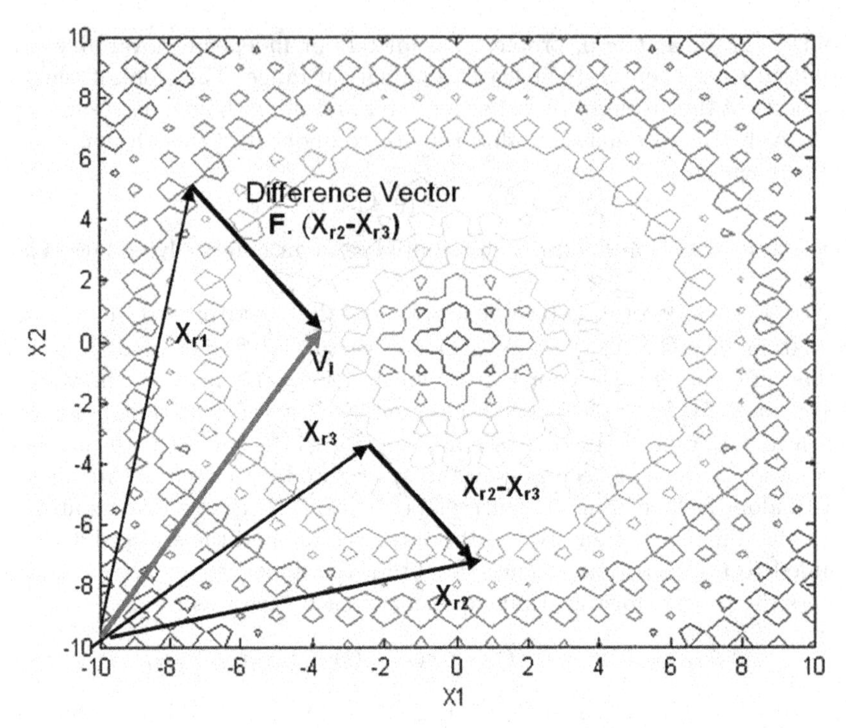

Fig. 5. Illustrating creation of the donor vector in 2-D parameter space (The constant cost contours are for two-dimensional Ackley Function)

```
L = 0;
do
{
  L=L+1;
} while (rand (0, 1) < CR) AND (L<D));
```

Hence in effect probability $(L > m) = (\mathrm{CR})^{\mathrm{m}-1}$ for any m > 0. CR is called "Crossover" constant and it appears as a control parameter of DE just like F. For each donor vector V, a new set of n and L must be chosen randomly as shown above. However, in "Binomial" crossover scheme, the crossover is performed on each of the D variables whenever a randomly picked number between 0 and 1 is within the CR value. The scheme may be outlined as

$$\begin{aligned} u_{i,j}(t) &= v_{i,j}(t) \quad \text{if rand}\,(0,1) < \mathrm{CR}, \\ &= x_{i,j}(t) \quad \text{else}\ldots\ldots \end{aligned} \tag{8}$$

In this way for each trial vector $\vec{X}_i(t)$ an offspring vector $\vec{U}_i(t)$ is created. To keep the population size constant over subsequent generations, the next step of the algorithm calls for "selection" to determine which one of the target vector and the trial vector will survive in the next generation, i.e., at time

$t = t + 1$. DE actually involves the Darwinian principle of "Survival of the fittest" in its selection process which may be outlined as

$$\vec{X}_i(t+1) = \vec{U}_i(t) \quad \text{if} \quad f(\vec{U}_i(t)) \leq f(\vec{X}_i(t)),$$
$$= \vec{X}_i(t) \quad \text{if} \quad f(\vec{X}_i(t)) < f(\vec{U}_i(t)), \ldots \tag{9}$$

where $f()$ is the function to be minimized. So if the new trial vector yields a better value of the fitness function, it replaces its target in the next generation; otherwise the target vector is retained in the population. Hence the population either gets better (w.r.t. the fitness function) or remains constant but never deteriorates. The DE/rand/1 algorithm is outlined below:

Procedure DE

Input: Randomly initialized position and velocity of the particles: $\vec{x}_i(0)$

Output: Position of the approximate global optima $\vec{X}^*$

Begin

Initialize population;
Evaluate fitness;
For i = 0 to max-iteration **do**
Begin
Create Difference-Offspring;
Evaluate fitness;
If an offspring is better than its parent
Then replace the parent by offspring in the next generation;
End If;
End For;
End.

Example 3. This example illustrates the complete searching on the fitness landscape of a two-dimensional sphere function by a simple DE. Sphere is perhaps one of the simplest two-dimensional functions and has been chosen to provide easy visual depiction of the search process. The function is given by

$$f(\vec{x}) = x_1^2 + x_2^2.$$

As can be easily perceived, the function has only one global minima $f^* = 0$ at $X^* = [0,\ 0]^T$. We start with a randomly initialized population of five vectors in the search range $[-10, 10]$. Initially, these vectors are given by

$$X_1(0) = [5, -9]^T$$
$$X_2(0) = [6, 1]^T$$
$$X_3(0) = [-3, 5]^T$$
$$X_4(0) = [-7, 4]^T$$
$$X_5(0) = [6, 7]^T$$

Figures 6–9 illustrate the initial orientation of the search variable vectors in the two-dimensional $X_1 - X_2$ space. The concentric circular lines are the *constant cost contours* of the function, i.e., locus in the $X_1 - X_2$ plane. Now following the mutation and recombination schemes as presented in expressions (7) and

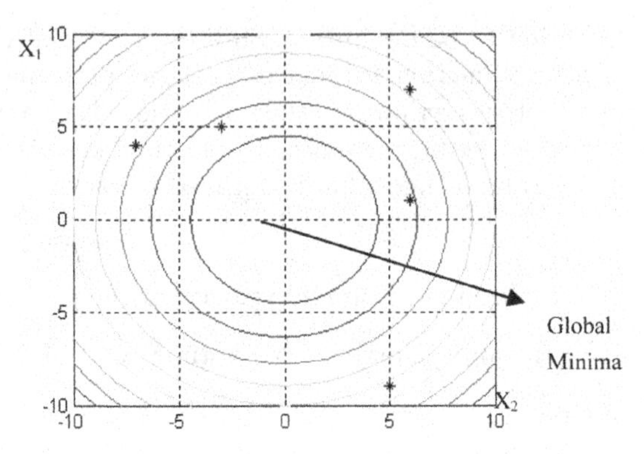

Fig. 6. Orientation of the initial solutions in the two-dimensional search space

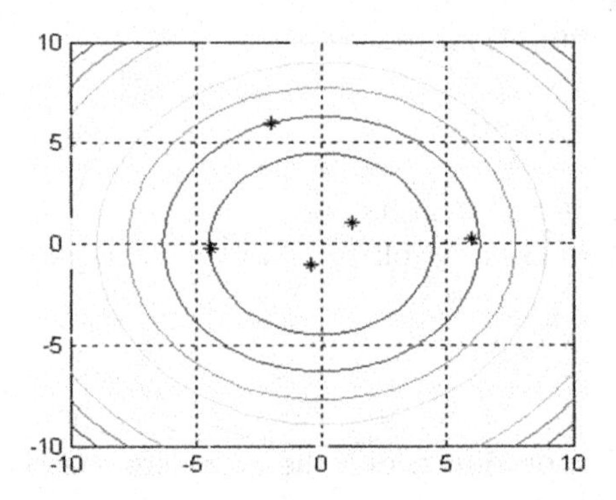

Fig. 7. Orientation of the population members after five iterations

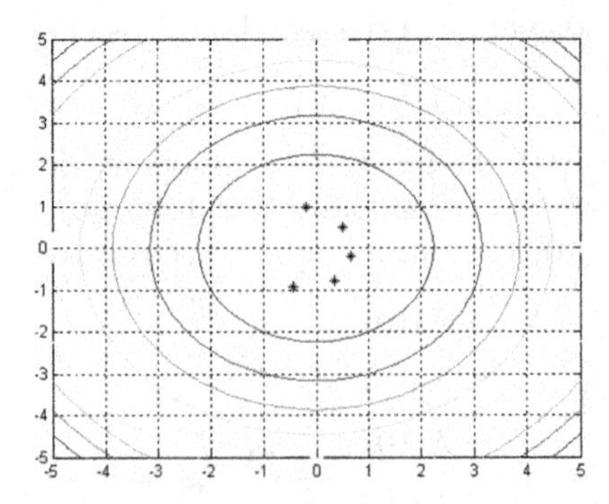

Fig. 8. Orientation of the population members after 10 iterations

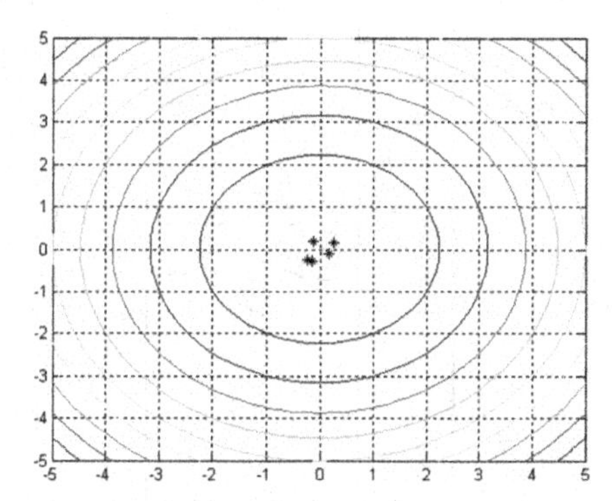

Fig. 9. Orientation of the population members after 20 iterations

(8), we form five donor vectors and then create five offspring vectors for time $t = 1$. Next we apply the selection method described by (9) and evolve the entire population at time $t = 1$. These steps have been summarized in Table 1.

7.2 The Complete DE Family of Storn and Price

Actually, it is the process of mutation, which demarcates one DE scheme from another. In the former section, we have illustrated the basic steps of a simple DE. The mutation scheme in (5) uses a randomly selected vector $\vec{X}_{r1}$ and only one weighted difference vector $F \cdot (\vec{X}_{r2} - \vec{X}_{r3})$ is used to perturb it. Hence, in literature the particular mutation scheme is referred to as DE/rand/1. We can now have an idea of how different DE schemes are named. The general convention used, is DE/x/y. DE stands for DE, x represents a string denoting the type of the vector to be perturbed (whether it is randomly selected or it is the best vector in the population with respect to fitness value) and y is the number of difference vectors considered for perturbation of x. Below we outline the other four different mutation schemes, suggested by Price et al. [33].

Scheme DE/rand to best/1

DE/rand to best/1 follows the same procedure as that of the simple DE scheme illustrated earlier. The only difference being that, now the donor vector, used to perturb each population member, is created using any two randomly selected member of the population as well as the best vector of the current generation (i.e., the vector yielding best suited objective function value at $t = t$). This can be expressed for the ith donor vector at time $t = t + 1$ as

$$\vec{V}_i(t+1) = \vec{X}_i(t) + \lambda \cdot (\vec{X}_{best}(t) - \vec{X}_i(t)) + F \cdot (\vec{X}_{r_2}(t) - \vec{X}_{r3}(t)) \qquad (10)$$

where λ is another control parameter of DE in [0, 2], $X_i(t)$ is the target vector and $\vec{X}_{best}(t)$ is the best member of the population regarding fitness at current

Table 1. Evolution of the population from $t = 0$ to $t = 1$ in Example 3

Population at $t = 0$	Fitness at $t = 0$	Donor vector at $t = 1$	Offspring vector at $t = 1$	Fitness of offspring at $t = 1$	Evolved population at $t = 1$
$X_1(0) = [2, -1]$	5	$V_1(1) = [-0.4, 10.4]$	$T_1(1) = [-0.4, -1]$	1.16	$X_1(1) = [-0.4, -1]$
$X_2(0) = [6, 1]$	37	$V_2(1)X_1(0) = [1.2, -0.2]$	$T_2(1) = [1.2, 1]$	2.44	$X_2(1) = [1.2, 1]$
$X_3(0) = [-3, 5]$	34	$V_3(1) = [-4.4, -0.2]$	$T_3(1) = [-4.4, -0.2]$	19.4	$X_3(1) = [-4.4, -0.2]$
$X_4(0) = [-2, 6]$	40	$V_4(1) = [9.2, -4.2]$	$T_4(1) = [9.2, 6]$	120.64	$X_4(1) = [-2, 6]$
$X_5(0) = [6, 7]$	85	$V_5(1) = [5.2, 0.2]$	$T_5(1) = [6, 0.2]$	36.04	$X_5(1) = [6, 0.2]$

time step $t = t$. To reduce the number of control parameters a usual choice is to put $\lambda = F$.

Scheme DE/best/1

In this scheme everything is identical to DE/rand/1 except the fact that the trial vector is formed as

$$\vec{V}_i(t+1) = \vec{X}_{best}(t) + F \cdot (\vec{X}_{r1}(t) - \vec{X}_{r2}(t)), \tag{11}$$

here the vector to be perturbed is the best vector of the current population and the perturbation is caused by using a single difference vector.

Scheme DE/best/2

Under this method, the donor vector is formed by using two difference vectors as shown below:

$$\vec{V}_i(t+1) = \vec{X}_{best}(t) + F \cdot (\vec{X}_{r1}(t) + \vec{X}_{r2}(t) - \vec{X}_{r3}(t) - \vec{X}_{r4}(t)). \tag{12}$$

Owing to the central limit theorem the random variations in the parameter vector seems to shift slightly into the Gaussian direction which seems to be beneficial for many functions.

Scheme DE/rand/2

Here the vector to be perturbed is selected randomly and two weighted difference vectors are added to the same to produce the donor vector. Thus for each target vector, a totality of five other distinct vectors are selected from the rest of the population. The process can be expressed in the form of an equation as

$$\vec{V}_i(t+1) = \vec{X}_{r1}(t) + F_1 \cdot (\vec{X}_{r2}(t) - \vec{X}_{r3}(t)) + F_2 \cdot (\vec{X}_{r4}(t) - \vec{X}_{r5}(t)) \tag{13}$$

Here F_1 and F_2 are two weighing factors selected in the range from 0 to 1. To reduce the number of parameters we may choose $F_1 = F_2 = F$.

Summary of all Schemes

In 2001 Storn and Price [2] suggested total ten different working strategies of DE and some guidelines in applying these strategies to any given problem. These strategies were derived from the five different DE mutation schemes outlined above. Each mutation strategy was combined with either the "exponential" type crossover or the "binomial" type crossover. This yielded $5 \times 2 = 10$ DE strategies, which are listed below.

1. DE/best/1/exp
2. DE/rand/1/exp
3. DE/rand-to-best/1/exp
4. DE/best/2/exp
5. DE/rand/2/exp
6. DE/best/1/bin
7. DE/rand/1/bin
8. DE/rand-to-best/1/bin
9. DE/best/2/bin
10. DE/rand/2/bin

The general convention used above is again DE/x/y/z, where DE stands for DE, x represents a string denoting the vector to be perturbed, y is the number of difference vectors considered for perturbation of x, and z stands for the type of crossover being used (exp: exponential; bin: binomial).

7.3 More Recent Variants of DE

DE is a stochastic, population-based, evolutionary search algorithm. The strength of the algorithm lies in its simplicity, speed (how fast an algorithm can find the optimal or suboptimal points of the search space) and robustness (producing nearly same results over repeated runs). The rate of convergence of DE as well as its accuracy can be improved largely by applying different mutation and selection strategies. A judicious control of the two key parameters namely the scale factor F and the crossover rate CR can considerably alter the performance of DE. In what follows we will illustrate some recent modifications in DE to make it suitable for tackling the most difficult optimization problems.

DE with Trigonometric Mutation

Recently, Lampinen and Fan [34] has proposed a trigonometric mutation operator for DE to speed up its performance. To implement the scheme, for each target vector, three distinct vectors are randomly selected from the DE population. Suppose for the ith target vector $\vec{X}_i(t)$, the selected population members are $\vec{X}_{r1}(t)$, $\vec{X}_{r2}(t)$ and $\vec{X}_{r3}(t)$. The indices r_1, r_2 and r_3 are mutually different and selected from $[1, 2, \ldots, N]$ where N denotes the population size. Suppose the objective function values of these three vectors are given by, $f(\vec{X}_{r1}(t))$, $f(\vec{X}_{r2}(t))$ and $f(\vec{X}_{r3}(t))$. Now three weighing coefficients are formed according to the following equations:

$$p' = \left|f(\vec{X}_{r1})\right| + \left|f(\vec{X}_{r2})\right| + \left|f(\vec{X}_{r3})\right|, \tag{14}$$

$$p_1 = \left|f(\vec{X}_{r1})\right| \Big/ p', \tag{15}$$

$$p_2 = \left|f(\vec{X}_{r2})\right| \Big/ p', \tag{16}$$

$$p_3 = \left|f(\vec{X}_{r3})\right| \Big/ p'. \tag{17}$$

Let rand (0, 1) be a uniformly distributed random number in (0, 1) and Γ be the trigonometric mutation rate in the same interval (0, 1). The trigonometric mutation scheme may now be expressed as

$$\begin{aligned}\vec{V}_i(t+1) &= (\vec{X}_{r1} + \vec{X}_{r2} + \vec{X}_{r3})/3 + (p_2 - p_1) \cdot (\vec{X}_{r1} - \vec{X}_{r2}) \\ &\quad + (p_3 - p_2) \cdot (\vec{X}_{r2} - \vec{X}_{r3}) + (p_1 - p_3) \cdot (\vec{X}_{r3} - \vec{X}_{r1}) \\ &\qquad\qquad\qquad\qquad\qquad\qquad \text{if rand}\,(0,1) < \Gamma \\ \vec{V}_i(t+1) &= \vec{X}_{r1} + F \cdot (\vec{X}_{r2} + \vec{X}_{r3}) \qquad \text{else.}\end{aligned} \tag{18}$$

Thus, we find that the scheme proposed by Lampinen et al. uses trigonometric mutation with a probability of Γ and the mutation scheme of DE/rand/1 with a probability of $(1 - \Gamma)$.

DERANDSF (DE with Random Scale Factor)

In the original DE [9] the difference vector $(\vec{X}_{r1}(t) - \vec{X}_{r2}(t))$ is scaled by a constant factor "F". The usual choice for this control parameter is a number between 0.4 and 1. We propose to vary this scale factor in a random manner in the range (0.5, 1) by using the relation

$$F = 0.5^*(1 + \text{rand}\ (0,1)), \tag{19}$$

where rand (0, 1) is a uniformly distributed random number within the range [0, 1]. We call this scheme DERANDSF (DE with Random Scale Factor) [35]. The mean value of the scale factor is 0.75. This allows for stochastic variations in the amplification of the difference vector and thus helps retain population diversity as the search progresses. Even when the tips of most of the population vectors point to locations clustered near a local optimum due to the randomly scaled difference vector, a new trial vector has fair chances of pointing at an even better location on the multimodal functional surface. Therefore, the fitness of the best vector in a population is much less likely to get stagnant until a truly global optimum is reached.

DETVSF (DE with Time Varying Scale Factor)

In most population-based optimization methods (except perhaps some hybrid global-local methods) it is generally believed to be a good idea to encourage

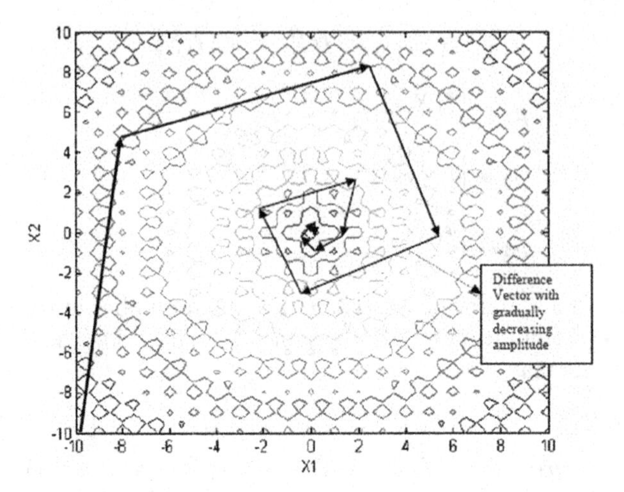

Fig. 10. Illustrating DETVSF scheme on two-dimensional cost contours of Ackley function

the individuals (here, the tips of the trial vectors) to sample diverse zones of the search space during the early stages of the search. During the later stages it is important to adjust the movements of trial solutions finely so that they can explore the interior of a relatively small space in which the suspected global optimum lies. To meet this objective we reduce the value of the scale factor linearly with time from a (predetermined) maximum to a (predetermined) minimum value:

$$R = (R_{\max} - R_{\min})^{*}(MAXIT - iter)/MAXIT \tag{20}$$

where $F_{\max}$ and $F_{\min}$ are the maximum and minimum values of scale factor F, *iter* is the current iteration number and *MAXIT* is the maximum number of allowable iterations. The locus of the tip of the best vector in the population under this scheme may be illustrated as in Fig. 10. The resulting algorithm is referred as DETVSF (DE with a time varying scale factor) [35].

DE with Local Neighborhood

Only in 2006, a new DE-variant, based on the neighborhood topology of the parameter vectors was developed [36] to overcome some of the disadvantages of the classical DE versions. The authors in [36] proposed a neighborhood-based local mutation operator that draws inspiration from PSO. Suppose we have a DE population $P = [\vec{X}_1, \vec{X}_2, \ldots, \vec{X}_{N_p}]$ where each $\vec{X}_i$ $(i = 1, 2, \ldots, N_p)$ is a D-dimensional vector. Now for every vector $\vec{X}_i$ we define a neighborhood of radius k, consisting of vectors $\vec{X}_{i-k}, \ldots \vec{X}_i, \ldots \vec{X}_{i+k}$. We assume the vectors to be organized in a circular fashion such that two immediate neighbors of vector $\vec{X}_1$ are $\vec{X}_{N_p}$ and $\vec{X}_2$. For each member of the population a local mutation is created by employing the fittest vector in the neighborhood of that member and two other vectors chosen from the same neighborhood. The model may be expressed as:

$$\vec{L}_i(t) = \vec{X}_i(t) + \lambda' \cdot (\vec{X}_{nbest}(t) - \vec{X}_i(t)) + F' \cdot (\vec{X}_p(t) - \vec{X}_q(t)) \qquad (21)$$

where the subscript *nbest* indicates the best vector in the neighborhood of $\vec{X}_i$ and $p, q \in (i-k, i+k)$. Apart from this, we also use a global mutation expressed as:

$$\vec{G}_i(t) = \vec{X}_i(t) + \lambda \cdot (\vec{X}_{best}(t) - \vec{X}_i(t)) + F \cdot (\vec{X}_r(t) - \vec{X}_s(t)), \qquad (22)$$

where the subscript *best* indicates the best vector in the entire population, and $r, s \in (1, N_p)$. Global mutation encourages exploitation, since all members (vectors) of a population are biased by the same individual (the population best); local mutation, in contrast, favors exploration, since in general different members of the population are likely to be biased by different individuals. Now we combine these two models using a time-varying scalar weight $w \in (0, 1)$ to form the actual mutation of the new DE as a weighted mean of the local and the global components:

$$\vec{V}_i(t) = w \cdot \vec{G}_i(t) + (1-w) \cdot \vec{L}_i(t). \qquad (23)$$

The weight factor varies linearly with time as folows:

$$w = w_{\min} + (w_{\max} - w_{\min}) \cdot \left(\frac{iter}{MAXIT}\right), \qquad (24)$$

where *iter* is the current iteration number, *MAXIT* is the maximum number of iterations allowed and $w_{\max}$, $w_{\min}$ denote, respectively, the maximum and minimum value of the weight, with $w_{\max}, w_{\min} \in (0, 1)$. Thus the algorithm starts at $iter = 0$ with $w = w_{\min}$ but as *iter* increases towards *MAXIT*, w increases gradually and ultimately when $iter = MAXIT$ w reaches $w_{\max}$. Therefore at the beginning, emphasis is laid on the local mutation scheme, but with time, contribution from the global model increases. In the local model attraction towards a single point of the search space is reduced, helping DE avoid local optima. This feature is essential at the beginning of the search process when the candidate vectors are expected to explore the search space vigorously. Clearly, a judicious choice of $w_{\max}$ and $w_{\min}$ is necessary to strike a balance between the exploration and exploitation abilities of the algorithm. After some experimenting, it was found that $w_{\max} = 0.8$ and $w_{\min} = 0.4$ seem to improve the performance of the algorithm over a number of benchmark functions.

8 A Synergism of PSO and DE – Towards a New Hybrid Evolutionary Algorithm

Das et al. proposed a new scheme of adjusting the velocities of the particles in PSO with a vector differential operator borrowed from the DE family [37]. The canonical PSO updates the velocity of a particle using three terms. These include a previous velocity term that provides the particle with the necessary momentum, a *social* term that indicates how the particle is stochastically drawn towards the globally best position found so far by the entire swarm, and finally a *cognitive* term that reflects the personal thinking of the particle, i.e., how much it is drawn towards the best position so far encountered in its own course. In the proposed scheme the cognitive term is omitted; instead the particle velocities are perturbed by a new term containing the weighted difference of the position vectors of any two distinct particles randomly chosen from the swarm. This differential velocity term is inspired by the DE mutation scheme and hence the authors name this algorithm as PSO-DV (particle swarm with differentially perturbed velocity). A *survival of the fittest* mechanism has also been incorporated in the swarm following the greedy selection scheme of DE.

8.1 The PSO-DV Algorithm

PSO-DV introduces a differential operator (borrowed from DE) in the velocity-update scheme of PSO. The operator is invoked on the position vectors of two randomly chosen particles (population-members), not on their individual best positions. Further, unlike the PSO scheme, a particle is actually shifted to a new location only if the new location yields a better fitness value, i.e., a selection strategy has been incorporated into the swarm dynamics. In the proposed algorithm, for each particle i in the swarm two other distinct particles, say j and k $(i \neq j \neq k)$, are selected randomly. The difference between their positional coordinates is taken as a difference vector:

$$\vec{\delta} = \vec{X}_k - \vec{X}_j.$$

Then the dth velocity component $(1 < d < n)$ of the target particle i is updated as

$$\left.\begin{aligned} V_{\mathrm{id}}(t+1) &= \omega \cdot V_{\mathrm{id}}(t) + \beta \cdot \delta_{\mathrm{d}} + C_2 \cdot \varphi_2 \cdot (P_{\mathrm{gd}} - jX_{\mathrm{id}}(t)), && \text{if } \mathrm{rand_d}\,(0,1) \leq \mathrm{CR}. \\ &= V_{\mathrm{id}}(t), && \text{otherwise,} \end{aligned}\right\} \tag{25}$$

where CR is the crossover probability, δ_{d} is the dth component of the difference vector defined earlier, and β is a scale factor in [0, 1]. In essence the cognitive part of the velocity update formula in (1) is replaced with the vector differential operator to produce some additional exploration capability. Clearly, for CR ≤ 1, some of the velocity components will retain their old values. Now, a new trial location Tr_i is created for the particle by adding the updated velocity to the previous position X_i:

$$\vec{Tr}_i = \vec{X}_i(t) + \vec{V}_i(t+1). \tag{26}$$

The particle is placed at this new location only if the coordinates of the location yield a better fitness value. Thus if we are seeking the minimum of an n-dimensional function $f(\vec{X})$, then the target particle is relocated as follows:

$$\begin{aligned} &\vec{X}_i(t+1) = \vec{T}r_i && \text{if } f(\vec{T}r_i) \leq f(\vec{X}_i(t)), \\ &\vec{X}_i(t+1) = \vec{X}_i(t) && \text{Otherwise,} \end{aligned} \tag{27}$$

Therefore, every time its velocity changes, the particle either moves to a better position in the search space or sticks to its previous location. The current location of the particle is thus the best location it has ever found. In other words, unlike the classical PSO, in the present scheme, P_{lid} always equals X_{id}. So the cognitive part involving $|P_{lid} - X_{id}|$ is automatically eliminated in our algorithm. If a particle gets stagnant at any point in the search space (i.e., if its location does not change for a predetermined number of iterations), then the particle is shifted by a random mutation (explained below) to a new location. This technique helps escape local minima and also keeps the swarm "moving":

If $(\vec{X}_i(t) = \vec{X}_i(t+1) = \vec{X}_i(t+2) = \ldots. = \vec{X}_i(t+N))$ and $f(\vec{X}_i(t+N))$ **then**

for (r = 1 to n)

$$X_{\mathrm{ir}}(t+N+1) = X_{\min} + \mathrm{rand_r}\,(0,1)^*(X_{\max} - X_{\min}), \tag{28}$$

where f* is the global minimum of the fitness function, N is the maximum number of iterations up to which stagnation can be tolerated and $(X_{\max}, X_{\min})$ define the permissible bounds of the search space. The scheme is conceptually outlined in Fig. 11. The pseudo-code for this algorithm is illustrated as follows:

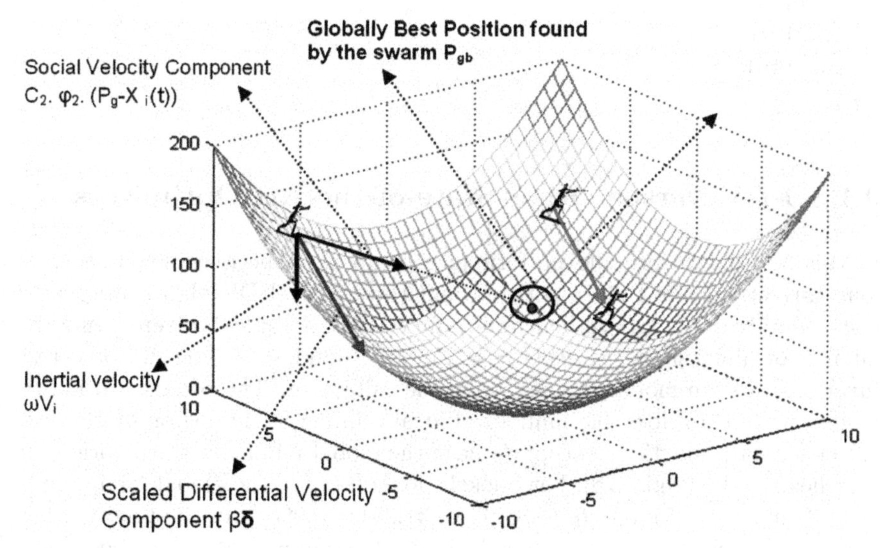

Fig. 11. Illustrating PSO-DV on a two-dimensional function surface

Procedure PSO-DV

begin
initialize population;
while stopping condition not satisfied do
for $i = 1$ to no_of_particles
evaluate fitness of particle;
update P_{gd};
select two other particles j and k ($i \neq j \neq k$) randomly;
construct the difference vector as $\vec{\delta} = \vec{X}_k - \vec{X}_j$;
for $d = 1$ to no_of_dimensions
if $\text{rand}_d\ (0, 1) \leq \text{CR}$
$V_{id}(t+1) = \omega \cdot V_{id}(t) + \beta \cdot \delta_d + C_2 \cdot \varphi_2 \cdot (P_{gd} - X_{id}(t))$;
else $V_{id}(t+1) = V_{id}(t)$;
endif
endfor
create trial location as $\vec{Tr}_i = \vec{X}_i(t) + \vec{V}_i(t+1)$;
if $(f(\vec{Tr}_i) \leq f(\vec{X}_i(t)))$ **then** $\vec{X}_i(t+1) = \vec{Tr}$;
else $\vec{X}_i(t+1) = \vec{X}_i(t)$;
endif
endfor
for $i = 1$ to no_of_particles
if X_i stagnates for N successive generations
for $r = 1$ to no_of_dimensions
$X_{ir}(t+1) = X_{min} + \text{rand}_r\ (0,1)^*(X_{max} - X_{min})$
end for
end if
end for
end while
end

9 PSO-DV Versus Other State-of-the-Art Optimizers

In this section we provide performance comparison between PSO-DV and four variants of the PSO algorithm and the classical DE algorithm over a test-bed of five well known benchmark functions. In Table 2, n represents the number of dimensions (we used n = 25, 50, 75 and 100). The first two test functions are uni-modal, having only one minimum. The others are multi-modal, with a considerable number of local minima in the region of interest. All benchmark functions except f_6 have the global minimum at the origin or very near to the origin [28]. For Shekel's foxholes (f_6), the global minimum is at (−31.95, −31.95) and f_6 (−31.95, −31.95) $\approx$ 0.998, and the function has only two dimensions. An asymmetrical initialization procedure has been used here following the work reported in [29].

Table 2. Benchmark functions used

Function	Mathematical representation
Sphere function	$f_1(x) = \sum_{i=1}^{n} x_i^2$
Rosenbrock	$f_2(x) = \sum_{i=1}^{n-1} [100(x_{i+1} - x_i^2)^2 + (x_i - 1)^2]$
Rastrigin	$f_3(x) = \sum_{i=1}^{n} [x_i^2 - 10\cos(2\pi x_i) + 10]$
Griewank	$f_4(x) = \frac{1}{4000} \sum_{i=1}^{n} x_i^2 - \prod_{i=1}^{n} \cos\left(\frac{x_i}{\sqrt{i}}\right) + 1$
Ackley	$f_5(x) = -20\exp\left(-0.2\sqrt{(\frac{1}{n}\sum_{i=1}^{n} x_i^2)} - \exp\left(\frac{1}{n}\sum_{i=1}^{n}\cos 2\pi x_i\right) + 20 + e\right.$
Shekel's Foxholes	$f_6(x) = \left[\frac{1}{500} + \sum_{j=1}^{25} \frac{1}{j + \sum_{i=1}^{2} (x_i - a_{ij})^6}\right]^{-1}$

Simulations were carried out to obtain a comparative performance analysis of the new method with respect to: (a) canonical PSO, (b) PSO-TVIW [18,37], (c) MPSO-TVAC [20], (d) HPSO-TVAC [20], and (e) classical DE. Thus a total of six algorithms were considered – one new, the other five existing in the literature. For a given function of a given dimension, 50 independent runs of each of the six algorithms were executed, and the average best-of-run value and the standard deviation were obtained. Different maximum generations ($G_{\max}$) were used according to the complexity of the problem. For all benchmarks (excluding Schaffer's f_6 and f_7) the stopping criterion was set as reaching a fitness of 0.001. However, for Shekel's Foxholes function (f_7) it was fixed at 0.998. Table 3 compares the algorithms on the quality of the best solution. The mean and the standard deviation (within parentheses) of the best-of-run solution for 50 independent runs of each of the six algorithms are presented in Table 3. Missing values of standard deviation in this table indicate a zero standard deviation. The best solution in each case has been shown in bold. Table 4 shows results of unpaired t-tests between the best algorithm and the second best in each case (standard error of difference of the two means, 95% confidence interval of this difference, the t value, and the two-tailed P value). For all cases in Table 4, sample size = 50 and degrees of freedom = 98. It is interesting to see from Tables 3 and 4 that the proposed method performs equally or better than other algorithms in a statistically meaningful way.

In Fig. 12, we have graphically presented the rate of convergence of all the methods for two most difficult functions Rosenbrock (f_2) and Ackley (f_5) functions (in 30 dimensions). We do not provide convergence graphs for all the functions in order to save space.

Table 3. Average and standard deviation of the best-of-run solution obtained for 50 runs of each of the six different methods

F	Dim	N	G_{max}	Average (standard deviation)					
				PSO	PSO-TVIW	MPSO-TVAC	HPSO-TVAC	DE	PSO-DV
F_1	10	50	1000	**0.001**	**0.001**	**0.001**	**0.001**	**0.001**	**0.001**
	20	100	2000	**0.001**	**0.001**	**0.001**	**0.001**	**0.001**	**0.001**
	30	150	4500	**0.001**	**0.001**	**0.001**	**0.0001**	**0.001**	**0.001**
F_2	10	75	3000	21.705 (40.162)	16.21 (14.917)	1.234 (4.3232)	1.921 (4.330)	2.263 (4.487)	**0.0063 (0.0561)**
	20	150	4000	52.21 (148.32)	42.73 (72.612)	22.732 (30.638)	20.749 (12.775)	18.934 (9.453)	**0.0187 (0.554)**
	30	250	5000	77.61 (81.172)	61.78 (48.933)	34.229 (36.424)	10.414 (44.845)	6.876 (1.688)	**0.0227 (0.182)**
F_3	10	50	3000	2.334 (2.297)	2.1184 (1.563)	1.78 (2.793)	0.039 (0.061)	0.006 (0.0091)	**0.0014 (0.0039)**
	20	100	4000	13.812 (3.491)	16.36 (4.418)	11.131 (0.91)	0.2351 (0.1261)	0.0053 (0.0032)	**0.0028 (0.0017)**
	30	150	5000	6.652 (21.811)	24.346 (6.317)	50.065 (21.139)	1.903 (0.894)	0.099 (0.112)	**0.0016 (0.277)**
F_4	10	50	2500	0.1613 (0.097)	0.092 (0.021)	**0.00561 (0.047)**	0.057 (0.045)	0.054 (0.0287)	0.024 (0.180)
	20	100	3500	0.2583 (0.1232)	0.1212 (0.5234)	0.0348 (0.127)	0.018 (0.0053)	0.019 (0.0113)	**0.0032 (0.0343)**
	30	150	5000	0.0678 (0.236)	0.1486 (0.124)	0.0169 (0.116)	0.023 (0.0045)	0.005 (0.0035)	**0.0016 (0.0022)**
F_5	10	50	2500	0.406 (1.422)	0.238 (1.812)	0.169 (0.772)	0.0926 (0.0142)	**0.00312 (0.0154)**	0.00417 (0.1032)
	20	100	3500	0.572 (3.094)	0.318 (1.118)	0.537 (0.2301)	0.117 (0.025)	0.029 (0.0067)	**0.0018 (0.028)**
	30	150	5000	1.898 (2.598)	0.632 (2.0651)	0.369 (2.735)	0.068 (0.014)	0.0078 (0.0085)	**0.0016 (0.0078)**
F_6	2	40	1000	1.235 (2.215)	1.239 (1.468)	1.321 (2.581)	1.328 (1.452)	1.032 (0.074)	**0.9991 (0.0002)**

Table 4. Results of unpaired t-tests on the data of Table 3

Fn, dim	Std. err.	t	95% Conf. intvl	Two-tailed P	Significance
f_2, 10	0.612	3.1264	(−3.129, −0.699)	0.0023	Very significant
f_2, 20	1.339	14.1249	(−21.573, −16.258)	<0.0001	**Extremely significant**
f_2, 30	0.054	1.3981	(−0.184, 0.032)	0.1653	Not significant
f_3, 10	0.004	6.3954	(−0.037, −0.019)	<0.0001	**Extremely significant**
f_3, 20	0.000	0.000	(−0.000421, 0.000421)	1.0000	Not significant
f_3, 30	0.042	2.3051	(−0.181, −0.014)	0.0233	Significant
f_4, 10	0.026	0.6990	(−0.071, 0.034)	0.4862	Not significant
f_4, 20	0.005	2.6781	(−0.023, −0.003)	0.0087	Very significant
f_4, 30	0.000	5.3112	(−0.0027, −0.0012)	<0.0001	**Extremely significant**
f_5, 10	0.015	0.0712	(−0.030, 0.028)	0.9434	Not significant
f_5, 20	0.004	9.3541	(−0.045, −0.03)	<0.0001	**Extremely significant**
f_5, 30	0.001	4.0532	(−0.009, −0.003)	<0.0001	**Extremely significant**
f_7, 2	2.615	0.0241	(−5.251, 5.125)	0.9809	Not significant

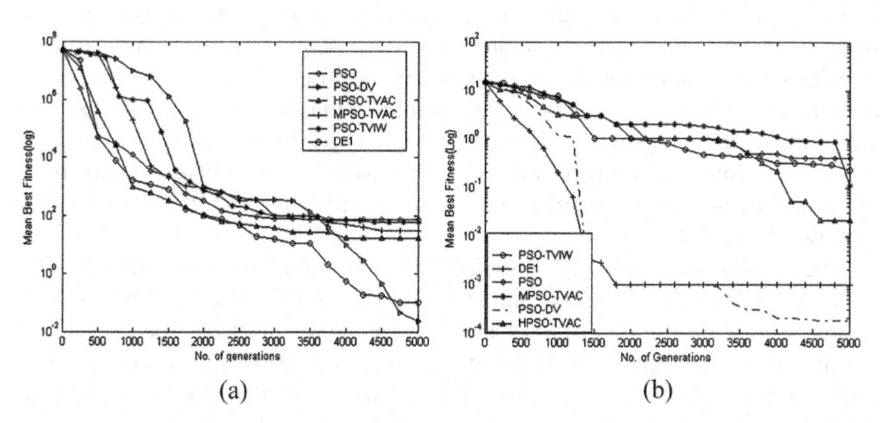

Fig. 12. Variation of the mean best value with time (all the graphs are for dimension = 30 except for Shekel's Foxholes (f_6) which is 2D) (**a**) Rosenbrock Function (f_2) (**b**) Ackley Function (f_5)

10 Applications

PSO has been applied to solve many interesting problems including a lot of practical application problems. It has been applied to evolve weights and structure of neural networks [38–40], analyze human tremor [41], register 3D-to-3D biomedical image [42], play games [43], control reactive power and voltage [44, 45], pattern recognition [46], etc. Generally speaking, PSO can be applied to solve most optimization problems and problems that can be converted to search or optimization problems.

Differential evolution (DE) has successfully been applied to many artificial and real optimization problems such as aerodynamic shape optimization

[47], automated mirror design [48], optimization of radial active magnetic bearings [49], and optimization of fermentation by using a high ethanol-tolerance yeast [50]. A DE based neural network-training algorithm was first introduced in [51]. In [52] the method's characteristics as a global optimizer were compared to other neural network training methods. Das et al. in [53] have compared the performance of some variants of the DE with other common optimization algorithms like PSO, GA, etc. in context to the partitional clustering problem and concluded in their study that DE rather than GAs should receive primary attention in such partitional cluster algorithms.

In this section we describe a simple application of the aforementioned algorithms to the design of two dimensional IIR filters [54]. In signal processing, the function of a filter is to remove unwanted parts of the signal, such as random noise, or to extract useful parts of the signal, such as the components lying within a certain frequency range. There are two main kinds of filter, *analog* and *digital.* They are quite different in their physical makeup and in how they work. An analog filter uses analog electronic circuits made up from components such as resistors, capacitors and op-amps to produce the required filtering effect. Such filter circuits are widely used in such applications as noise reduction, video signal enhancement, graphic equalisers in hi-fi systems, and many other areas. A digital filter uses a digital processor to perform numerical calculations on sampled values of the signal. The processor may be a general-purpose computer such as a PC, or a specialised DSP (digital signal processor) chip.

Digital filters are broadly classified into two main categories namely, FIR (*finite impulse response*) filters and IIR (*infinite impulse response*) filters. The impulse response of a digital filter is the output sequence from the filter when a unit impulse is applied at its input. (A unit impulse is a very simple input sequence consisting of a single value of 1 at time $t = 0$, followed by zeros at all subsequent sampling instants). An FIR filter is one whose impulse response is of finite duration. The output of such a filter is calculated solely from the current and previous input values. This type of filter is hence said to be *non-recursive.* On the other hand, an IIR filter is one whose impulse response (theoretically) continues for ever in time. They are also termed as *recursive* filters. The current output of such a filter depends upon previous output values. These, like the previous input values, are stored in the processor's memory. The word recursive literally means "running back", and refers to the fact that previously-calculated output values go back into the calculation of the latest output. The recursive (previous output) terms feed back energy into the filter input and keep it going.

In our work [54, 55] the filter design is mainly considered from a frequency domain perspective. Frequency domain filtering consists in first, taking the fourier transform of the two-dimensional signal (which may be the pixel intensity value in case of a gray-scale image), then multiplying the frequency domain signal by the transfer function of the filter and finally inverse trans-

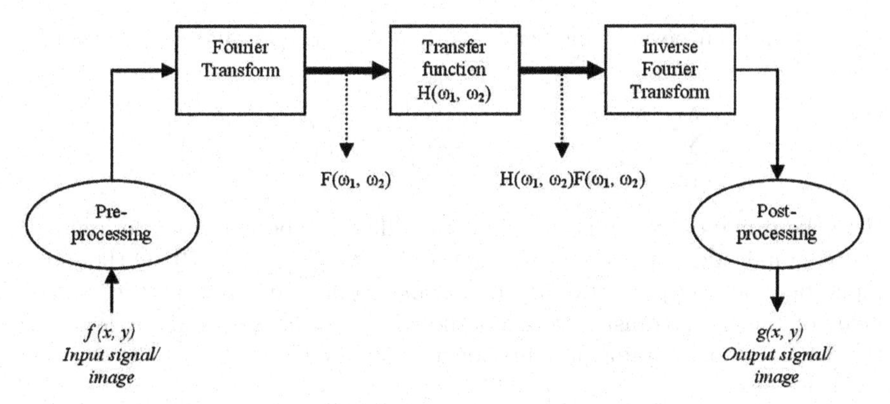

Fig. 13. Scheme of filtering with 2-D digital filter

forming the product in order to get the output response of the filter. The scheme is illustrated in Fig. 13.

Let the general prototype 2-D transfer function for the digital filter be

$$H(z_1, z_2) = H_0 \frac{\sum\limits_{i=0}^{N} \sum\limits_{j=0}^{N} p_{ij} z^i z^j}{\prod\limits_{k=1}^{N} (1 + q_k z_1 + r_k z_2 + s_k z_1 \cdot z_2)} \tag{29}$$

It is a general practice to take $p_{00} = 1$ (by normalizing p_{ij}'s with respect to the value of p_{00}). Also, let us assume that the user-specified amplitude response of the filter to be designed is M_{d} which is obviously a function of digital frequencies ω_1 and ω_2 (ω_1, $\omega_2 \in [0, \pi]$). Now the main design problem is to determine the coefficients in the numerator and denominator of (29) in such a fashion that $H(z_1, z_2)$ follows the desired response M_{d} (ω_1, ω_2) as closely as possible. Such an approximation of the desired response can be achieved by minimizing

$$J(p_{ij}, q_k, r_k, s_k, H_0) = \sum_{n_1=0}^{N_1} \sum_{n_2=0}^{N_2} [|M(\omega_1, \omega_2)| - M_d(\omega_1, \omega_2)]^b, \tag{30}$$

where

$$M(\omega_1, \omega_2) = H(z_1, z_2) \Big|_{\substack{z_1 = e^{j\omega_1} \\ z_2 = e^{j\omega_2}}} \tag{31}$$

and

$$\omega_1 = (\pi/N_1)_{n_1};$$
$$\omega_2 = (\pi/N_2)_{n_2};$$

and b is an even positive integer (usually $b = 2$ or 4). Equation (30) can be restated as

$$J = \sum_{n_1=0}^{N_1} \sum_{n_2=0}^{N_2} \left[\left| M\left(\frac{\pi n_1}{N_1}, \frac{\pi n_2}{N_2}\right) \right| - M_d\left(\frac{\pi n_1}{N_1}, \frac{\pi n_2}{N_2}\right) \right]^b . \tag{32}$$

Here the prime objective is to reduce the difference between the desired and actual amplitude responses of the filter at $N_1 \cdot N_2$ points. For BIBO (bounded input bounded output) stability the prime requirement is that the z-plane poles of the filter transfer function should lie within the unit circle. Since the denominator contains only first degree factors, we can assert the stability conditions as

$$|q_k + r_k| - 1 < s_k < 1 - |q_k - r_k|, \tag{33}$$

where $k = 1, 2, \ldots, N$.

We solve the above constrained minimization problems using a binary coded GA [56], PSO, DE and PSO-DV and observe that the synergistic PSO-DV performs best as compared to the other competitive algorithms over this problem.

To judge the accuracy of the algorithms, we run all of them for a long duration upto 100,000 FEs. Each algorithm is run independently (with a different seed for the random number generator in every run) for 30 times and the mean best J value obtained along with the standard deviations have been repored for the design problem (32) in Table 5. Figures 14 and 15 illustrate the frequency responses of the filters. The notation J_b has been used to denote four sets of experiments performed with the value of J obtained using exponent $b = 1$, 2, 4 and 8. Table 6 summarizes the results of the unpaired t-test on the J values (standard error of difference of the two means, 95% confidence interval of this difference, the t value, and the two-tailed P value) between the best and next-to-best results in Table 4. For all cases in Table 6, sample size = 30 and number of degrees of freedom = 58.

The frequency response of the various filters deigned by the above mentioned algorithms are shown below.

Table 5. Mean value and standard deviations of the final results (J values) with exponent $p = 1$, 2, 4, 8 after 100,000 FEs (mean of 20 independent runs of each of the competitor algorithms)

Value of J for different exponents	PSO-DV	PSO	DE	Binary GA in [56]
J_1	**61.7113 ± 0.0054**	98.5513 ± 0.0327	95.7113 ± 0.0382	96.7635 ± 0.8742
J_2	**9.0215 ± 0.0323**	11.9078 ± 0.583	10.4252 ± 0.0989	10.0342 ± 0.0663
J_4	**0.5613 ± 0.00054**	0.9613 ± 0.0344	0.5732 ± 0.0024	0.6346 ± 0.0154
J_8	0.0024 ± 0.001	0.2903 ± 0.0755	**0.0018 ± 0.0006**	0.0091 ± 0.0014

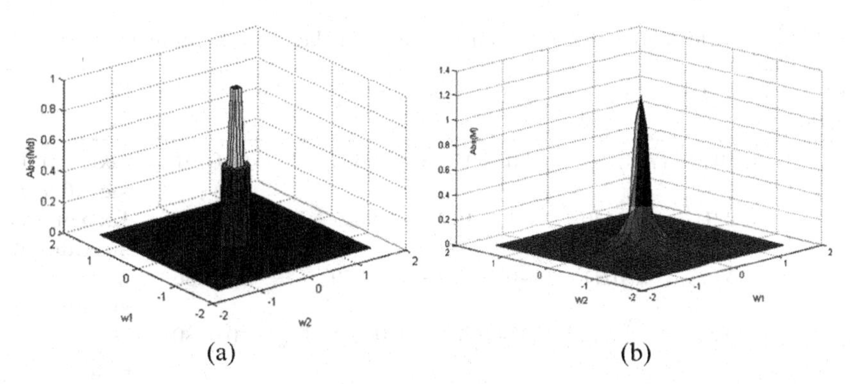

Fig. 14. (**a**) Ideal desired filter response (**b**) Frequency response of the filter designed by PSO-DV

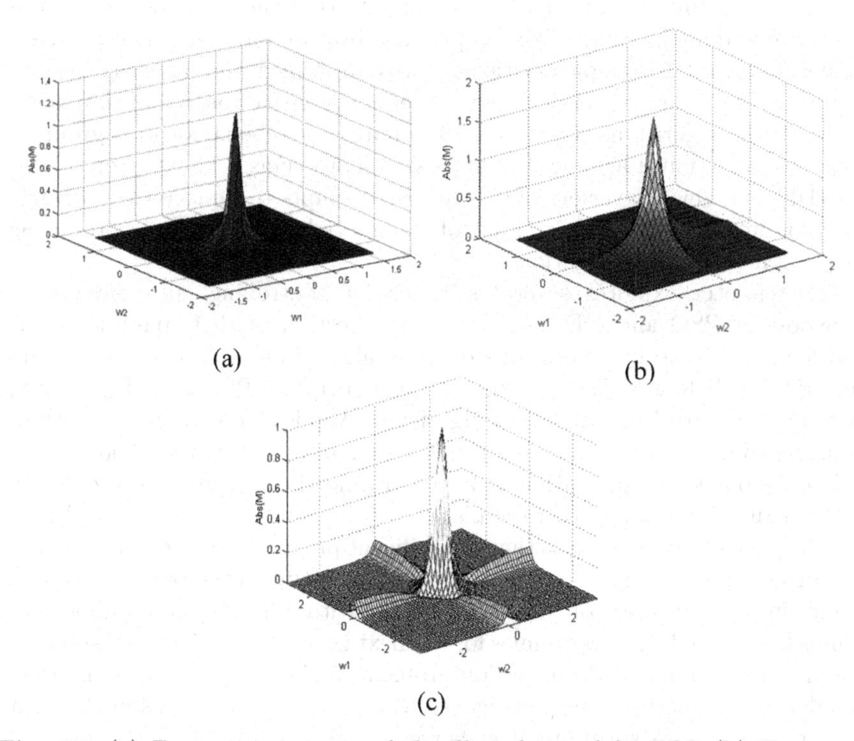

Fig. 15. (**a**) Frequency response of the filter designed by PSO (**b**) Frequency response of the filter designed by DE (**c**) Frequency response of the filter designed by a binary GAv [56]

Table 6. Results of unpaired t-tests on the data of Table 5

Cost function	Std. err	t	95% Conf. intvl	Two-tailed P	Significance
J_1	0.195	179.313	−35.447 to −34.656	<0.0001	**Extremely significant**
J_2	0.221	6.3440	−1.8516 to −0.9557	<0.0001	**Extremely significant**
J_4	0.030	3.1041	−0.1518 to −0.0319	0.0036	**Very significant**
J_8	0.000	1.9551	−0.0000213 to 0.0012213	0.0580	Not significant

11 Conclusions

Search and optimization problems are ubiquitous through the various realms of science and engineering. This chapter has provided a comprehensive overview of two promising optimization algorithms, which are currently gaining popularity for their greater accuracy, faster convergence speed and simplicity. One of these algorithms, known as PSO mimics the behavior of a group of social insects in multi-agent cooperative search problems. The latter one called DE (DE) is a deviant variety of GA, which attempts to replace the crossover operator in GA by a special type of differential operator for reproducing offspring in the next generation.

The chapter explores several schemes for controlling the convergence behaviors of PSO and DE by a judicious selection of their parameters. It also focuses on the hybridizations of these algorithms with other soft computing tools. It finally discusses the mutual synergy of PSO with DE leading to a more powerful global search algorithm. Applications of the algorithms to diverse domains of engineering problems have been surveyed. The chapter elaborates one such application of PSO, DE and their variants to the design of IIR digital filters in greater details.

The present article reveals that a significant progress has been made in the field of swarm intelligence and evolutionary computing in the past 10 years. In recent times, a symbiosis of swarm intelligence with other soft computing algorithms has opened up new avenues for the next generation computing systems. Engineering search and optimization problems including pattern recognition, bioinformatics and machine intelligence will find new dimensions in the light of hybridization of swarm intelligence with other algorithms.

References

1. Konar A (2005), Computational Intelligence: Principles, Techniques and Applications, Springer, Berlin Heidelberg New York.
2. Holland JH (1975), Adaptation in Natural and Artificial Systems, University of Michigan Press, Ann Arbor.

3. Goldberg DE (1975), Genetic Algorithms in Search, Optimization and Machine Learning, Addison-Wesley, Reading, MA.
4. Kennedy J, Eberhart R and Shi Y (2001), Swarm Intelligence, Morgan Kaufmann, Los Altos, CA.
5. Kennedy J and Eberhart R (1995), Particle Swarm Optimization, In Proceedings of IEEE International Conference on Neural Networks, pp. 1942–1948.
6. Storn R and Price K (1997), Differential Evolution – A Simple and Efficient Heuristic for Global Optimization Over Continuous Spaces, Journal of Global Optimization, 11(4), 341–359.
7. Venter G and Sobieszczanski-Sobieski J (2003), Particle Swarm Optimization, AIAA Journal, 41(8), 1583–1589.
8. Yao X, Liu Y, and Lin G (1999), Evolutionary Programming Made Faster, IEEE Transactions on Evolutionary Computation, 3(2), 82–102.
9. Shi Y and Eberhart RC (1998), Parameter Selection in Particle Swarm Optimization, Evolutionary Programming VII, Springer, Lecture Notes in Computer Science 1447, 591–600.
10. Shi Y and Eberhart RC (1999), Empirical Study of Particle Swarm Optimization, In Proceedings of the 1999 Congress of Evolutionary Computation, vol. 3, IEEE Press, New York, pp. 1945–1950.
11. Angeline PJ (1998), Evolutionary Optimization Versus Particle Swarm Optimization: Philosophy and Performance Differences, Evolutionary Programming VII, Lecture Notes in Computer Science 1447, Springer, Berlin Heidelberg New York, pp. 601–610.
12. Shi Y and Eberhart RC (1998), A Modified Particle Swarm Optimiser, IEEE International Conference on Evolutionary Computation, Anchorage, Alaska, May 4–9.
13. Shi Y and Eberhart RC (2001), Fuzzy Adaptive Particle Swarm Optimization, In Proceedings of the Congress on Evolutionary Computation 2001, Seoul, Korea, IEEE Service Center, IEEE (2001), pp. 101–106.
14. Clerc M and Kennedy J (2002), The Particle Swarm – Explosion, Stability, and Convergence in a Multidimensional Complex Space, IEEE Transactions on Evolutionary Computation, 6(1), 58–73.
15. Eberhart RC and Shi Y (2000), Comparing Inertia Weights and Constriction Factors in Particle Swarm Optimization, In Proceedings of IEEE International Congress on Evolutionary Computation, vol. 1, pp. 84–88.
16. van den Bergh F and Engelbrecht PA (2001), Effects of Swarm Size on Cooperative Particle Swarm Optimizers, In Proceedings of GECCO-2001, San Francisco, CA, pp. 892–899.
17. Ratnaweera A, Halgamuge SK, and Watson HC (2004), Self-Organizing Hierarchical Particle Swarm Optimizer with Time-Varying Acceleration Coefficients, IEEE Transactions on Evolutionary Computation, 8(3), 240–255.
18. Kennedy J (1999), Small Worlds and Mega-Minds: Effects of Neighborhood Topology on Particle Swarm Performance, In Proceedings of the 1999 Congress of Evolutionary Computation, vol. 3, IEEE Press, New York, pp. 1931–1938.
19. Kennedy J and Eberhart RC (1997), A Discrete Binary Version of the Particle Swarm Algorithm, In Proceedings of the 1997 Conference on Systems, Man, and Cybernetics, IEEE Service Center, Piscataway, NJ, pp. 4104–4109.
20. Løvbjerg M, Rasmussen TK, and Krink T (2001), Hybrid Particle Swarm Optimizer with Breeding and Subpopulations, In Proceedings of the Third Genetic and Evolutionary Computation Conference (GECCO-2001).

21. Krink T, Vesterstrøm J, and Riget J (2002), Particle Swarm Optimization with Spatial Particle Extension, In Proceedings of the IEEE Congress on Evolutionary Computation (CEC-2002).
22. Løvbjerg M and Krink T (2002), Extending Particle Swarms with Self-Organized Criticality, In Proceedings of the Fourth Congress on Evolutionary Computation (CEC-2002).
23. Miranda V and Fonseca N (2002), EPSO – Evolutionary Particle Swarm Optimization, a New Algorithm with Applications in Power Systems, In Proceedings of IEEE T&D AsiaPacific 2002 – IEEE/PES Transmission and Distribution Conference and Exhibition 2002: Asia Pacific, Yokohama, Japan, vol. 2, pp. 745–750.
24. Blackwell T and Bentley PJ (2002), Improvised Music with Swarms. In Proceedings of IEEE Congress on Evolutionary Computation 2002.
25. Robinson J, Sinton S, and Rahmat-Samii Y (2002), Particle Swarm, Genetic Algorithm, and Their Hybrids: Optimization of a Profiled Corrugated Horn Antenna, In Antennas and Propagation Society International Symposium, 2002, vol. 1, IEEE Press, New York, pp. 314–317.
26. Krink T and Løvbjerg M (2002), The Lifecycle Model: Combining Particle Swarm Optimization, Genetic Algorithms and Hill Climbers, In Proceedings of PPSN 2002, pp. 621–630.
27. Hendtlass T and Randall M (2001), A Survey of Ant Colony and Particle Swarm Meta-Heuristics and Their Application to Discrete Optimization Problems, In Proceedings of the Inaugural Workshop on Artificial Life, pp. 15–25.
28. vandenBergh F and Engelbrecht A (2004), A Cooperative Approach to Particle Swarm Optimization, IEEE Transactions on Evolutionary Computation 8(3), 225–239.
29. Parsopoulos KE and Vrahatis MN (2004), On the Computation of All Global Minimizers Through Particle Swarm Optimization, IEEE Transactions on Evolutionary Computation, 8(3), 211–224.
30. Price K, Storn R, and Lampinen J (2005), Differential Evolution – A Practical Approach to Global Optimization, Springer, Berlin Heidelberg New York.
31. Fan HY, Lampinen J (2003), A Trigonometric Mutation Operation to Differential Evolution, International Journal of Global Optimization, 27(1), 105–129.
32. Das S, Konar A, Chakraborty UK (2005), Two Improved Differential Evolution Schemes for Faster Global Search, ACM-SIGEVO Proceedings of GECCO' 05, Washington D.C., pp. 991–998.
33. Chakraborty UK, Das S and Konar A (2006), DE with Local Neighborhood, In Proceedings of Congress on Evolutionary Computation (CEC 2006), Vancouver, BC, Canada, IEEE Press, New York.
34. Das S, Konar A, Chakraborty UK (2005), Particle Swarm Optimization with a Differentially Perturbed Velocity, ACM-SIGEVO Proceedings of GECCO' 05, Washington D.C., pp. 991–998.
35. Salerno J (1997), Using the Particle Swarm Optimization Technique to Train a Recurrent Neural Model, IEEE International Conference on Tools with Artificial Intelligence, pp. 45–49.
36. van den Bergh F (1999), Particle Swarm Weight Initialization in Multi-Layer Perceptron Artificial Neural Networks, Development and Practice of Artificial Intelligence Techniques, Durban, South Africa, pp. 41–45.

37. He Z, Wei C, Yang L, Gao X, Yao S, Eberhart RC, and Shi Y (1998), Extracting Rules from Fuzzy Neural Network by Particle Swarm Optimization, In Proceedings of IEEE Congress on Evolutionary Computation (CEC 1998), Anchorage, Alaska, USA.
38. Eberhart RC and Hu X (1999), Human Tremor Analysis Using Particle Swarm Optimization, In Proceedings of the IEEE Congress on Evolutionary Computation (CEC 1999), Washington D.C., pp. 1927–1930.
39. Wachowiak MP, Smolíková R, Zheng Y, Zurada MJ, and Elmaghraby AS (2004), An Approach to Multimodal Biomedical Image Registration Utilizing Particle Swarm Optimization, IEEE Transactions on Evolutionary Computation, 8(3), 289–301.
40. Messerschmidt L and Engelbrecht AP (2004), Learning to Play Games Using a PSO-Based Competitive Learning Approach, IEEE Transactions on Evolutionary Computation 8(3), 280–288.
41. Yoshida H, Kawata K, Fukuyama Y, Takayama S, and Nakanishi Y (2000), A Particle Swarm Optimization for Reactive Power and Voltage Control Considering Voltage Security Assessment, IEEE Transactions on Power Systems, 15(4), 1232–1239.
42. Abido MA (2002), Optimal Design of Power System Stabilizers Using Particle Swarm Optimization, IEEE Transactions on Energy Conversion, 17(3), 406–413.
43. Paterlini S and Krink T (2006), Differential Evolution and Particle Swarm Optimization in Partitional Clustering, Computational Statistics and Data Analysis, vol. 50, 1220–1247.
44. Rogalsky T, Kocabiyik S and Derksen R (2000), Differential Evolution in Aerodynamic Optimization, Canadian Aeronautics and Space Journal, 46(4), 183–190.
45. Doyle S, Corcoran D, and Connell J (1999), Automated Mirror Design Using an Evolution Strategy, Optical Engineering, 38(2), 323–333.
46. Stumberger G, Dolinar D, Pahner U, and Hameyer K (2000), Optimization of Radial Active Magnetic Bearings Using the Finite Element Technique and Differential Evolution Algorithm, IEEE Transactions on Magnetics, 36(4), 1009–1013.
47. Wang FS and Sheu JW (2000), Multi-Objective Parameter Estimation Problems of Fermentation Processes Using High Ethanol Tolerance Yeast, Chemical Engineering Science, 55(18), 3685–3695.
48. Masters T and Land W (1997), A New Training Algorithm for the General Regression Neural Network," Proceedings of Computational Cybernetics and Simulation, Organized by IEEE Systems, Man, and Cybernetics Society, 3, 1990–1994.
49. Zelinka I and Lampinen J (1999), An Evolutionary Learning Algorithms for Neural Networks, In Proceedings of Fifth International Conference on Soft Computing, MENDEL'99, pp. 410–414.
50. Das S, Abraham A, and Konar A (2007), Adaptive Clustering Using Improved Differential Evolution Algorithm, IEEE Transactions on Systems, Man and Cybernetics – Part A, IEEE Press, New York, USA.
51. Das S and Konar A (2007), A Swarm Intelligence Approach to the Synthesis of Two-Dimensional IIR Filters, Engineering Applications of Artificial Intelligence, 20(8), 1086–1096. http://dx.doi.org/10.1016/j.engappai.2007.02.004

52. Das S and Konar A (2006), Two-Dimensional IIR Filter Design with Modern Search Heuristics: A Comparative Study, International Journal of Computational Intelligence and Applications, 6(3), Imperial College Press.
53. Mastorakis N, Gonos IF, Swamy MNS (2003), Design of Two-Dimensional Recursive Filters Using Genetic Algorithms, IEEE Transactions on Circuits and Systems, 50, 634–639.
54. Liu H, Abraham A, and Clerc M (2007), Chaotic Dynamic Characteristics in Swarm Intelligence, Applied Soft Computing Journal, Elsevier Science, 7(3), 1019–1026.
55. Abraham A, Liu H, and Chang TG (2006), Variable Neighborhood Particle Swarm Optimization Algorithm, Genetic and Evolutionary Computation Conference (GECCO-2006), Seattle, USA, Late Breaking Papers, CD Proceedings, Jörn Grahl (Ed.).
56. Abraham A, Das S, and Konar A (2007), Kernel Based Automatic Clustering Using Modified Particle Swarm Optimization Algorithm, 2007 Genetic and Evolutionary Computation Conference, GECCO 2007, ACM Press, Dirk Thierens et al. (Eds.), ISBN 978-1-59593-698-1, pp. 2–9.

Web Services, Policies, and Context: Concepts and Solutions

Zakaria Maamar[1], Quan Z. Sheng[2], Djamal Benslimane[3], and Philippe Thiran[4]

[1] Zayed University, U.A.E., zakaria.maamar@zu.ac.ae
[2] The University of Adelaide, Australia, qsheng@cs.adelaide.edu.au
[3] Claude Bernard Lyon 1 University, France, djamal.benslimane@liris.cnrs.fr
[4] Louvain School of Management & University of Namur, Belgium, pthiran@fundp.ac.be

1 Introduction

Despite the extensive adoption of Web services by IT system developers, they still lack the capabilities that could enable them to match and eventually surpass the acceptance level of traditional integration middleware (e.g., CORBA, Java RMI). This lack of capabilities is to a certain extent due to the *trigger-response* interaction pattern that frames the exchanges of Web services with third parties. Adhering to this interaction pattern means that a Web service only performs the requests it receives without considering its internal execution state, or even questioning if it would be rewarded for performing these requests (e.g., to be favored over similar Web services during selection). There exist, however, several situations that insist on Web services self-management so that *scalability*, *flexibility*, and *stability* requirements are satisfied.

The objective of this chapter is to discuss the value-added of integrating *context* and *policies* into a *Web services* composition approach.

Web services offer new opportunities to deploy B2B applications, which tend to crosscut companies' boundaries. Web services are independent from specific platforms and computing paradigms, and have the capacity to form high-level business processes referred to as composite Web services [3].

Policies are considered as external, dynamically modifiable rules and parameters that are used as input to a system [13]. This permits to the system to adjust to administrative decisions and changes in the execution environment. In the field of Web services, policies are intended to specify different aspects of the behavior of a Web service, so this one can align its capabilities to users' requirements and resources' constraints.

Context "*... is not simply the state of a predefined environment with a fixed set of interaction resources. It is part of a process of interacting with an ever-changing environment composed of reconfigurable, migratory, distributed,*

Z. Maamar et al.: *Web Services, Policies, and Context: Concepts and Solutions*, Studies in Computational Intelligence (SCI) **116**, 39–55 (2008)
www.springerlink.com

and multiscale resources" [4]. In the field of Web services, context is used to facilitate the development and deployment of flexible Web services. Flexibility refers to a Web service that selects appropriate operations based on the requirements of the business scenario that this Web service implements.

While context and policies are separately used for different needs of Web services, this chapter discusses their role in framing the composition process of Web services. We propose a three-level approach to compose Web services. This approach does not only make Web services bind to each other, but emphasizes the cornerstone of refining this binding at the following levels: *component* level ($\mathcal{W}$-level) to deal with Web services' definitions and capabilities, *composite* level ($\mathcal{C}$-level) to address how Web services are discovered and combined (semantic mediation is discarded), and finally, *resource* level ($\mathcal{R}$-level) to focus on the performance of Web services.

The role of policies and context to support the composition of Web services is depicted as follows:

- Policies manage the transitions between the three levels. Going from one level to another direct level requires policy activation. Two types of policies are put forward: *engagement* policies manage the participations of Web services in compositions, and *deployment* policies manage the interactions of Web services with computing resources.
- Context provides information on the environment wherein the composition of Web services occurs. Because of the three levels in the approach, three types of context are defined, namely $\mathcal{W}/\mathcal{C}/\mathcal{R}$-context standing for context of $\mathcal{W}$eb service, context of $\mathcal{C}$omposite Web service, and context of $\mathcal{R}$esource.

The rest of this chapter is organized as follows. Section 2 presents the approach to compose Web services. Section 3 discusses the impact of policies on Web services and specifies the policies for the behavior of Web services. Section 4 is about exception handling. Section 5 reviews some related works. Finally, Sect. 5 concludes the chapter.

2 The Proposed Composition Approach

2.1 Presentation

Figure 1 illustrates our approach to compose Web services. The figure reads as follows: bottom-up during normal progress of composition and top-down during exception in composition. The underlying idea is that context coupled to policies handle the specification and execution of a composite Web service. Three levels form this approach. The component level is concerned with the definition of Web services and their announcements to composite Web services that are located in the composition level. This level is concerned with the way component Web services are discovered (based on their capabilities)

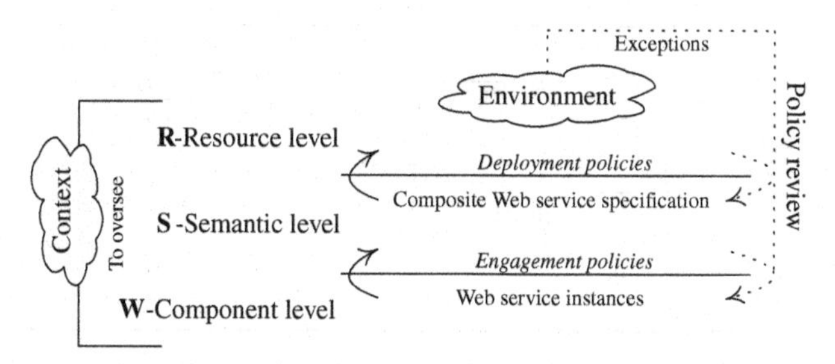

Fig. 1. Approach to compose Web services

and assembled (according to particular collaboration patterns). Finally, the resource level focuses on the performance of a composite Web service. This performance is subject to the availability of resources and the commitments of these resources to other concurrent composite Web services.

In Fig. 1, policies support level transitions. Transiting from one level to a higher direct-level is subject to policy execution. This execution uses the information that context provides on different elements like number of Web services that are currently running, composite Web services that are under preparation, etc. Two types of policies are shown in Fig. 1: engagement and deployment. While policies support level transitions (i.e., how a composite Web service specification progresses), context oversees first, the actions that occur within and across these levels and second, the elements like Web service, composite Web service, or resource, that are involved by these actions. Prior to triggering any policy and executing any action, the information that context provides is validated.

Any deviation from the specification of a composite Web service like execution delays, raises exceptions. This leads to reviewing the different actions of the policies that were carried out within and across the levels. The review process occurs in a descending way commencing with resource, composite, and then, component levels (Fig. 1). Transiting to a lower level means that the current level is not faulty and extra investigations of the exception are deemed appropriate at this lower level. To keep track of the actions of policies that happened, are happening, and might happen, hence corrective measures (e.g., compensation policies) are scheduled and undertaken as part of the review process, context monitors the whole composition of Web services. Additional details on exception handling are discussed in Sect. 4.

2.2 Description of the Three Levels

The W-Component level shows the Web services that participate in a composite Web service. This participation is upon approval and happens in accordance with the Web services *instantiation* principle that was introduced in [15]. This principle emphasizes the simultaneous participation of a Web service in several compositions.

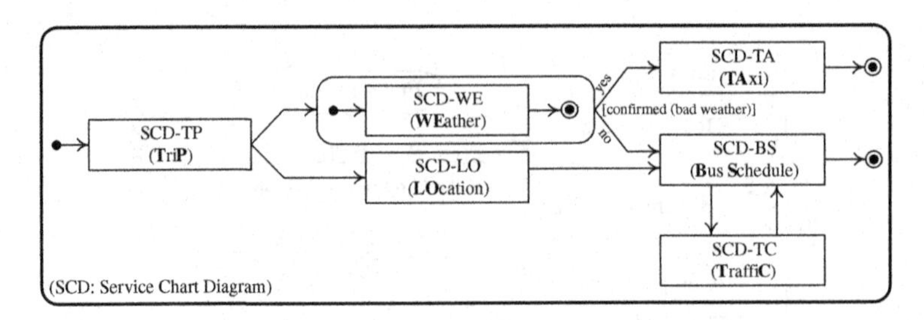

Fig. 2. Sample of a composite Web service specification

Before a Web service accepts the invitation to participate in a composite Web service, it consults its $\mathcal{W}$-context so that it can verify the number of current participations vs. the maximum number of participations, the expected completion time of its other participations, and the features of the newly received invitation of participation like when the new Web service instance is required. It happens that a Web service refuses an invitation of participation for various reasons like risk of overload.

The $\mathcal{C}$-Composite level is about the specification of a composite Web service in term of business logic. Figure 2 shows a simple specification that combines service chart diagrams and state chart diagrams [12]. The component Web services of this specification are: trip (`TP`), weather (`WE`), location (`LO`), taxi (`TA`), bus schedule (`BS`), and traffic (`TC`). The respective service chart diagrams of these components are connected through transitions. Some of these transitions have constraints to satisfy like [confirmed(bad weather)]. A composite Web service consults its $\mathcal{C}$-context so that it can follow up the execution progress of its specification. This enables, for example, identifying the next Web services to be subject to invitation of participation and the status of each component Web service of the composite Web service (e.g., initiated, running, suspended, resumed). Figure 3 illustrates the editor we developed for specifying composite Web services. The editor provides means for directly manipulating service chart diagrams, states, and transitions using drag and drop actions.

The $\mathcal{R}$-Resource level represents the computing facilities upon which Web services operate. Scheduling the execution requests of Web services is prioritized when sufficient resources are unavailable to satisfy all these requests at once. A Web service requires resources for different operations like self-assessment of current state before accepting/rejecting participations in compositions, and satisfying users' needs upon request. Before a resource accepts supporting the execution of an additional Web service, it consults its $\mathcal{R}$-context so that it can verify the number of Web services currently executed vs. the maximum number of Web services under execution, the approximate completion time of the ongoing executions of Web services, and the features of the newly received request like requested time of execution. Like with Web

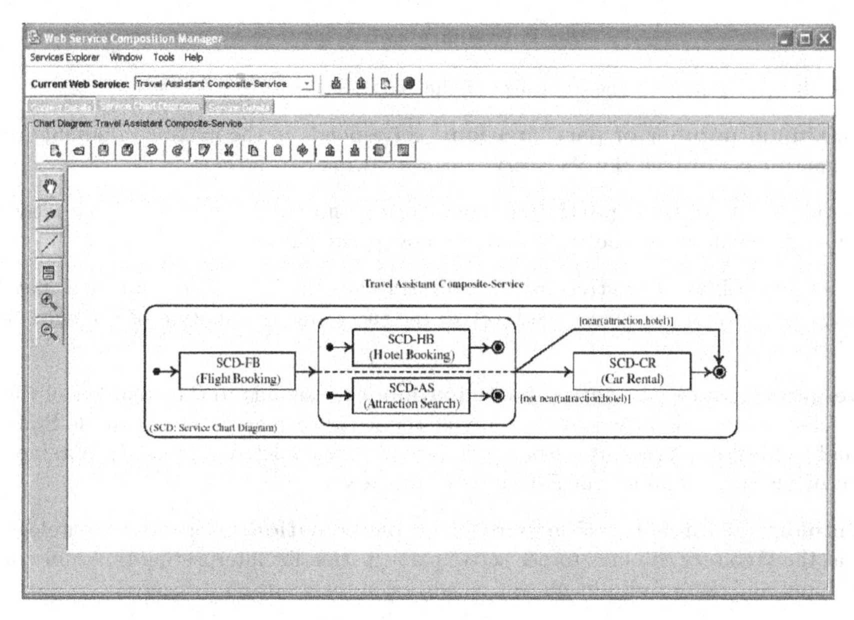

Fig. 3. Editor for composite Web service specification

services, similar considerations apply to resources when it comes to turning down execution requests of Web services.

2.3 Description of the Three Contexts

To comply with the three levels of the composition approach, three types of context are defined: $\mathcal{W}$-context, $\mathcal{C}$-context, and $\mathcal{R}$-context (Fig. 1).

The $\mathcal{W}$-*context* of a Web service returns information on the participations of the Web service in different compositions. These participations are made possible because of the Web services instantiation principle. $\mathcal{W}$-context has the following arguments (Table 1): label, maximum number of participations, number of active participations, next possibility of participation, resource&state per active participation, local ontology per active participation, previous Web services per active participation, current Web services per active participation, next Web services per active participation, regular actions, reasons of failure per active participation, corrective actions per failure type and per active participation, and date. Figure 4a, b shows how context arguments are instantiated at run-time.

The $\mathcal{C}$-*context* of a composite Web service is built upon the $\mathcal{W}$-contexts of its component Web services and permits overseeing the progress of a composition.

The $\mathcal{R}$-*context* of a resource oversees the execution of the Web services that operate on top of this resource prior it accepts additional Web services for execution. A resource has computation capabilities that continuously change

Table 1. Arguments of $\mathcal{W}$-context

Label: corresponds to the identifier of the Web service

Maximum number of participations: corresponds to the maximum number of compositions in which the Web service can participate at a time

Number of active participations: corresponds to the number of active compositions in which the Web service is now participating

Next possibility of participation: indicates when the Web service can participate in a new composition. This is subject to the successful termination of the current active participations

Resource&State per active participation: corresponds to the identifier of the selected resource and the state of the Web service in each active composition. State can be of types in-progress, suspended, aborted, or completed, and will be obtained out of the state argument of $\mathcal{R}$-context of this resource

Ontology of interpretation per active participation: refers to the ontology that the Web service binds to per active participation for interpreting the values of its input and output arguments when conversion functions are triggered

Previous Web services per active participation: indicates the Web services that were successfully completed before the Web service per active composition (null if there are no predecessors)

Current Web services per active participation: indicates the Web services that are concurrently being performed with the Web service per active composition (null if there is no concurrent processing)

Next Web services per active participation: indicates the Web services that will be executed after the Web service successfully completes its execution per active composition (null if there are no successors)

Regular actions: illustrates the actions that the Web service normally performs

Reasons of failure per active participation: informs about the reasons that are behind the failure of the execution of the Web service per active composition

Corrective actions per failure type and per active participation: illustrates the actions that the Web service has performed due to execution failure per active composition

Date: identifies the time of updating the arguments above

depending on the number[1] of Web services that are now under execution and the execution duration per Web service. Due to lack of space, the structures of $\mathcal{C}$-context and $\mathcal{R}$-context are not presented.

[1] Number is used for illustration purposes. Additional criteria could be execution load of Web services.

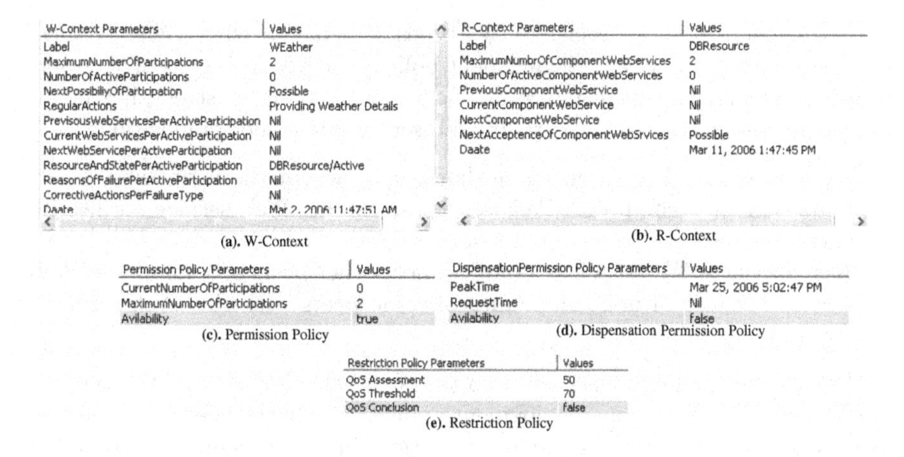

W-Context Parameters	Values
Label	WEather
MaximumNumberOfParticipations	2
NumberOfActiveParticipations	0
NextPossibiliyOfParticipation	Possible
RegularActions	Providing Weather Details
PrevisousWebServicesPerActiveParticipation	Nil
CurrentWebServicesPerActiveParticipation	Nil
NextWebServicePerActiveParticipation	Nil
ResourceAndStatePerActiveParticipation	DBResource/Active
ReasonsOfFailurePerActiveParticipation	Nil
CorrectiveActionsPerFailureType	Nil
Daate	Mar 2, 2006 11:47:51 AM

(a). W-Context

R-Context Parameters	Values
Label	DBResource
MaximumNumbrOfComponentWebServices	2
NumberOfActiveComponentWebServices	0
PreviousComponentWebService	Nil
CurrentComponentWebService	Nil
NextComponentWebService	Nil
NextAcceptenceOfComponentWebSrvices	Possible
Daate	Mar 11, 2006 1:47:45 PM

(b). R-Context

Permission Policy Parameters	Values
CurrentNumberOfParticipations	0
MaximumNumberOfParticipations	2
Avilability	true

(c). Permission Policy

DispensationPermission Policy Parameters	Values
PeakTime	Mar 25, 2006 5:02:47 PM
RequestTime	Nil
Avilability	false

(d). Dispensation Permission Policy

Restriction Policy Parameters	Values
QoS Assessment	50
QoS Threshold	70
QoS Conclusion	false

(e). Restriction Policy

Fig. 4. Illustration of context and policy specification

2.4 Description of the Two Policies

Policies permit during a normal progress scenario to move a composition from one level to a higher-direct level (Fig. 1). Policies' outcomes are also subject to review in case of exceptions, i.e., abnormal progress.

Engagement policies frame the participation of component Web services in composite Web services. This highlights the opportunity of a Web service to take part in several compositions at a time. Since Web services are context-aware, they assess their participations and use the engagement policies to back their either acceptance or denial of participation in additional compositions.

Deployment policies frame the interactions between component Web services and computing resources. We earlier argued that resources have limited computation capabilities, and scheduling execution requests of component Web services is deemed appropriate. Since resources are context-aware, they assess their commitments and use the deployment policies to back their decisions of either accepting or denying a component Web service's execution request.

3 Role of Policies

3.1 Behavioral Web Services

In the composition approach of Fig. 1, policies have an impact on first, the behavior that Web services expose towards composite Web services and second, the behavior that composite Web services (through their component Web services) expose towards computing resources. We identify this behavior with the following attributes: permission and restriction.

Engagement policies associate two types of behavior with a Web service: permission and restriction. Engagement policies satisfy the needs of Web services' providers who have interests and/or obligations in strengthening or restricting the participation of their Web services in some compositions.

- *Permission.* A Web service is authorized to accept the invitation of participation in a composite Web service once this Web service verifies its current status.
- *Restriction.* A Web service cannot connect other peers of a composite Web service because of non-compliance of some peers with this Web service's requirements.

Deployment policies associate two types of behavior with a Web service: permission and restriction. Deployment policies satisfy the needs of resources' providers who have interests and/or obligations in strengthening or restricting the execution of Web services on these resources.

- *Permission.* A Web service receives an authorization of execution from a resource once this resource checks its current state.
- *Restriction.* A Web service is not accepted for execution on a resource because of non-compliance of this Web service with the resource's requirements.

3.2 Specification of Policies

In what follows, the policies that define the behavior of Web services are specified. To this end, a policy definition language is required. In this chapter, the Web Services Policy Language (WSPL) is used [1]. The syntax of WSPL is strictly based on the OASIS eXtensible Access Control Markup Language (XACML) standard[2]. Figure 4c–e shows how policies are instantiated at run-time.

Engagement Policy

Figure 5 illustrates the way the different behaviors of a Web service are connected together based on the execution outcome of the engagement policy per type of behavior. For instance, a positive permission (i.e., yes) is not confirmed until no restrictions are identified (i.e., no). More details on this figure are given in the description of each type of behavior. The dashed lines in Fig. 5 represent potential cases of exceptions to be discussed in Sect. 4.

1. An engagement policy for permission consists of authorizing a Web service to be part of a composite Web service. The authorization is based

[2] www.oasis-open.org/committees/download.php/2406/oasis-xacml-1.0.pdf.

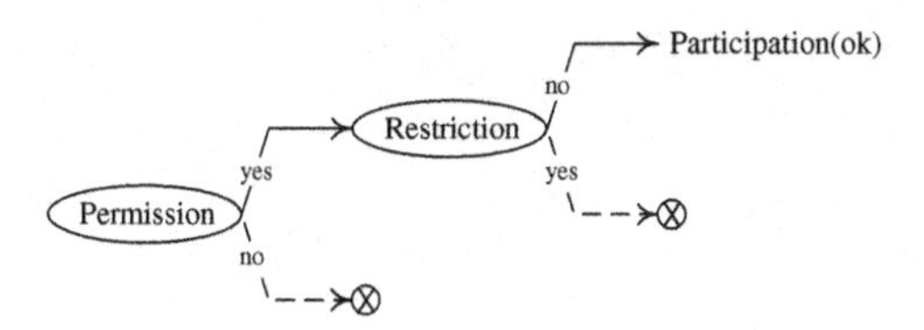

Fig. 5. Behaviors of a Web service in an engagement scenario

on the status of the Web service that is known through $\mathcal{W}$-context. The following illustrates an engagement policy for permission in WSPL. It states that a Web service participates in a composition subject to evaluating `<Condition>` to true. This latter refers to some arguments like the number of current active participations of the Web service in compositions and the next possibility of participation of the Web service in additional compositions. These arguments are known as vocabulary items in WSPL. In the policy, `<TrueConclusion>` shows the permission of participation, whereas `<FalseConclusion>` shows the contrary. In case of positive permission, `yes-permission-participation` procedure is executed, which results in updating the following arguments in $\mathcal{W}$-context of the Web service: *number of active participations*, *previous Web services*, *current Web services*, and *next Web services*. Figure 6 shows the outcome of executing an engagement policy of type permission. This policy grants a Web service the permission to take part in a composite Web service.

```
Policy (Aspect="PermissionParticipation") {
<Rule xmlns="urn:oasis:names:tc:xacml:3.0:permission:policy:schema:wd:01"
xmlns:proc="permission-participation" RuleId="PermissionParticipationWS">
  <Condition>
    <Apply FunctionId="and">
    <Apply FunctionId="integer-less-than" DataType="boolean">
    <SubjectAttributeDesignator AttributeId="CurrentNumberOfParticipations"
    DataType="integer"/>
    <SubjectAttributeDesignator AttributeId="MaximumNumberOfParticipations"
    DataType="integer"/>
    </Apply>
    <SubjectAttributeDesignator AttributeId="Availability" DataType="boolean"/>
    </Apply>
  </Condition>
  <Conclusions>
    <TrueConclusion>
    <proc:do> yes-permission-participation </proc:do>
    </TrueConclusion>
    <FalseConclusion>
    <proc:do> no-permission-participation </proc:do>
    </FalseConclusion>
  </Conclusions>
</Rule>}
```

2. An engagement policy for restriction consists of preventing a Web service from taking part in a composite Web service. A restriction does not implement a negative permission of participation. However, it follows a positive permission of participation (Fig. 5). Restrictions could be geared towards

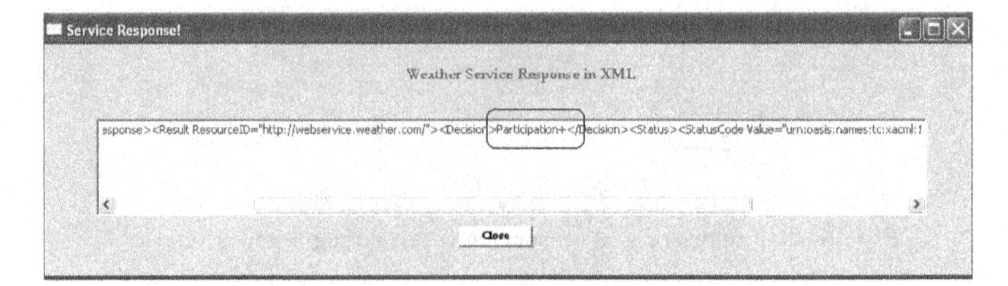

Fig. 6. Outcome of policy execution

the quality issue (e.g., time of response, reputation) of the component Web services with whom a Web service will interact. For example, a Web service's provider has only interest in the Web services that have a "good" QoS record [16]. The following illustrates a restriction policy in WSPL. It states that a Web service can be restricted from participation subject to evaluating `<Condition>` to true. This latter checks that a positive permission of participation exists and the assessment level of the QoS of the Web services is low. These Web services are identified using *previous Web services per active participation* argument of $\mathcal{W}$-context of the Web service. In this policy, `QoSAssessment` is an integer value that is the result of evaluating the QoS of a Web service, and `QoSThreshold` is the minimum QoS assessment value that is acceptable for composition.

```
Policy (Aspect="RestrictionPermission") {
    <Rule xmlns="urn:oasis:names:tc:xacml:3.0:generalization:policy:schema:wd:01"
    RuleId="RestrictionParticipationWS">
      <Condition>
        <Apply FunctionId="and">
        <SubjectAttributeDesignator AttributeId="YesPermissionParticipation">
        DataType="boolean"/>
        <Apply FunctionId="integer-great-than-or-equal">
        <SubjectAttributeDesignator AttributeId="QoSAssessment" DataType="integer"/>
        <SubjectAttributeDesignator AttributeId="QoSThreshold" DataType="integer"/>
        </Apply>
        </Apply>
      </Condition>
      <Conclusions>
        <TrueConclusion RestrictionParticipation = "No"/>
        <FalseConclusion RestrictionParticipation = "Yes"/>
      </Conclusions>
</Rule>}
```

Deployment Policy

Figure 7 illustrates the way the different behaviors of a Web service are connected together based on the execution outcome of the deployment policy per type of behavior. For instance, a positive permission for execution (i.e., yes) is confirmed if there are no restrictions that could deny this permission at run time. More details about this figure are given in the description of each type

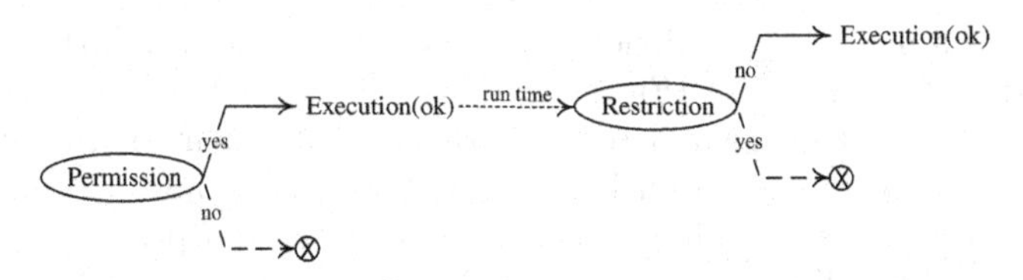

Fig. 7. Behaviors of a Web service in a deployment scenario

of behavior. The dashed lines in Fig. 7 represent potential cases of exceptions and are discussed in Sect. 4.

1. A deployment policy for permission is about a Web service that receives the necessary execution authorizations from a resource. These authorizations are based on the state of the resource, which is reflected through $\mathcal{R}$-context. The following illustrates a deployment policy for permission in WSPL. It states that a resource accepts the execution request of a Web service subject to evaluating `<Condition>` to true. This latter refers to some arguments like the number of active component Web services that the resource supports their execution and the next acceptance of the resource for additional component Web services. In the policy, `<TrueConclusion>` shows the permission of execution, whereas `<FalseConclusion>` shows the contrary. In case of positive permission of execution, `yes-permission-deployment` procedure is executed, which results in updating the following arguments: *resource&state per active participation* of $\mathcal{W}$-context of the Web service and *number of active component Web services* of $\mathcal{R}$-context of the resource.

```
Policy (Aspect="PermissionDeployment") {
<Rule xmlns="urn:oasis:names:tc:xacml:3.0:permission:policy:schema:wd:01"
xmlns:proc="permission-deployment" RuleId="PermissionDeploymentWS">
  <Condition>
    <Apply FunctionId="and">
    <Apply FunctionId="integer-less-than" DataType="boolean">
    <SubjectAttributeDesignator AttributeId="NumberofActiveComponentWebServices"
    DataType="integer"/>
    <SubjectAttributeDesignator AttributeId="MaximumNumberofComponentWebServices"
    DataType="integer"/>
    </Apply>
    <SubjectAttributeDesignator AttributeId="NextAcceptanceofComponentWebServices"
    DataType="boolean"/>
    </Apply>
  </Condition>
  <Conclusions>
    <TrueConclusion>
    <proc:do> yes-permission-deployment </proc:do>
    </TrueConclusion>
    <FalseConclusion>
    <proc:do> no-permission-deployment </proc:do>
    </FalseConclusion>
  </Conclusions>
</Rule>}
```

2. A deployment policy for restriction consists of preventing a Web service form being engaged in an execution over a resource. A restriction does not

implement a negative permission of deployment, rather it follows a positive permission of deployment (Fig. 5). Besides the example of resource failure, restrictions could be geared towards the reinforcement of the execution features that were agreed on between a Web service and a resource. For example a Web service binds a resource for execution prior to the scheduled time. The following illustrates a deployment policy for restriction in WSPL. It states that a Web service can be restricted from execution subject to evaluating `<Condition>` to true. This latter checks that a positive permission of execution has been issued and the agreed execution time is valid. The execution time of a Web service is identified using *next component Web services per active participation* argument of $\mathcal{R}$-context of the resource.

```
Policy (Aspect="RestrictionDeployment") {
<Rule xmlns="urn:oasis:names:tc:xacml:3.0:generalization:policy:schema:wd:01"
RuleId="RestrictionDeploymentWS">
  <Condition>
    <Apply FunctionId="and">
    <SubjectAttributeDesignator AttributeId="YesPermissionDeployment"
    DataType="boolean"/>
    <Apply FunctionId="equal" DataType="boolean">
    <SubjectAttributeDesignator AttributeId="ExecutionTime" DataType="String"/>
    </Apply>
    </Apply>
  </Condition>
  <Conclusions>
    <TrueConclusion RestrictionDeployment = "No"/>
    <FalseConclusion RestrictionDeployment = "Yes"/>
  </Conclusions>
</Rule>}
```

4 Exception Handling

4.1 Rationale

In [22], Yu et al. adopt the example of a programmer who writes a BPEL business process. The complexity of some business domains results sometimes in processes with errors. When a process is put into operation, exception handling is to occur otherwise, angry customers and loss of revenues arise. In [19], Russell et al. looked into the range of issues that may result in exceptions during workflow execution and the various ways these issues can be addressed. This work classifies exceptions using patterns. Some of the exception types that are inline with what is handled in the proposed composition approach include work item failure, deadline expiry, and resource unavailability. In the composition approach, exception handling means reviewing policies' outcomes and taking appropriate actions in a top-down manner by consulting the three levels shown in Fig. 1. An exception could take place at one of the following levels:

- *Resource level.* There is no guarantee that a particular resource is up at the execution time of a Web service. A resource could be down due to power failure.
- *Composite level.* There is no guarantee that the specification of a composite Web service is error-free. Conflicting actions like concurrent accept and reject and deadlocks may occur during this specification execution.
- *Component level.* There is no guarantee that a particular Web service is still available at execution-time. A provider could withdraw its Web service from a composition without prior notice.

4.2 Exception Types per Policy Type

Deployment policy. Permission and restriction show the behaviors of a Web service following deployment-policy triggering. Potential origins of exception are as follows. First, an execution permission for a Web service over a resource was not issued but there was tentative of execution over this resource. Second, an execution restriction on a Web service was detected but that was not properly reported.

Exceptions in case of permission could arise because of the inaccurate assessment of some arguments in $\mathcal{R}$-context of a resource. These arguments are as follows:

- *Number of active component Web services* argument. A resource did not correctly assess the exact number of Web services that currently run on top of it. This number should not exceed the maximum number of authorized Web services for execution.
- *Next acceptance of component Web services* argument. The execution of a Web service got disrupted, which has negatively affected the execution scheduling of the forthcoming Web services over a resource.

Exceptions in case of restriction could arise because of the inaccurate assessment of some arguments in a resource's $\mathcal{R}$-context like *next component Web services per active participation.* The execution of a Web service should not occur prior to the time that was agreed upon between the resource and the Web service.

Engagement policy. Permission and restriction show the behaviors of a Web service following the triggering of an engagement policy. The dashed lines in Fig. 5 represent the potential origins of exceptions. First, a participation permission for a Web service was not issued, but there was tentative of participation. Second, a participation restriction on a Web service was detected, but the restriction has not been taken effectively.

Exceptions in case of permission could arise because of the inaccurate assessment of some arguments in a Web service's $\mathcal{W}$-context:

- *Number of active participations* argument. The Web service did not exactly assess its current participations in compositions. This number should not exceed the maximum number of authorized compositions.

In the engagement policy for permission given in Sect. 3.2, the condition of participation is shown with the following statements.

```
<Apply FunctionId="integer-less-than" ... >
<SubjectAttributeDesignatorAttributeId="CurrentNumberOfParticipations" .../>
<SubjectAttributeDesignator AttributeId="MaximumNumberOfParticipations" .../>
</Apply>
```

Since there is no permission of participation the invocation of the Web service fails due to unavailability. A policy needs to be triggered so, actions are taken such as reviewing the current number of active participations of the Web service and notifying the relevant composite Web service about the disengagement of this Web service.

– *Next possibility of participation* argument. The execution of a Web service got delayed, which has negatively affected the participation scheduling of the Web service in forthcoming compositions.

Exceptions in case of restriction could arise because of the inaccurate assessment of some arguments in a Web service's $\mathcal{W}$-context like *next Web services per active participation.* A Web service cannot connect other peers if they are not listed in this argument.

5 Related Work

This related work is presented from two perspectives: context for Web services and policies for Web services.

In literature the first time "context-aware" appeared was in [21]. Schilit and Theimer describe context as location, identities of nearby people, objects and changes in those objects. Dey defines context as the user's emotional state, focus of attention, location and orientation, date and time, as well as objects and people in the user's environment [7]. These kinds of definitions are often too wide. One of the most accurate definitions is given by Dey and Abowd [6]. They refer to context as "*any information that can be used to characterize the situation of entities (i.e., whether a person, place or object) that are considered relevant to the interaction between a user and an application, including the user and the application themselves*".

Making Web services context-aware is not straightforward; many issues are to be addressed (adapted from [20]): how is context structured, how does a Web service bind context, how are changes detected and assessed for context update purposes, and what is the overload on a Web service of taking context into account? Responses to some of these questions have guided Maamar et al. to organize a Web service's context along three interconnected perspectives [14]. The participation perspective is about overseeing the multiple composition scenarios in which a Web service concurrently takes part. The execution perspective is about looking for the computing resources upon which a Web service operates, and monitoring the capabilities of these computing resources so that the Web service's requirements are constantly satisfied. Finally, the preference perspective is about ensuring that user preferences (e.g., execution time at 2 pm) are integrated into the specification of a composite Web service.

Some research projects have focused on how to facilitate the development and deployment of context-aware and adaptable Web services. Standard Web services descriptions are augmented with context information (e.g., location, time, user profile) and new frameworks are developed to support this. The approach proposed in [17] is intended to provide an enhancement of WSDL language with context aware features. The proposed Context-based Web Service Description Language (CWSDL) adds to WSDL a new part called Context Function which is used in order to select the best service among several ones. This function represents the sensitivity of the service to context information. Another interesting approach was proposed in [11] to deal with context in Web services. The approach consists of two parts: a context infrastructure and a context type set. The context infrastructure allows context information to be transmitted as a SOAP header-block within the SOAP messages.

Kagal et al. argue that policies should be part of the representation of Web services, and in particular of semantic Web services [10]. Policies provide the specification of who can use a service under what conditions, how information should be provided to the service, and how provided information will be used later. In [2], Baresi et al. proposed a policy-based approach to monitor Web services' functional (e.g., constraints on exchanged data) and non-functional (e.g., security, reliability) requirements. In this approach Baresi et al. report on the different types of policies that can be defined along the life cycle of a Web service [18]. These types are service policies, server policies, supported policies, and requested policies.

Desai et al. have recognized the importance of separation of concerns from a software engineering perspective [5]. Indeed, they split a business process into two parts: protocol and policy. The protocol part is a modular, public specification of an interaction among different roles that collaborate towards achieving a desired goal. The policy part is a private description of a participant's business logic that controls how this participant takes part in a protocol.

Policies have also been used to access and use resources in other fields. In grid computing, Dumitrescu et al. describe two strategies that a virtual organization must deploy to ensure grid-wide resource policy enforcement in a decentralized manner [8]. In the semantic Web, Gavriloaie et al. present an approach to manage access to sensitive resources using PeerTrust, a language for expressing access control policies [9]. PeerTrust is based on guarded distributed logic programs and has been demonstrated for automated trust negotiation.

6 Conclusion

In this chapter we proposed an approach to compose Web services. The role of policies and context in this composition has been depicted as follows.

- Policies manage the transitions between component, composite, and resource levels. Transiting from one level to another requires activating policies. Two types of policies were developed: engagement policies assess a Web service participation in a given composition, and deployment policies manage the interactions between component Web services and computing resources.
- Context provides useful information concerning the environment wherein the composition of Web services occurs. Context caters the necessary information that enables tracking the whole composition process by enabling for instance to trigger the appropriate policies and to regulate the interactions between Web services according to the current status of the environment. Three types of context were defined, namely $\mathcal{W}/\mathcal{C}/\mathcal{R}$-context standing for context of Web service, context of composite Web service, and context of resource.

References

1. A. H. Anderson. Predicates for Boolean Web Service Policy Language (SWPL). In *Proceedings of the International Workshop on Policy Management for the Web (PM4W'2005) Held in Conjunction with The Fourteenth International World Wide Web Conference (WWW'2005)*, Chiba, Japan, 2005.
2. L. Baresi, S. Guinea, and P. Plebani. WS-Policy for Service Monitoring. In *Proceedings of the Sixth VLDB Workshop on Technologies for E-Services (TES'2005) Held in Conjunction with the 31st International Conference on Very Large Data Bases (VLDB'2005)*, Trondheim, Norway, 2005.
3. I. Budak Arpinar, B. Aleman-Meza, R. Zhang, and A. Maduko. Ontology-driven Web services composition platform. In *Proceedings of the IEEE International Conference on E-Commerce Technology (CEC'2004)*, San-Diego, USA, 2004.
4. J. Coutaz, J. L. Crowley, S. Dobson, and D. Garlan. Context is Key. *Communications of the ACM*, 48(3), March 2005.
5. N. Desai, A. U. Mallya, A. K. Chopra, and M. P. Singh. Processes = Protocols + Policies: A Methodology for Business Process Development. In *Proceedings of the 14th International World Wide Web Conference (WWW'2005)*, Chiba, Japan, 2005.
6. A. K. Dey and G. D. Abowd. Towards a Better Understanding of Context and Context-Awareness. In *Proceedings of the Workshop on the What, Who, Where, When, and How of Context-Awareness Held in Conjuction with CHI'2000*, The Hague, The Netherlands, 2000.
7. Anind K. Dey. Context-Aware Computing: The CyberDesk Project. In *Proceedings of the AAAI'98 Spring Symposium on Intelligent Environments (AAAI'1998)*, Menlo Park, CA, USA, 1998.

8. C. Dumitrescu, M. Wilde, and I. T. Foster. A model for usage policy-based resource allocation in grids. In *Proceedings of the Sixth IEEE International Workshop on Policies for Distributed Systems and Networks (POLICY'05)*, New-York, USA, 2005.
9. R. Gavriloaie, W. Nejdl, D. Olmedilla, K. Seamons, and M. Winslett. No registration needed: how to use declarative policies and negotiation to access sensitive resources on the semantic Web. In *Proceedings of The First European Semantic Web Symposium on The Semantic Web: Research and Applications (ESWS'2004)*, Heraklion, Crete, Greece, 2004.
10. L. Kagal, M. Paolucci, N. Srinivasan, G. Denker, and K. Finin, T. Sycara. Authorization and privacy for semantic Web services. *IEEE Intelligent Systems*, 19(4), July/August 2004.
11. M. Keidl and A. Kemper. A framework for context-aware adaptable web services. In *EDBT*, pages 826–829, 2004.
12. Z. Maamar, B. Benatallah, and W. Mansoor. Service chart diagrams – description & application. In *Proceedings of the Alternate Tracks of The 12th International World Wide Web Conference (WWW'2003)*, Budapest, Hungary, 2003.
13. Z. Maamar, D. Benslimane, and A. Anderson. Using policies to manage Composite Web Services? *IEEE IT Professional*, 8(5), Sept./Oct. 2006.
14. Z. Maamar, D. Benslimane, and Nanjangud C. Narendra. What Can Context do for Web Services? *Communications of the ACM*, Volume 49(12), December 2006.
15. Z. Maamar, S. Kouadri Mostéfaoui, and H. Yahyaoui. Towards an agent-based and context-oriented approach for Web services composition. *IEEE Transactions on Knowledge and Data Engineering*, 17(5), May 2005.
16. D. A. Menascé. QoS Issues in Web Services. *IEEE Internet Computing*, 6(6), November/December 2002.
17. S. Kouadri Mostéfaoui and B. Hirsbrunner. Towards a context-based service composition framework. In *ICWS*, pages 42–45, 2003.
18. N. Mukhi and P. Plebani. Supporting policy-driven behaviors in web services: experiences and issues. In *Proceedings of the Second International Conference on Service-Oriented Computing (ICSOC'2004)*, New York City, NY, USA, 2004.
19. N. Russell, W. M. P. van der Aalst, and A. H. M. ter Hofstede. Exception handling patterns in process-aware information systems. Technical Report, BPM Center Report BPM-06-04, BPMcenter.org, 2006.
20. M. Satyanarayanan. Pervasive computing: vision and challenges. *IEEE Personal Communications*, 8(4), August 2001.
21. B. Schilit and M. Theimer. Disseminating active map information to mobile hosts. *IEEE Network*, 8(5), 1994.
22. X. Yu, L. Zhang, Y. Li, and Y. Chen. WSCE: A flexible web service composition environment. In *Proceedings of The IEEE International Conference on Web Services (ICWS'2004)*, San-Diego, California, USA, 2004.

Data Mining with Privacy Preserving in Industrial Systems

Kevin Chiew

School of Informatics and Engineering, Flinders University, Adelaide, SA 5001, Australia, kevin.chiew.au@gmail.com

1 Introduction

1.1 Background and Motivation

The pervasive impact of business computing has made information technology an indispensable part of daily operations and the key to success for enterprises. Data mining, as one of the IT services most needed by enterprises, has been realized as an important way for discovering knowledge from the data and converting "data rich" to "knowledge rich" so as to assist strategic decision making. The benefits of using data mining for decision making have been demonstrated in various industries and governmental sectors, e.g., banking, insurance, direct-mail marketing, telecommunications, retails, and health care [8, 14, 22]. Among all the available data mining methods, the discovery of associations between business events or transactions is one of the most commonly used data mining techniques. Association rule mining has been an important application in decision support and marketing strategy [19] for an enterprise.

At the same time, many enterprises have accumulated large amount of data from various channels in today's digitalized age. It is important to make these data available for decision making. Enterprise data mining provides such a technique for the exploration and analysis of data so as to reveal hidden information and knowledge. These processes involve extensive collaborations (e.g., exchange or sharing of business data) across different divisions of an enterprise or even enterprises themselves. However, there is also a security concern of potential risk of exposing privacy (and losing business intelligence) of an enterprise during the practice [20]. This is because either the data or the revealed information may contain the privacy of an enterprise. During the data analysis process, e.g., data mining, data transferring, and data sharing, it involves some elements containing sensitive information from which an adversary can decipher the privacy of an enterprise. Without proper security policy and technology, enterprise privacy could be very vulnerable to security

K. Chiew: *Data Mining with Privacy Preserving in Industrial Systems*, Studies in Computational Intelligence (SCI) **116**, 57–79 (2008)
www.springerlink.com

breaches. Therefore, it is urgent and critical to provide solutions to protecting enterprise privacy for data mining in different application scenarios.

Consider a typical application scenario as follows. In an organization (e.g., an enterprise or a governmental sector), there are several divisions including an IT division which provides IT services for the whole organization. A functional division may have to delegate its data mining tasks to the IT division because of two reasons: lack of IT expertise and lack of powerful computing resources which are usually centrally managed by the IT division. The data used for the data mining usually involve privacy that the functional division may not want to disclose to anyone outside the division. To protect the privacy, this division should first convert (or encrypt) the source data to another format of presentation before transferring to the IT division. Therefore, there are two factors which are important for enabling a functional division to delegate data mining tasks to the IT division: (1) the computational time of data conversion is less than that of data mining; otherwise it is not at all worthwhile to do so; and (2) the storage space of converted data should be acceptable (the less the better though, several times more is still acceptable and practical).

This scenario can be extended to a more general circumstance in which all divisions are individually independent organizations or enterprises. This is because in today's fast-paced business environment, it is impossible for any single enterprise to understand, develop, and implement every information technology needed. It can also be extended to online scenarios, e.g., a distributed computing environment in which some edge servers undertaking delegated mining tasks may be intruded by hacking activities and thus may not be fully trusted.

When delegating mining tasks[1], we should protect the following three elements which may expose data privacy: (1) the source data which is the database of all transactions; (2) the mining requests which are itemsets of interests; and (3) the mining results which are frequent itemsets and association rules.

People have proposed various methods to preserve customer privacy in data mining for some scenarios, such as a distributed environment. However, those existing methods cannot protect all three elements simultaneously. This is because when a first party[2] delegates its mining tasks to a second party[3], it has to provide the source database (which might be someway encrypted) together with some additional information (e.g., plain text of mining requests) without which this second party may not be able to carry out the mining tasks. Given this situation, those proposed methods are unable to efficiently prevent

[1] Without further specification, we always refer to association rule mining tasks.

[2] This is the party that delegates its data mining tasks. It may be a functional division in the scenario discussed above or a center server in a distributed environment with client-server architecture.

[3] This is the party that is authorized by the first party to undertake the delegated data mining tasks. It may be the IT division of an organization or an edge server in a distributed environment with client-server architecture.

the exposure of private information to the second party, or unable to prevent the second party from deciphering further information from the mining results (which would be sent back to the first party) with the additional information.

1.2 Our Solution

In this chapter[4], we present a Bloom filter based approach which provides an algorithm for privacy preserving association rule mining with computation efficiency and predictable (controllable) analysis precision. The Bloom filter [11] is a stream (or a vector) of binary bits. It is a computationally efficient and irreversible coding scheme that can represent a set of objects while preserving privacy of the objects (technical details will be presented in Sect. 3.1).

With our approach, firstly the source data are converted to Bloom filter representation and handed over to a second party (e.g., the IT division of the organization) together with mining algorithms. Then the first party sends its mining requests to the second party. Mining requests are actually candidates of frequent itemsets which are also represented by Bloom filters. Lastly, the second party runs the mining algorithms with source data and mining requests, and comes out the mining results which are frequent itemsets or association rules represented by Bloom filters. In the above mining process, what the first party exposes to the second party does not violate privacy [17]; that is, the second party would not be able to distill down private information from Bloom filters. Therefore all the three elements mentioned above are fully protected by Bloom filters.

The goal of privacy protection can be achieved by Bloom filter because it satisfies simultaneously the following three conditions. First, transactions containing different numbers of items are mapped to Bloom filters with the same length. This prevents an adversary from deciphering the compositions of transactions by analyzing the lengths of transactions. Second, Bloom filters support membership queries. This allows an authorized second party to carry out data mining tasks with only Bloom filters (i.e., Bloom filters of either transactions or candidates of frequent itemsets). Third, without knowing all possible individual items in the transactions, it is difficult to identify what items are included in the Bloom filter of a transaction by counting the numbers of 1's and 0's. This is because the probability of a bit in a Bloom filter being 1 or 0 is 0.5 given that the parameters of the Bloom filter are optimally chosen (see a formal description in Appendix 1 and the detailed mathematical analysis in [24, 25]).

The experimental results show that (1) the data conversion time is much less than mining time, which supports the worthiness to delegate mining tasks; (2) there is a tradeoff between storage space and mining precision; (3) there is a positive relationship between privacy security level and mining precision;

[4] A preliminary version of this chapter can found found in [26].

(4) the converted data do not require more storage space compared with the original storage format.

1.3 Organization of the Chapter

The remaining sections of this chapter are organized as follows. Firstly in Sect. 2 we review the related work on privacy preserving data mining. After that in Sect. 3 we present our solution which uses a technique of keyed Bloom filters to encode the raw data, the data mining requests, and also the results of data analysis during the data exchanges for privacy protection. Next, we demonstrate in Sect. 4 the implementation of the proposed solution over a point-of-sale dataset and two web clickstream datasets. We present experiments which investigate the tradeoffs among the level of privacy control, analysis precisions, computational requirements, and storage requirements of our solution with comparisons over other mining methods. Lastly in Sect. 5, we conclude the chapter with discussions of different application scenarios made possible by the solution, and point out some directions for further study.

2 Literature Review

Association rule mining has been an active research area since its introduction [2]. Various algorithms have been proposed to improve the performance of mining association rules and frequent itemsets. An interesting direction is the development of techniques that incorporate privacy concerns. Rayardo and Srikant [27] provided a detailed summary and classification for the existing solutions. They classified the solutions into the following categories: privacy policy encoding methods (e.g., P3P and EPAL), Hippocratic database [3], privacy-preserving data mining methods [6,10,31], secured information sharing across private repositories (e.g., cryptographic protocols and secure coprocessors) [32], and privacy-preserving searching [13].

Privacy preserving data mining techniques can be classified into two major types, namely data perturbation based and secure multiparty computation based (see [10] for a review of existing algorithms). With data perturbation methods, the raw data is modified by randomization or adding noise so that it no longer represents real value, while preserving the statistical property of the data [7]. An early work of Agrawal and Srikant [6] proposed a perturbation based approach for decision tree learning. Some recent work [9, 15, 21, 28, 29] investigates the tradeoff between the extent of private information leakage and the degree of data mining accuracy. One problem of perturbation based approach is that it may introduce some false association rules. Another drawback of this approach is that it cannot always fully preserve privacy of data while achieving precision of mining results [18], the effect of the amount of perturbation of the data on the accuracy of mining results is unpredictable.

The distributed privacy preserving data mining [16, 23, 30] is based on secure multiparty computation [1]. This approach is only applicable when there are multiple parties among which each possesses partial data for the overall mining process and wants to obtain any overall mining results without disclosing their own data source. Moreover, this method needs sophisticated protocols (secure multi-party computation based). These make it infeasible for our scenario.

Both types of techniques are designed to protect privacy by masquerading the original data. They are not designed to protect data privacy from the mining requests or the mining results, which are accessible by data miners.

Recently, Agrawal et al. presented an order-preserving encryption scheme for numeric data that allows comparison operations to be directly applied on encrypted data [4]. However, encryption is time consuming and it may require auxiliary indices. It is only designed for certain type of queries and may not be suitable for complex tasks such as association rule mining.

3 Our Solution: Bloom Filter-Based Approach

From the literature, there is no single method that can enable industrial enterprises to delegate data mining tasks while protecting all three elements involving the disclosure of privacy in the process of delegating data mining tasks. Our main objective in this chapter is to propose a computationally feasible and efficient solution for the scenario.

A large number of examples from different industries (such as financial, medical, insurance, and retails) can be used for the study of the thread of privacy and business knowledge disclosure. In this chapter, we consider the well-known association rule mining [2] which is also known as market basket analysis [12] in business analytics. Association rule mining can be carried out by two steps: (1) mining of frequent itemsets, followed by (2) mining of association rules from frequent itemsets. Currently a well-known algorithm for mining of frequent itemsets is Apriori algorithm proposed by Agrawal and Srikant in [5]. Based on Apriori algorithm, in our early study [25] we investigated the feasibility of using a Bloom filter based approach for mining of frequent itemsets with privacy concerns. In this chapter, we propose a solution with concerns of privacy protection by extending the Bloom filter-based approach to the delegating scenario and the whole process of association rule mining.

In what follows, we first introduce the mechanisms of constructing Bloom filters and membership queries over Bloom filters with discussion on the feature of privacy protection. We then present algorithms of frequent itemset mining and association rule mining.

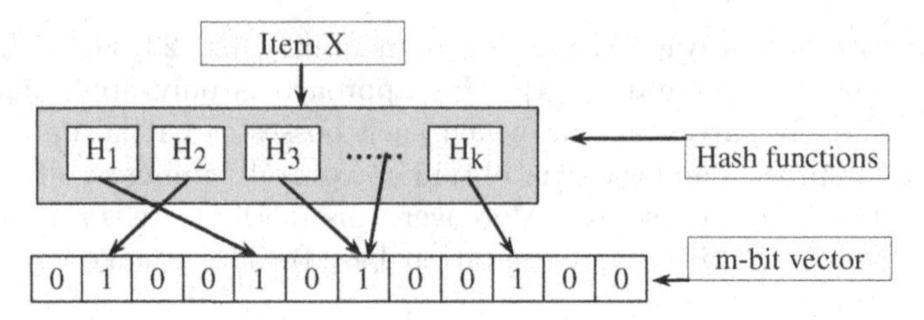

Fig. 1. Constructing a Bloom filter of an item

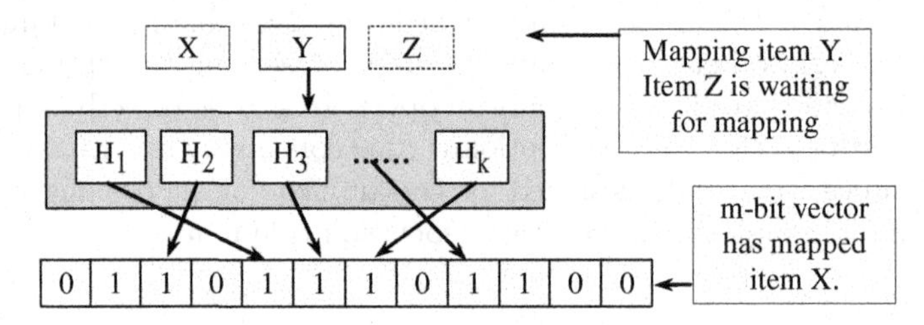

Fig. 2. Constructing a Bloom filter of a transaction

3.1 Bloom Filters

The Bloom filter [11] is a computationally efficient hash-based probabilistic scheme that can represent a set of objects with minimal memory requirements. It can be used to answer membership queries with *zero* false negatives (i.e., without missing of useful information) and low false positives (i.e., with incurring of some extra results that are not of interests). A brief revisit of Bloom filter with formal definitions is given in Appendix 1. We provide an intuitive description in what follows.

The mechanism of a Bloom filter contains (1) a binary vector (or stream) with length m and (2) k hash functions h_1, h_2, $\ldots$, h_k of range from 1 to m. Given an item x, the Bloom filter of x, denoted as $B(x)$, is constructed by the following steps: (1) initialize by setting all bits of the vector with 0 and (2) set bit $h_i(x)$ (where $1 \leqslant i \leqslant k$) of the vector with 1. For example, if $h_2(x) = 7$, then bit 7 (i.e., the 7th bit) of the vector is set with 1. It is possible that several hash functions set the same bit of the vector, i.e., $h_i(x) = h_j(x)$. Thus, after conversion, the number of 1's in $B(x)$ is not greater than k. Figure 1 illustrates how to construct a Bloom filter of an item.

It is similar to construct a Bloom filter of a transaction $T = \{X, Y, Z\}$ (or an itemset). Figure 2 illustrates such a process in which item Y (or Z) is mapped to the binary vector to which item X (or items X and Y) has already been mapped. This process can be presented as $B(T) = B(X) \oplus B(Y) \oplus B(Z)$ where operator $\oplus$ stands for bitwise OR[5].

[5] The bitwise OR operation is defined as: $0 \oplus a = a$ and $1 \oplus a = 1$ where a is a binary variable 0 or 1.

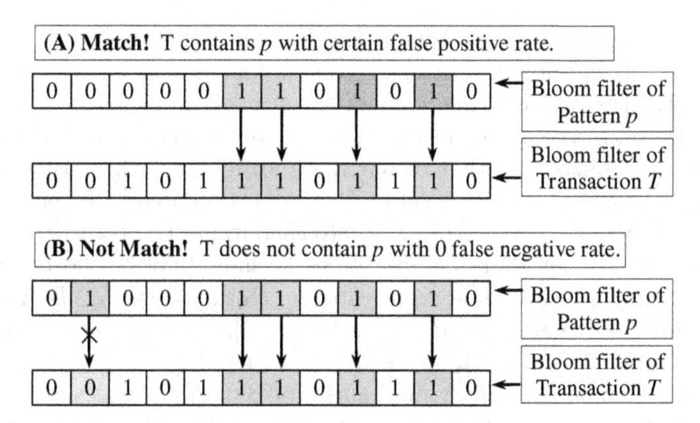

Fig. 3. Membership query over Bloom filters

It should be pointed out that converting the data into Bloom filters is an irreversible process. Any unauthorized party accessing to Bloom filters will have no way to know/infer the original value of the data represented by Bloom filters unless they have the access to the original data, the hash functions, and the secrete keys (introduced later in this subsection).

To check whether a pattern p is contained by a transaction T, we examine whether $B(p) \otimes B(T) = B(p)$ holds true, where operator $\otimes$ stands for bitwise *AND*[6]. If $B(p) \otimes B(T) \neq B(p)$, then p is definitely not contained by T; otherwise, p is a member of T with very low probability of false (i.e., false positive rate). Figure 3 shows the process of membership query with Bloom filters.

A Bloom filter does not incur any false negative, meaning that it will not suggest that a pattern is not in T if it is; but it may yield a false positive, meaning that it may suggest that a pattern is in T even though it is not. In our application, the false positive rate is upper-bounded by 0.5^k, where k is the number of hash functions, and the optimal value of k is given by $k = \frac{m}{n} \ln 2$, where m is the length of Bloom filters (i.e., the number of bits in the binary vectors), and n is the average length of transactions (i.e., the average number of items in transactions). Lemma 1 in Appendix 1 gives the technical details for deriving the optimal value of k. Therefore, the false positive rate decreases exponentially with linear increase of the number of hash functions or the length of Bloom filters. For many applications, this is acceptable as long as the false positive rate is sufficiently small.

The privacy of data can be preserved by Bloom filters due to the irreversible feature. Given the above parameters of a Bloom filter, there are m^k possible mappings (for example, if we set the length of a Bloom filter $m = 80$ and the number of hash functions $k = 25$, then there are totally 80^{25} possible

[6] The bitwise *AND* operation is defined as: $0 \otimes a = 0$ and $1 \otimes a = a$ where a is a binary variable 0 or 1.

mappings in constructing the Bloom filter). Thus a Bloom filter is effective against some straightforward attacks (e.g., unknown-text attack and brute-force attack). It is possible that some other encryption algorithms (e.g., DES or RSA) are more secure against attacks; however, the computational cost is much more higher than that of our method. The length of Bloom filters is a tradeoff between security level and computational cost. To enhance the security level, we insert a secret key K into each itemset or transaction before constructing its Bloom filter. The secret key K should not be chosen from the items. This amendment can be represented as $B_K(T) = B(T) \oplus B(K)$, where $B_K(T)$ is referred to as a keyed Bloom filter [25]. Without further mention, we always assume that Bloom filters are constructed with a secret key.

With the membership query mechanism of Bloom filters described above, we are able to conduct association rule mining with access to only Bloom filters. Given the irreversible feature of Bloom filters, a first party can convert all data involving disclosure of privacy to Bloom Filters and safely delegate the mining tasks to a second party without disclosing any value of the data in the database, the mining requests, and the mining results. We do not need to worry about missing of useful information (i.e., frequent itemsets and strong association rules in our application) due to *zero* false negative rate; but we may get some extra information (which may confuse data hacker while not affecting the quality of mining results) with low probability of false positive rate. However, performing mining tasks over Bloom filters differs from mining over original datasets. This is because the frequency learnt from Bloom filters may be larger than its real frequency learnt from the original datasets. Appendix 2 gives the detailed analysis and estimate of false positive and false negative.

3.2 Mining Processes and Algorithms

The procedure of mining frequent itemsets is the process of membership queries over Bloom filters. Based on Apriori algorithm, a frame work of our method is shown in Algorithm 1. Algorithm 1 can be divided into three phases: *counting phase* (lines 3–5), *pruning phase* (lines 6–8), and *candidates generating phase* (lines 9–10) in each round ℓ, where ℓ indicates the size of each candidate itemset dealt with. In the counting phase, each candidate filter is checked against all transaction filters[7] and the candidate's count is updated. In the pruning phase, any Bloom filter is eliminated from the candidate set if its count (i.e., Support(x)) is less than the given threshold $N \cdot \tau$. Finally, in the candidates generating phase, new candidate Bloom filters are generated from the Bloom filters discovered in the current round. The new candidates will be used for data mining in the next round. With the results of frequent itemsets, the mining of association rules is relatively simple, which is shown in Algorithm 2.

[7] All transactions are organized in a tree hierarchy so as to minimize the times of membership queries. See details in [25].

Algorithm 1 Mining of frequent itemsets from Bloom filters

1: $C_1 = \{B(I_1), \ldots, B(I_d)\}$ // $B(I_i)$ is the Bloom filter of item I_i
2: **for** ($\ell = 1$; $C_\ell \neq \varnothing$; ℓ++) **do**
3: **for** each $B(S) \in C_\ell$ and each transaction filter $B(T_i)$ **do**
4: **if** $B(S) \otimes B(T_i) = B(S)$ **then**
Support(S)++ // S is a candidate frequent ℓ-itemset
5: **end for**
6: **for** each $B(S) \in C_\ell$ **do**
7: **if** Support(S)$< N \cdot \tau$ **then** // N is transaction number in the database,
delete $B(S)$ from C_ℓ // and τ the threshold of minimum support
8: **end for**
9: $F_\ell = C_\ell$ // F_ℓ is the collection of Bloom filters of all "frequent" ℓ-itemsets
10: $C_{\ell+1} = can\_gen(F_\ell)$
// generate filters of candidate itemsets for the next round
11: **end for**
12: Answer $= \bigcup_\ell F_\ell$ // all filters of frequent itemsets

Algorithm 2 Mining of association rules from Bloom filters of frequent itemsets

1: $AR = \varnothing$;
2: $F = \{B(F_1^1), \ldots, B(F_{d_1}^1), B(F_1^2), \ldots, B(F_{d_2}^2), \ldots, B(F_1^k)), \ldots, B(F_{d_k}^k)\}$
// $B(F_i^s)$ is the Bloom filter of frequent s-itemset F_i^s
// where $1 \leqslant s \leqslant k$ and $1 \leqslant i \leqslant d_s$
3: **for** ($s = 1$; $s < k$; s++) **do**
4: **for** ($t = s + 1$; $t \leqslant k$; t++) **do**
5: **for** each $B(F_i^s)$ and each $B(F_j^t)$ **do**
6: **if** $B(F_i^s) \otimes B(F_j^t) = B(F_i^s)$ and Support(F_j^t)/Support$(F_i^s) \geqslant \xi$
// ξ is the minimum threshold of confidence
7: **then** $F_i^s \Rightarrow F_j^t - F_i^s$ is a strong association rule and is added to AR
8: **end for**
9: **end for**
10: **end for**
11: **return** AR // All strong association rules

The complete process of association rule mining is given as follows. (1) The first party hands over to the second party the application software that performs frequent itemset mining together with the database of transactions represented by Bloom filters. (2) The first party sends to the second party mining requests which include candidate itemsets and the threshold of minimum support. The generation of candidates is done at the first party side by running Apriori_gen [5] which is the critical step of Apriori algorithm. This step has to be done in the first party side because it involves data privacy [25]. (3) The second party carries out mining tasks with the data received and finally returns the mining results which are Bloom filters of frequent itemsets together with their supports. (4) With frequent itemsets and their supports returned from the second party, it is easy to generate strong association rules with thresholds of minimum confidence. This job can be performed by the first party itself, or by the second party. If it is performed by the first party, there is no need to convert frequent itemsets to Bloom filters. If it is performed by the second party, for privacy considerations all data has to be converted to Bloom filters.

4 Experiments

4.1 Experimental Settings

We implement the solution and evaluate it with experiments on three real datasets BMS-POS, BMS-WebView-1, and BMS-WebView-2 which are publicly available for research communities[8]. Dataset BMS-POS contains several years of point-of-sale data from a large electronic retailer; whereas datasets BMS-WebView-1 and BMS-WebView-2 contains several months of clickstream data from an e-commerce website. Table 1 shows the number of items, the average size of transactions, and the number of transactions included in these datasets. Figure 4 shows the distribution of transaction sizes of the datasets. For dataset BMS-POS, a transaction is a list of items purchased in a basket; whereas for dataset BMS-WebView-1 or BMS-WebView-2, a transaction is a browsing session which contains a list of webpages visited by a customer.

Table 1. Characteristics of real datasets

Dataset	Distinct items	Max-size	Average size	Number of transactions
BMS-POS	1,657	164	6.53	515,597
BMS-WebView-1	497	267	2.51	59,602
BMS-WebView-2	3,340	161	4.62	77,512

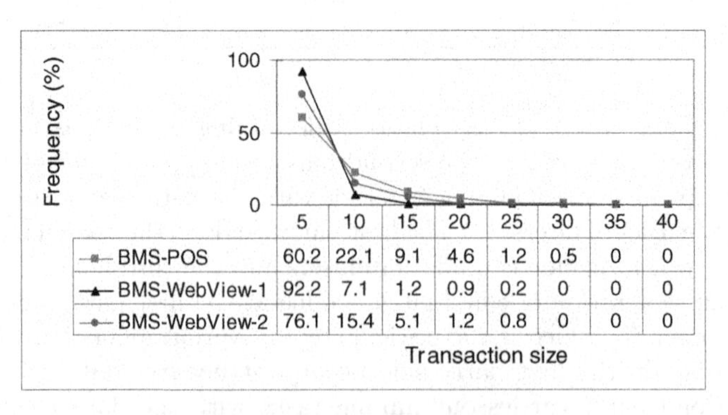

	5	10	15	20	25	30	35	40
BMS-POS	60.2	22.1	9.1	4.6	1.2	0.5	0	0
BMS-WebView-1	92.2	7.1	1.2	0.9	0.2	0	0	0
BMS-WebView-2	76.1	15.4	5.1	1.2	0.8	0	0	0

Fig. 4. Distribution of transaction sizes

[8] Downloadable at http://www.ecn.purdue.edu/KDDCUP.

The experiments are run on a Compaq desktop computer with Pentium-4 CPU clock rate of 3.00 GHz, 3.25 GB of RAM and 150 GB harddisk, with Microsoft Windows XP Professional SP2 as the operating system.

We have qualitatively analyzed the privacy preserving feature of Bloom filters in Sect. 3.1 (further theoretical analysis and discussions can be found in [24, 25]). Therefore the emphasis of this set of experiments is to investigate the relationship among the level of privacy protection (determined by the number of hash functions), storage requirement, computation time, and analysis precision. In the experiments, we set the threshold of minimum support $\tau = 1\%$ and cluster the transactions in each dataset into 4 groups based on their transaction sizes (the technical details of grouping is given in Appendix 2). We change k the number of hash functions used for Bloom filters from 25 to 40 in the experiments.

4.2 Experimental Results

Figures 5 to 7 show that the time of mining frequent itemsets is much more than the time of converting data to Bloom filter presentations, meaning that the mining process takes the major part of running time. This result verifies the worthiness in terms of running time for data format conversion before delegating mining tasks (satisfying the first factor enabling to delegate mining tasks as mentioned in Sect. 1).

Figure 8 shows the mining precisions with the change of k. There is a globally decreasing trend of false positive rates for each real dataset. For dataset BMS-POS, the false positive rate is less than 1% for $k \geqslant 25$. For datasets BMS-WebView-1 and BMS-WebView-2 the false positive rate is less than 4% for $k \geqslant 30$.

Figure 9 shows that the running time changes slightly with hash function number k. The running time is around 8 minutes for dataset BMS-POS and within 0.5 minute for datasets BMS-WebView-1 and BMS-WebView-2, because comparatively dataset BMS-POS contains 7 to 9 times as many as transactions.

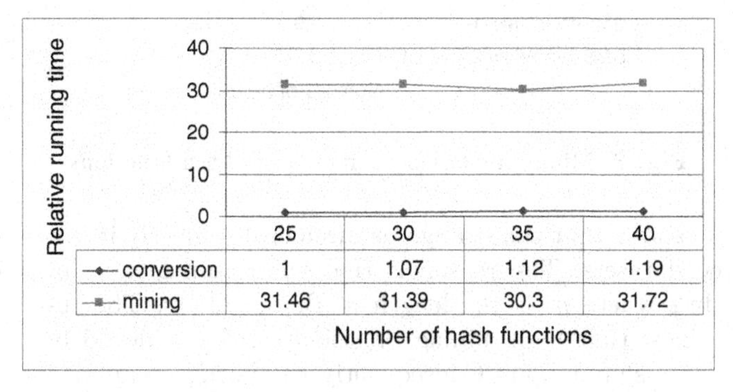

Fig. 5. Data conversion time vs. mining time on dataset BMS-POS

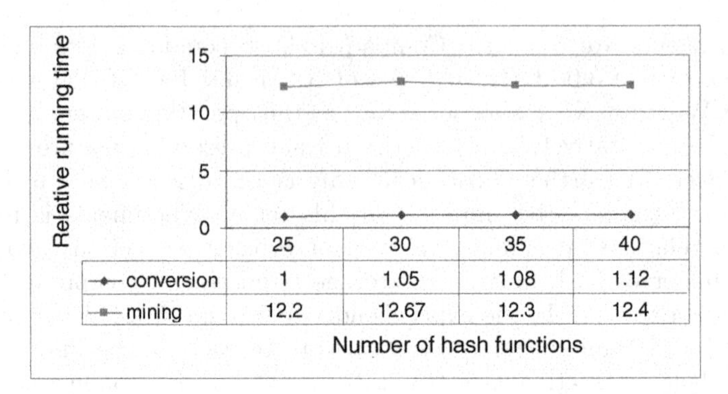

Fig. 6. Data conversion time vs. mining time on dataset BMS-WebView-1

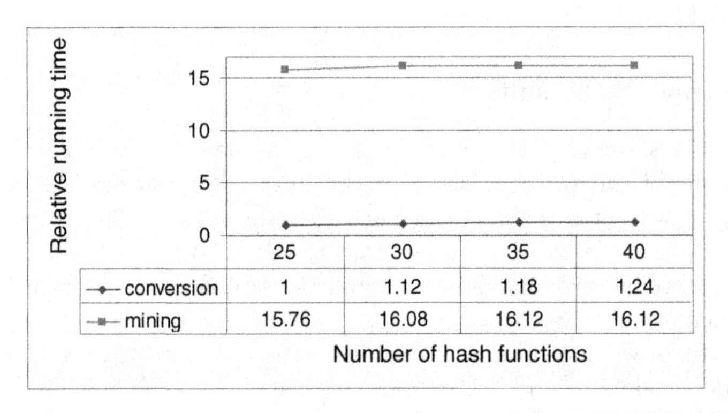

Fig. 7. Data conversion time vs. mining time on dataset BMS-WebView-2

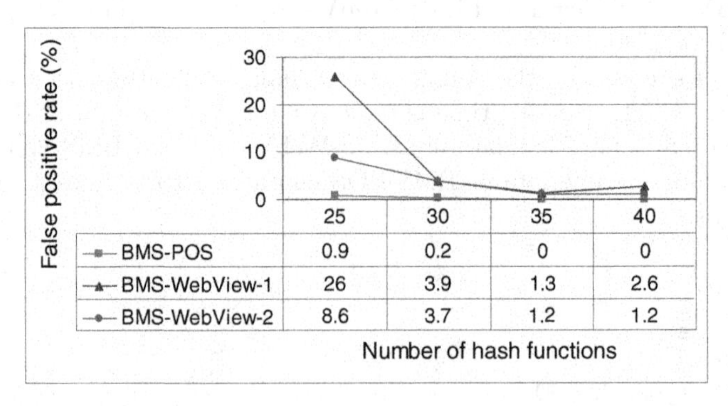

Fig. 8. Mining precision vs. number of hash functions

Figure 10 shows that the storage requirement is linearly increasing with k for all three datasets. The reason is that the optimal value of k is given by $k = \frac{m}{n} \ln 2$ where m is the length of Bloom filters. The results of this experiment show that high mining precision can be achieved by increasing the number of hash functions. Consequently, the storage requirement increases linearly due to the use of longer Bloom filters.

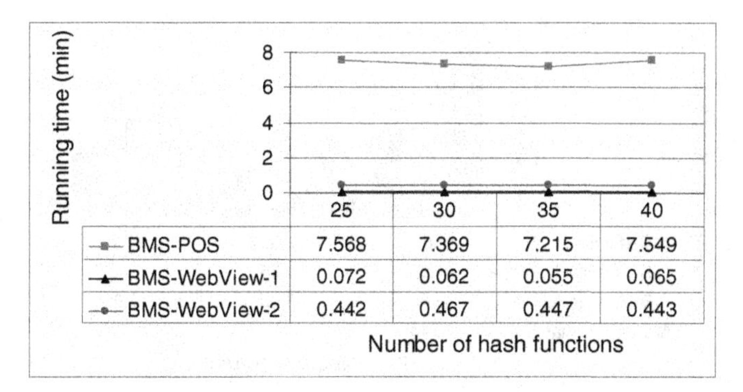

Fig. 9. Running time vs. number of hash functions

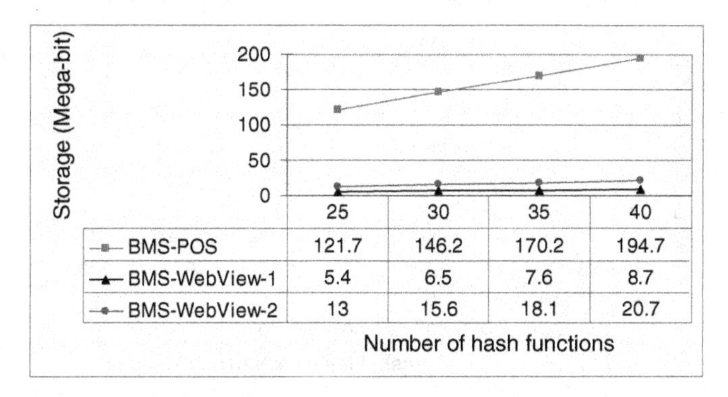

Storage (Mega-bit)
200
150
100
50
0
25 30 35 40
BMS-POS 121.7 146.2 170.2 194.7
BMS-WebView-1 5.4 6.5 7.6 8.7
BMS-WebView-2 13 15.6 18.1 20.7
Number of hash functions

Fig. 10. Storage requirement vs. number of hash functions

Figure 11 shows a comparison of the average storage space required by a transaction under difference storage formats. The results show that the storage space of Bloom filter format is practical, i.e., it is less than text format but a bit more than binary format depending on the precision requirement (adjustable by k). This satisfies the second factor that it is worthwhile in term of storage space to adopt Bloom filter presentations as mentioned in Sect. 1. We can achieve further saving of storage space without decreasing mining precision with grouping technique (see Appendix 2).

With the mining results of frequent itemsets returning from the second party, we continue the mining of association rules with given threshold of minimum confidence. In this experiment, we let $k = 30$ under which the false positive rates of frequent itemsets are lower than 5% for all three datasets. We vary the threshold of minimum confidence from 55 to 75%. The false negative rates are *zero* for all datasets, meaning that our approach does not miss useful information. As shown in Fig. 12, the false positive rate is *zero* for datasets BMS-WebView-1 and BMS-WebView-2 and lower than 0.2% for dataset BMS-POS. The running time for any dataset is less than 0.1 s.

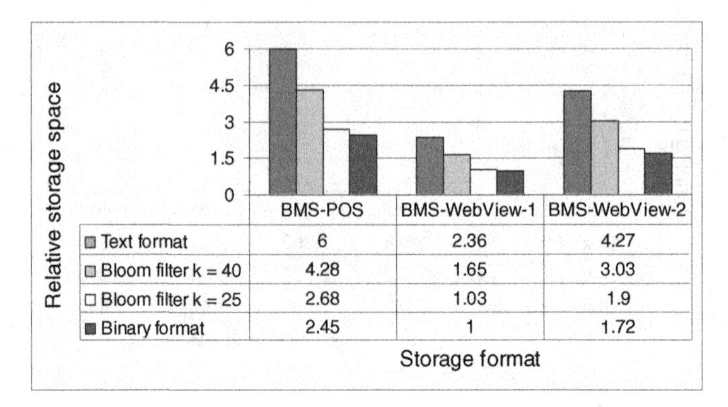

Fig. 11. Average storage space of a transaction

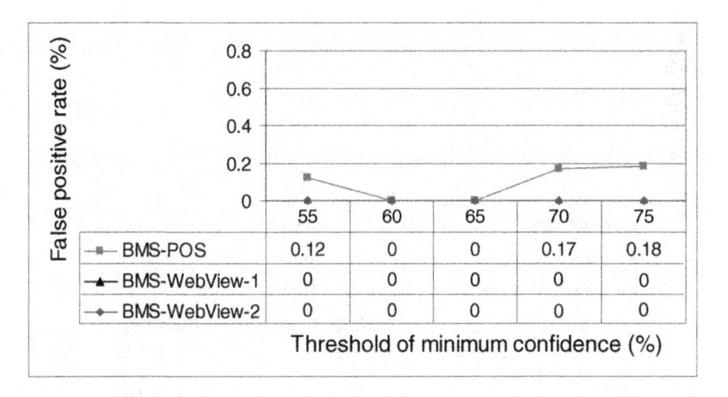

Fig. 12. Precision of mining association rules

5 Conclusions

In this chapter, we have discussed and identified the risks of exposing data privacy in the scenario of delegating data mining tasks. We have also identified the factors that enable us to delegate mining tasks. We have proposed a method for enterprise data mining with privacy protection and applied it to association rule mining in this delegation scenario. As compared with other existing methods, the metrics of our method include: (1) our approach is effective in protecting of three elements that can expose data privacy in the process of delegating mining tasks; (2) there is a positive relationship between the privacy security level and the analysis precision; (3) to increase the privacy security level, we only need to sacrifice data storage space; and (4) the solution is scalable, i.e., the storage space increases linearly with the privacy protection level or the analysis precision.

In our current study, we have developed a privacy protection method for association rule mining with a single (centralized) database. We can also apply our method to other mining tasks (e.g., mining of some other rules that are of interest to researchers). Further study to investigate the feasibility and implementation of the proposed solution in a multiple (distributed) databases environment is needed. Another future research direction could be investigating the feasibility of using the keyed Bloom filter approach in other tasks of business analytics.

Appendix: 1

Bloom Filter Revisited

A Bloom filter is a simple, space-efficient, randomized data structure for representing a set of objects so as to support membership queries.

Definition 1. *Given an n-element set $S = \{s_1, \ldots, s_n\}$ and k hash functions $h_1, \ldots, h_k$ of range m, the Bloom filter of S, denoted as $B(S)$, is a binary vector of length m that is constructed by the following steps: (a) every bit is initially set to 0; (b) every element $s \in S$ is hashed into the bit vector through the k hash functions, and the corresponding bits $h_i(s)$ are set*[9] *to 1. A Bloom filter function, denoted as $B(\cdot)$, is a mapping from a set (not necessarily n-element set) to its Bloom filter.*

For membership queries, i.e., whether an item $x \in S$, we hash x to the Bloom filter of S (through those hash functions) and check whether all $h_i(x)$ are 1's. If not, then clearly x is not a member of S. If yes, we say x is in S although this could be wrong with some probability.

Definition 2. *For an element s and a set S, define $s \in_B S$ if s hashes to all 1's in the Bloom filter of S, and $s \notin_B S$ otherwise. The false positive rate of the Bloom filter of S is defined as the probability of $s \in_B S$ while $s \notin S$, or $\Pr(s \in_B S \mid s \notin S)$.*

Assuming that all hash functions are perfectly random, we have the following:

Lemma 1. *Given an n-element set $S = \{s_1, \ldots, s_n\}$ and its Bloom filter $B(S)$ of length m constructed from k hash functions, the probability for a specific bit in $B(S)$ being 0 is*

$$p_0 = \left(1 - \frac{1}{m}\right)^{kn} \approx e^{-kn/m}$$

and the probability for a specific bit being 1 is

$$p_1 = 1 - p_0 \approx 1 - e^{-kn/m}.$$

[9] A location can be set to one multiple times, but only the first change has an effect.

Then the false positive rate of $B(S)$ *is*

$$f = p_1^k \approx \left(1 - e^{-kn/m}\right)^k \tag{1}$$

Given n *and* m, f *is minimized when* $p_0 = p_1 = 0.5$ *and* $k = \frac{m}{n}\ln 2$, *in which case* $f = 1/2^k = (0.6185)^{m/n}$.

Appendix: 2

Analysis

In this section, we analyze the possible error rates introduced by mining the frequent itemsets based on Bloom filters instead of the original dataset. For any given itemset, the frequency learnt from Bloom filters may be larger than its real frequency learnt from original transactions due to the false positive of Bloom filters. We make this clear in the following analysis. By default, we assume that for any itemset, there is a Bloom filter function $B(\cdot)$ which produces a binary vector of length m through k hash functions.

Preliminaries

The false positive of a Bloom filter was defined for checking an element from a Bloom filter. Now we extend the concept of false positive to checking an itemset from a Bloom filter.

Definition 3. *Given an itemset* S *and a transaction* T_i, *define* $S \subseteq_B T_i$ *if for all items* $s \in S$, $s \in_B T_i$, *and define* $S \nsubseteq_B T_i$ *otherwise. The false positive rate for checking* S *from the Bloom filter of* T_i, *denoted as* f_i, *is defined as the probability of* $S \subseteq_B T_i$ *while* $S \nsubseteq T_i$, *or* $\Pr(S \subseteq_B T_i \mid S \nsubseteq T_i)$.

Due to the false positive of checking an itemset from a Bloom filter, the support or frequency learnt from a collection of Bloom filters is different from that learnt from original transactions. Regarding such support and frequency, we have the following:

Definition 4. *Given a collection of* N *Bloom filters* $\{B(T_1), \ldots, B(T_N)\}$ *for transaction database* $\mathcal{D}$ *over* $\mathcal{I}$, *the support of an itemset* $S \in 2^{\mathcal{I}}$ *that is learnt from the collection of filters, denoted as* $Bsup(S)$, *is defined as the number of filters* $B(T_i)$ *that satisfy* $S \subseteq_B T_i$. *The frequency of* S *that is learnt from the collection of filters, denoted as* $Bfq(S)$, *is defined as* $Bsup(S)/N$.

Lemma 2. *In the setting of Definition 4, the following statements hold: (a)* $S \subseteq_B T_i$ *iff* $B(S) \wedge B(T_i) = B(S)$, *where* $\wedge$ *is bitwise AND. (b) If* $S \subseteq T_i$, *then* $S \subseteq_B T_i$. *(c) If* $S \nsubseteq T_i$, *then* $S \subseteq_B T_i$ *with probability* f_i, *and* $S \nsubseteq_B T_i$ *with probability* $1 - f_i$. *(d)* $Bfq(S) \geqslant fq(S)$.

Theorem 1. *Given itemset S and transaction T_i, the false positive rate of checking S from the Bloom filter of T_i is*

$$f_i = \left(1 - e^{-kn_i/m}\right)^{||B(S-T_i)||}$$

where $n_i = |T_i|$ is the length of transaction T_i in terms of the number of items, and $||\cdot||$ indicates the number of 1's in a binary vector.

Proof. From (1), one can derive that the false positive rate for checking any single item $s \in S - T_i$ is p_1^k, where $p_1 = \left(1 - e^{-kn_i/m}\right)$ is the probability that a specific bit is 1 in $B(T_i)$ and k is the number of bits to which the item is hashed. From Lemma 2, we know that $S \cap T_i \subseteq_B T_i$. Since the items in $S - T_i$ are hashed to $||B(S - T_i)||$ bits all together, the false positive rate f_i for checking S from $B(T_i)$ is $p_1^{||B(S-T_i)||} = \left(1 - e^{-kn_i/m}\right)^{||B(S-T_i)||}$. □

Corollary 1. *Given an itemset S and a transaction T_i, the false positive rate f_i of checking S from the Bloom filter of T_i is bounded:*

$$\left(1 - e^{-kn_i/m}\right)^{||B(S)||} \leqslant f_i \leqslant \left(1 - e^{-kn_i/m}\right)^k$$

Proof. According to the definition of false positive, we have $S \nsubseteq T_i$ and thus $1 \leqslant ||B(S - T_i)|| \leqslant ||B(S)||$. Combining this with Theorem 1, we have $\left(1 - e^{-kn_i/m}\right)^{||B(S)||} \leqslant f_i \leqslant \left(1 - e^{-kn_i/m}\right)^k$. □

False Positive and False Negative

In our data mining problem, an outsourced server has access to the Bloom filters of all transactions. However, it has no access to the original data. Therefore, given a Bloom filter $B(S)$, the server cannot compute the frequency $fq(S)$ directly. The approaches to the traditional frequent itemset mining problem cannot be applied directly to our problem.

According to Lemma 2(a), the frequency $Bfq(S)$ can be derived from those Bloom filters. Our solution is to find all Bloom filters $B(S)$ such that $Bfq(S) \geqslant \tau'$ where τ' is a revised threshold. Note that $fq(S) \geqslant \tau$ is required in our data mining problem; thus, we must ensure that $\{B(S) : Bfq(S) \geqslant \tau'\} \approxeq \{B(S) : fq(S) \geqslant \tau\}$. Because the two sets are not necessarily the same, we need to define the false positive rate and false negative rate for checking an itemset from all Bloom filters.

Definition 5. *Given an itemset S and N Bloom filters $B(T_i)$, $i = 1, \ldots, N$, the false positive rate for checking S from all Bloom filters using revised threshold $\tau' \geqslant \tau$, denoted as $f_{\tau'}^{+}$, is defined as the probability of $Bfq(S) \geqslant \tau'$ while $fq(S) < \tau$, or $\Pr\left(Bfq(S) \geqslant \tau' \mid fq(S) < \tau\right)$. The false negative rate for checking S from all Bloom filters using τ', denoted as $f_{\tau'}^{-}$, is defined as the probability of $Bfq(S) < \tau'$ while $fq(S) \geqslant \tau$, or $\Pr\left(Bfq(S) < \tau' \mid fq(S) \geqslant \tau\right)$.*

If the revised threshold τ' is the same as the threshold τ, then the false negative rate will be zero due to the fact $Bfq(S) \geqslant fq(S)$ (Lemma 2 (d)); however, the false positive rate may be greater than zero in this case. In general, one may use $\tau' \geqslant \tau$ so as to balance false negative rate and false positive rate. The higher the τ', the higher the false negative rate, and the lower the false positive rate. If the false negative rate is of major concerns, one may choose $\tau' = \tau$ to zero out false negative rate. Note that choosing $\tau' < \tau$ is meaningless as it only increases the false positive rate without decreasing the false negative rate which is zero.

To formalize our analysis, we define a random variable for checking an itemset from a Bloom filter. Then we re-write in terms of the defined variables the frequency of an itemset learnt from Bloom filters and the false positive/negative rates for checking an itemset from all Bloom filters.

Definition 6. *For an itemset S and a transaction T_i such that $S \not\subseteq T_i$, define a random 0–1 variable e_i such that $e_i = 1$ if $S \subseteq_B T_i$ and $e_i = 0$ if $S \not\subseteq_B T_i$.*

The defined variable indicates whether $S \subseteq_B T_i$; that is, $e_i = 1$ with probability f_i and $e_i = 0$ with probability $1 - f_i$ (see Lemma 2). In other words, e_i represents a Bernoulli trial with probabilities f_i of success and $1 - f_i$ of failure. Without loss of generality, we can assume that $S \not\subseteq T_i$ for the first $N \cdot \big(1 - fq(S)\big)$ transactions T_i. Then we have

$$Bfq(S) = fq(S) + \frac{1}{N} \cdot \sum_{i=1}^{N(1-fq(S))} e_i \tag{2}$$

Lemma 3. *Let s_e be the sum of $N \cdot (1 - fq(S))$ random 0–1 variables e_i defined for an itemset S. The false positive and false negative rates for checking S from all Bloom filters using a revised threshold $\tau' \geqslant \tau$ are*

$$f^+_{\tau'} = \Pr\Big(s_e \geqslant N\big(\tau' - fq(S)\big) \mid fq(S) < \tau\Big)$$

$$f^-_{\tau'} = \Pr\Big(s_e < N\left(\tau' - fq(S)\right) \mid \tau \leqslant fq(S) < \tau'\Big)$$

Estimate of False Positive and False Negative

We estimate the false positive and false negative rates in the case of $n_i \cong n$, where n_i is the size of each transaction, and $n = \frac{1}{N} \sum_{i=1}^{N} n_i$ is the average size of N transactions. Note that if $n_i \ncong n$, the transaction data can be clustered into multiple groups such that in each group, the size of each transaction is equal or close to the average size of the transactions in the group. The data mining task can be easily extended to each group. For simplicity, we assume that $n_i = n$ in our analysis; we leave the multi-group case in the coming subsection.

Let $k = \frac{m}{n} \ln 2$ be the optimal number of hash functions that are used for generating the Bloom filter of length m for each transaction, which consists of

n items. According to Lemma 1 and Corollary 1, the false positive rate f_i for checking an itemset S from a Bloom filter has a lower bound and an upper bound:

$$2^{-||B(S)||} \leqslant f_i \leqslant 2^{-k} \tag{3}$$

First consider a special case where $f_i = \bar{f}$ for all $i = 1, \ldots, N$. In this case s_e is the sum of $N \cdot (1 - fq(S))$ independent Bernoulli trials with probabilities $\bar{f}$ for success and $1 - \bar{f}$ for failure. Let $b(\alpha, \beta, \bar{f}) = \frac{\beta!}{\alpha!(\beta-\alpha)!} \bar{f}^{\alpha}(1 - \bar{f})^{\beta-\alpha}$ be the probability that α Bernoulli trials with probabilities $\bar{f}$ for success and $(1 - \bar{f})$ for failure result in β successes and $\beta - \alpha$ failures $(\alpha \leqslant \beta)$. Let $\mathcal{C}(\alpha, \beta, \bar{f}) = \sum_{i=\alpha}^{\beta} b(i, \beta, \bar{f})$ be the cumulative binomial probability of having at least α successes in β trials[10]. Let $A = N \cdot \big(\tau' - fq(S)\big)$ and $B = N \cdot \big(1 - fq(S)\big)$, then the false positive and false negative rates for checking an itemset S from all Bloom filters are

$$f_{\tau'}^{+}(\bar{f}) = \mathcal{C}\big(A, B, \bar{f}\big), \qquad \text{where } fq(S) \in (0, \tau) \tag{4}$$

$$f_{\tau'}^{-}(\bar{f}) = 1 - \mathcal{C}\big(A, B, \bar{f}\big), \quad \text{where } fq(S) \in [\tau, \tau') \tag{5}$$

Since the cumulative binomial probability $\mathcal{C}(\alpha, \beta, \bar{f})$ is monotonic increasing with $\bar{f}$, from formulae (3) to (5) it is easy to know the lower bounds and upper bounds for the false positive rate and false negative rate in general case:

$$f_{\tau'}^{+}\left(2^{-||B(S)||}\right) \leqslant f_{\tau'}^{+} \leqslant f_{\tau'}^{+}\left(2^{-k}\right), \quad \text{where } fq(S) \in (0, \tau)$$

$$f_{\tau'}^{-}\left(2^{-k}\right) \leqslant f_{\tau'}^{-} \leqslant f_{\tau'}^{-}\left(2^{-||B(S)||}\right), \quad \text{where } fq(S) \in [\tau, \tau')$$

In a special case where $\tau' = \tau$, we have $f_{\tau}^{-} = 0$ and

$$\mathcal{C}\left(x,\ y,\ 2^{-||B(S)||}\right) \leqslant f_{\tau}^{+} \leqslant \mathcal{C}\left(x,\ y,\ 2^{-k}\right)$$

where $x = N \cdot \big(\tau - fq(S)\big)$ and $y = N \cdot \big(1 - fq(S)\big)$.

The above inequalities indicate that the greater the number k of hash functions, the smaller the false positive and false negative rates. In other words, the longer the Bloom filters, the smaller the false positive and false negative rates. To further understand this, we compare $Bfq(S)$ with $fq(S)$ in the average case. Let $E[\cdot]$ denote the mean of a random variable. From (2), we have

$$\begin{aligned} E\big[Bfq(S)\big] &= fq(S) + \frac{1}{N} \cdot E[s_e] \\ &= fq(S) + \frac{1}{N} \cdot \sum\nolimits_{i=1}^{N(1-fq(S))} f_i \end{aligned}$$

Recall the bounds for f_i (see (3)). In the false positive case (where $fq(S) \leqslant \tau$), we have

[10] Function $\mathcal{C}$ is a standard function provided in many commercial software packages such as Matlab.

$$N \cdot (1-\tau) \cdot 2^{-||B(S)||} \leqslant E[s_e] \leqslant N \cdot 2^{-k}$$

Similarly, in the false negative case (where $fq(S) \in [\tau, \tau')$), we have

$$N \cdot (1-\tau') \cdot 2^{-||B(S)||} \leqslant E[s_e] \leqslant N \cdot (1-\tau) \cdot 2^{-k}$$

The above three equations imply that the average value of $Bfq(S)$ is greater than $fq(S)$, but the difference is bounded. Note that $||B(S)|| \geqslant k$. We have

$$E\big[Bfq(S)\big] - fq(S) \leqslant (1-\tau) \cdot 2^{-k} < 2^{-k} \tag{6}$$

The longer the Bloom filters (i.e., the greater the k), the smaller the difference between $Bfq(S)$ and $fq(S)$. If the length of Bloom filters increases linearly, then the difference of frequencies decreases exponentially. For example, if $k \geqslant 20$, then the difference of frequencies will be less than 10^{-6}. This means that in the average case, the frequency of an itemset detected from Bloom filters will be greater than that detected from original data by at most 10^{-6}. Note that the above analysis is conducted for each itemset. The overall false positive and false negative rates depend on the distribution of itemsets and their frequencies.

Multiple Groups of Bloom Filters

The estimate of false positive and false negative rates is based on the assumption that the size of each transaction is equal or close to the average size of all transactions. This might not be true in many real datasets. A simple solution is to cluster the transactions into multiple groups such that in each group, the size of each transaction is equal or close to the average size of the transactions in the group. Across all groups, the same number k of hash functions are used for generating Bloom filters. In each group, $k = \frac{m}{n} \ln 2$ is optimized for the length m of Bloom filters and the average size n of the transactions in the group. This means that for different groups, the length of Bloom filters are different depending on the average size of transactions in the group. Roughly speaking, long Bloom filters will be used for long transactions, while short Bloom filters for short transactions.

Using different Bloom filters for different groups of transactions will not only save storage requirement but also increase the precision of data mining. Note that the bounds presented in the previous subsection are based on k. Since the same k is optimized across multiple groups, the bounds can be used to estimate the false positive and false negative rates in all groups.

In the case of multiple groups, the transactions are represented by the Bloom filters of different lengths. This requires that each candidate itemset be represented by Bloom filters of different lengths too. This is because in our data mining problem, the Bloom filter of each candidate itemset needs to be checked against the Bloom filters of the same length. To avoid transmitting multiple Bloom filters for each candidate itemset between client and server, only the longest Bloom filter of each candidate itemset is sent to a second party at one time. In data mining process, the mining program run by the second party needs to transform it into different lengths so as to check it against different groups of Bloom filters. We provide a simple solution called δ-folding to transform the Bloom filters.

Definition 7. *Given an m-bit Bloom filter B, δ-folding of B, denoted as B_δ, is defined as a δm-bit vector generated from*

$$B : bit(B_\delta, i) = bit(B, i) \bigvee_{\substack{0<j\leqslant m \\ j=i \mod \delta m}} bit(B, j)$$

for $0 < i \leqslant \delta m$, where $bit(B, i)$ denotes the ith bit of B, $\vee$ the bitwise OR, and $0 < \delta \leqslant 1$.

Example 1. Let B be a 10-bit Bloom filter. $B_{0.7}$ is defined as a 7-bit vector $B_{0.7}$: $bit(B_{0.7}, i) = bit(B, i) \vee bit(B, i + 7)$ for $i = 1, 2, 3$, and $bit(B_{0.7}, i) = bit(B, i)$ for $i = 4, 5, 6, 7$.

From the definition of Bloom filter, vector B_δ is actually a Bloom filter generated with the same k hash functions which are used for generating Bloom filter B.

Given a Bloom filter $B(S)$ of longest length m of candidate itemset S, the frequency $Bfq(S)$ is computed by checking $B(S)$ against the Bloom filters of transactions in all groups. For a particular group in which the Bloom filters have length $m' \leqslant m$ (note that m is the longest length for all groups), the mining program applies $\frac{m'}{m}$-folding to $B(S)$ such that the transformed Bloom filter has the length m'.

References

1. Agrawal, R., A. Evfimievski, and R. Srikant. Information sharing across private databases. In *Proceedings of the 2003 ACM SIGMOD International Conference on Management of Database*, pages 86–97, San Diego, California, 2003.
2. Agrawal, R., T. Imielinski, and A. Swami. Mining association rules between sets of items in large databases. In *Proceedings of the 1993 ACM SIGMOD International Conference on Management of Database*, pages 207–216, 1993.
3. Agrawal, R., J. Kiernan, R. Srikant, and Y. Xu. Hippocratic databases. In *Proceedings of the 28th International Conference on Very Large Data Bases (VLDB'02)*, pages 143–154, Hong Kong, China, August 20–23, 2002.
4. Agrawal, R., J. Kiernan, R. Srikant, and Y. Xu. Order preserving encryption for numeric data. In *Proceedings of the ACM SIGMOD ICMD*, pages 563–574, 2004.
5. Agrawal, R. and R. Srikant. Faster algorithms for mining association rules in large databases. In *Proceedings of the 20th International Conference on Very Large Data Bases (VLDB'94)*, pages 487–499, Santiago de Chile, Chile, September 12–15, 1994.

6. Agrawal, R. and R. Srikant. Privacy preserving data mining. In *Proceedings of the 2000 ACM SIGMOD International Conference on Management of Database*, pages 439–450, Texas, USA, May 16–18, 2000.
7. Agrawal, S. and J. R. Haritsa. A framework for high-accuracy privacy-preserving mining. In *Proceedings of the 21th IEEE International Conference on Data Engineering (ICDE'05)*, pages 193–204, Tokyo, Japan, 2005.
8. Apte, C., B. Liu, E. Pednault, and P. Smyth. Business applications of data mining. *Communications of the ACM*, 45(8):49–53, 2002.
9. Atallah, M., E. Bertino, A. K. Elmagarmid, M. Ibrahim, and V. S. Verykios. Disclosure limitation of sensitive rules. In *Proceedings of the IEEE KDEE*, pages 45–52, 1999.
10. Bertino, E. and I.-N. Fovino. A framework for evaluating privacy preserving data mining algorithms. *Data Mining and Knowledge Discovery*, 11(2):121–154, 2005.
11. Bloom, B. Space time tradeoffs in hash coding with allowable errors. *Communications of the ACM*, 7(13):422–426, 1970.
12. Chen, Y.-L., K. Tang, R.-J. Shena, and Y.-H. Hu. Market basket analysis in a multiple store environment. *Decision Support Systems*, 40(2):339–354, 2005.
13. Chor, B., O. Goldreich, E. Kushilevitz, and M. Sudan. Private information retrieval. *Journal of the ACM*, 45(6):965–982, 1998.
14. Culnan, M.-J. Gaint food and elensys: Looking out for customers or gross privacy invasion? *Communications of AIS*, 16:317–328, 2005.
15. Evfimievski, A., R. Srikant, R. Agrawal, and J. Gehrke. Privacy preserving mining of association rules. In *Proceedings of the 8th ACM SIGKDD KDD 2002*, pages 217–228, 2002.
16. Kantarcıoğlu, M. and C. Clifton. Privacy preserving distributed mining of association rules on horizontally partitioned data. In *Proceedings of the ACM SIGMOD Workshop on Research Issues on Data Mining and Knowledge Discovery*, pages 24–31, 2002.
17. Kantarcıoğlu, M., J. Jin, and C. Clifton. When do data mining results violate privacy? In *Proceedings of the 10th ACM SIGKDD KDD 2004*, pages 599–604, 2004.
18. Kargupta, H., S. Datta, Q. Wang, and K. Sivakumar. On the privacy preserving properties of random data perturbation techniques. In *Proceedings of the 3rd IEEE ICDM*, pages 99–106, 2003.
19. Lin, Q.-Y., Y.-L. Chen, J.-S. Chen, and Y.-C. Chen. Mining inter-organizational retailing knowledge for an alliance formed by competitive firms. *Information & Management*, 40(5):431–442, 2003.
20. Milne, G.-R. Privacy and ethical issues in database/interactive marketing and public policy: A research framework and overview of the special issue. *Journal of Public Policy and Marketing*, 19:1–6, 2000.
21. Oliveira, S. and O. Zaiane. Protecting sensitive knowledge by data sanitization. In *Proceedings of the 3rd IEEE ICDM*, pages 211–218, 2003.
22. Padmanabhan, B. and A. Tzhilin. On the use of optimization for data mining: Theoretical interactions and eCRM opportunities. *Management Science*, 49(10):1327–1343, 2003.
23. Pinkas, B. Cryptographic techniques for privacy preserving data mining. *ACM SIGKDD Explorations*, 4(2):12–19, 2002.

24. Qiu, L., Y. Li, and X. Wu. An approach to outsourcing data mining tasks while protecting business intelligence and customer privacy. In *Workshops Proceedings of the 6th IEEE International Conference on Data Mining (ICDM 2006)*, pages 551–558, Hong Kong, China, December 18–22, 2006.
25. Qiu, L., Y. Li, and X. Wu. Protecting business intelligence and customer privacy while outsourcing data mining tasks. *Knowledge and Information Systems: An International Journal*, Online first, November 17, 2007.
26. Qiu, L., K.-L. Ong, and S. M. Lui. Safely outsourcing data mining tasks. In *Proceedings of the 5th Australian Data Mining Conference (AusDM 2006)*, pages 1–8, Sydney, Australia, November 29–30, 2006.
27. Rayardo, R.-J. and R. Srikant. Technological solutions for protecting privacy. *IEEE Computer*, 36(9):115–118, 2003.
28. Rizvi, S. and J. Haritsa. Maintaining data privacy in association rule mining. In *Proceedings of VLDB'02*, pages 682–693, 2002.
29. Saygin, Y., V. S. Verykios, and C. Clifton. Using unknowns to prevent discovery of association rules. *Sigmod Record*, 30(4):45–54, 2001.
30. Vaidya, J. and C. Clifton. Privacy preserving association rule mining in vertically partitioned data. In *Proceedings of the 8th ACM SIGKDD KDD*, pages 639–644, 2002.
31. Vaidya, J. and C. Clifton. Privacy-preserving data mining: Why, how, and when. *IEEE Security and Privacy*, 2(6):19–27, 2004.
32. Yao, A. C.-C. How to generate and exchange secrets. In *Proceedings of the 27th IEEE Symposium on Foundations of Computer Science (FOCS'86)*, pages 162–167, Xi'an, China, 1986.

Kernels for Text Analysis

Evgeni Tsivtsivadze, Tapio Pahikkala, Jorma Boberg, and Tapio Salakoski

Turku Centre for Computer Science (TUCS), Department of Information Technology, University of Turku, Turku, Finland,
Evgeni.Tsivtsivadze@it.utu.fi, Tapio.Pahikkala@it.utu.fi,
Jorma.Boberg@it.utu.fi, Tapio.Salakoski@it.utu.fi

Summary. During past decade, kernel methods have proved to be successful in different text analysis tasks. There are several reasons that make kernel based methods applicable to many real world problems especially in domains where data is not naturally represented in a vector form. Firstly, instead of manual construction of the feature space for the learning task, kernel functions provide an alternative way to design useful features automatically, therefore, allowing very rich representations. Secondly, kernels can be designed to incorporate a prior knowledge about the domain. This property allows to notably improve performance of the general learning methods and their simple adaptation to the specific problem. Finally, kernel methods are naturally applicable in situations where data representation is not in a vectorial form, thus avoiding extensive preprocessing step. In this chapter, we present the main ideas behind kernel methods in general and kernels for text analysis in particular as well as provide an example of designing feature space for parse ranking problem with different kernel functions.

1 Introduction

Nowadays, when the amount of textual data available in digital form is growing rapidly, kernel based methods [1–4] are becoming increasingly important for automatic processing of natural language text documents. One key difference between traditional statistical and kernel based approach is that the former deals mainly with the data that is represented in vector form, whereas the latter can be naturally applied to non-vectorial types of data.

Recently, kernel methods have proved to be successful in many natural language processing tasks such as machine translation, document classification, parse reranking, etc. The approach used by kernel methods is to map the input data into the feature space by means of the kernel function and then use learning algorithm to discover relations in that space. This technique provides an advantage of detecting linear relationships in the feature space by using simple and efficient learning algorithms. The constructed feature space can possibly be of infinite dimensions to ensure necessary representational power. However,

E. Tsivtsivadze et al.: *Kernels for Text Analysis*, Studies in Computational Intelligence (SCI) **116**, 81–97 (2008)
www.springerlink.com

using a kernel function, learning algorithm can operate in that feature space without the need of explicitly computing coordinates of the data points, but rather using the inner products between all pairs of data points.

There are several reasons that make kernel based methods applicable to many real world problems especially in domains where data is not naturally represented in a vector form. Firstly, kernel methods allow an efficient computation of the mutual similarities of the data instances in a high dimensional feature space. Secondly, kernels can be designed to incorporate a prior knowledge about the domain. This property allows to notably improve performance of the general learning methods and their simple adaptation to the specific problem. Finally, kernel methods are naturally applicable in situations where data representation for the problem is not in a vectorial form, avoiding extensive preprocessing step. Kernels can be used on structured instances directly, decreasing the amount of preprocessing time and taking into account the structure of the data.

In this chapter, we aim at presenting main ideas behind kernel functions as well as describe some data representations for text analysis. We demonstrate how to construct a kernel function that is appropriate for the task in question and provide illustrative examples of different kernels. We start with basic principles for constructing kernels for natural language processing tasks and formulate "bag of words" kernels [5], kernel for strings [6], word-sequence [7] and other types of kernels applicable for sequential data. We also demonstrate how to incorporate prior knowledge about problem domain into the kernel function. We continue by presenting general convolution framework [8, 9] for constructing kernels and show usefulness of the approach. Finally, we conclude the chapter by providing in depth description and the application of graph kernels [10, 11] to the task of parse ranking.

2 Kernel Methods

A kernel based method consists of two main parts, namely a kernel function that computes the similarities of the data instances in a feature space and the actual leaning algorithm. The kernel function can be considered as an interface that provides the learning algorithm with the inner products of the feature vectors. Next, we go through the main properties of the kernel functions and continue by presenting examples of kernels for sequential data.

2.1 General Properties of Kernels

Below we give brief introduction into main properties of kernel function and some basic methods for their construction, for in depth review, we refer reader to [1–4]. The usage of the kernels in the learning algorithm enables us to take advantage of the high dimensional feature space without computational penalty that usually grows with the number of dimensions. This is done by evaluating the inner products of the feature vectors without explicitly computing their coordinates. Thus, the key properties we need to consider are the properties of inner product spaces.

A vector space H is an inner product space if there exists a real valued symmetric bilinear map $\langle \cdot, \cdot \rangle$ that satisfies

$$\langle z, z \rangle \geq 0, \forall z \in H \tag{1}$$

and

$$\langle z, z \rangle = 0 \text{ if and only if } z = 0. \tag{2}$$

Usually inner product space is also known as Hilbert space with additional properties of separability and completeness. In fact any feature space induced by a kernel function is a Hilbert space. Thus, properties of Hilbert spaces provide us with a formal basis of constructing and understanding new kernel functions. The map $\phi : X \longrightarrow H$ is called feature map from the input space X into the feature space H. We use a kernel function to evaluate the inner product in the feature space with some feature map ϕ. Therefore we can write following equality:

$$\langle \phi(x_i), \phi(x_j) \rangle = k(x_i, x_j). \tag{3}$$

One common way of constructing new valid kernel functions is by using closure properties of kernels. Recall that, if we can demonstrate that some function can be represented as an inner product in the feature space, it is indeed a valid kernel function. If k_1 and k_2 are kernels over $X \times X$, $\alpha \in \mathbb{R}_+$, $f(\cdot)$ is a real valued function on X, $\phi : X \longrightarrow H$ and k_3 is a kernel over $H \times H$, and B is a positive semidefinite matrix of dimension $n \times n$, then the following functions are kernels:

$$\begin{aligned}
k(x_i, x_j) &= k_1(x_i, x_j) + k_2(x_i, x_j) \\
k(x_i, x_j) &= \alpha k_1(x_i, x_j) \\
k(x_i, x_j) &= k_1(x_i, x_j) k_2(x_i, x_j) \\
k(x_i, x_j) &= f(x_i) f(x_j) \\
k(x_i, x_j) &= k_3(\phi(x_i), \phi(x_j)) \\
k(x_i, x_j) &= x_i^{\mathrm{T}} B x_j \text{ when } X = \mathbb{R}^n
\end{aligned}$$

Let us consider a simple example of the features vectors and their inner products.

2.2 Bag of Words Kernel

Let $S = \{x_1, \ldots, x_m\}$ be a set of documents. Consider that we also are given a dictionary W consisting of N words. The documents are similar if they share large amount of words. We define a function $\phi : S \longrightarrow \mathbb{R}^N$,

$$\phi(x) = (fr(w_1, x), fr(w_2, x), \ldots, fr(w_N, x)), w_i \in W, 1 \leq i \leq N, \tag{4}$$

where $fr(w_i, x)$, etc. represents frequency of the occurrence of the word w_i in the document x. This type of representation is usually called bag of words. One of the key aspects of this representation is that documents are described as vectors and thus, inner product can be used to compute their similarity. One of the drawbacks of this approach is that although documents are represented by the words they contain, their ordering, as well as punctuation, and other grammatical information in the document is lost. In addition, semantical information disappears as well, that is, the text in the document can not be understood anymore.

Bag of words representation is probably one of the simplest constructions used in text processing. Recently, researchers proposed kernel functions that are not only able to utilize syntactical information within the text but also take advantage of semantics (see e.g., [12]).

2.3 String Kernels

Strings of symbols naturally occur not only in various problems related to text analysis, but also in other areas such as bioinformatics, where data is represented as a sequence of amino acids, etc. Recently, several string kernels that compute inner product between the feature vectors of the sequences in high dimensional feature spaces were proposed.

Let us consider an alphabet Σ of symbols from which the strings are composed. In each case the kernel function is defined by an explicit feature map of all finite sequences from the alphabet Σ to a vector set indexed by a set of subsequences of length k from Σ. We call a k-length subsequence occurring in the input sequence a k-gram, and the set of all k-grams is defined by Σ^k.

One of the simplest of available string kernels is the spectrum kernel [13]. Informally, the spectrum kernel can be considered as a histogram of frequencies of all its contiguous substrings of length k. In other words, the k-spectrum kernel function is an inner product of the k-spectra of the two evaluated sequences. A straightforward generalization of the spectrum kernel is the possibility to allow mismatches between compared k-grams. A mismatch kernel [13] is defined via feature map $\Phi_{(k,m)}$ to the $|\Sigma|^k$ - dimensional vector space indexed by k-grams from Σ. For a k-gram $\alpha = \alpha_1 \dots \alpha_k$, where $\alpha_j \in \Sigma$, let $N_{k,m}(\alpha)$ be a set of all k-gram β from Σ that differ from α by at most m characters. For any k-gram α, the feature map is defined by

$$\Phi_{(k,m)}(\alpha) = (\phi_\beta(\alpha))_{\beta \in \Sigma^k},$$

where $\phi_\beta(\alpha) = 1$ if $\beta \in N_{k,m}(\alpha)$, and $\phi_\beta(\alpha) = 0$ otherwise. Then by extending the feature map to a sequence x of any length, we obtain

$$\Phi_{(k,m)}(x) = \sum_{k\text{-grams } \alpha \text{ in } x} \Phi_{(k,m)}(\alpha).$$

Each of the k-grams contributes to the coordinates in the neighborhood set, and thus $\Phi_{(k,m)}(x)$ represents a vector with the counts of all instances of k-grams with up to m mismatches in x. The (k,m)-mismatch kernel function represents the inner product between these vectors and is expressed as follows:

$$\kappa(x,y) = \left\langle \Phi_{(k,m)}(x), \Phi_{(k,m)}(y) \right\rangle .$$

The spectrum kernel is obtained from this by setting $m = 0$.

2.4 Gappy String Kernels

For the (g,k)-gappy string kernel [13], where $g \geq k$, we consider the same alphabet Σ and feature space as in mismatch kernel. The key difference between these kernels is the feature map that now is defined to adapt gappy matches of g-grams to k-grams. For a g-gram $\alpha = \alpha_1, \ldots, \alpha_g$, where $\alpha_j \in \Sigma$, let $G_{g,k}(\alpha)$ be a set of all g-grams β from Σ that occur in α by having at most $g - k$ gaps. Then the feature map on α is defined by

$$\Phi_{(g,k)}(\alpha) = (\phi_\beta(\alpha))_{\beta \in \Sigma^k},$$

where $\phi_\beta(\alpha) = 1$ if $\beta \in G_{k,m}(\alpha)$, and $\phi_\beta(\alpha) = 0$ otherwise. Then by extending the feature map on a sequence $x \in \Sigma$ of any length, we obtain

$$\Phi_{(g,k)}(x) = \sum_{g\text{-grams } \alpha \text{ in } x} \Phi_{(g,k)}(\alpha).$$

Again, similarly to the mismatch kernel, the gappy kernel is defined by taking inner product between feature vectors x and y.

On the other hand, we can also count the number of occurrences of each k-subsequence and weigh each occurrence by the number of gaps it contains. Then we can define for a k-gram α and g-gram β the weighting as follows:

$$\phi_\beta^\lambda(\alpha) = \frac{1}{\lambda^k} \sum_{1 \leq i_1 < i_2 < i_3 < \ldots < i_k \leq g \text{ where } \alpha_{i_j} = \beta_j \text{ for } j=1\ldots k} \lambda^{i_k - i_1 + 1},$$

where the weight parameter satisfies $0 < \lambda < 1$. Thus, we can obtain the weighted version of the gappy kernel from the feature map

$$\Phi_{(g,k,\lambda)}(x) = \sum_{g\text{-grams } \alpha \text{ in } x} (\phi_\beta^\lambda(\alpha))_{\beta \in \Sigma^k}. \tag{5}$$

This feature map was initially proposed by [13] and is related to the one proposed by [6]. The main difference of these feature maps is that in (5) only subsequences of length k that occur with most $k - g$ gaps contribute to the corresponding k-gram feature, whereas in [6] all gappy occurrences are considered. On the other hand, it is easy to approximate feature map proposed by [6] by slightly modifying (5) so that the first character of the k-gram occurs in the first position of the g-gram [13]

$$\hat{\phi}_{\beta}^{\lambda}(\alpha) = \frac{1}{\lambda^k} \sum_{1=i_1<i_2<i_3<\ldots<i_k \leq g \text{ where } \alpha_{i_j}=\beta_j \text{ for } j=1\ldots k} \lambda^{i_k - i_1 + 1}$$

and $0 < \lambda < 1$. This feature map is extended to the sequences of any length in exactly the same way as described above.

2.5 Convolution Kernels

Convolution kernels are defined between the input objects by applying convolution sub-kernels for the parts of the objects. Following [8], we briefly describe the convolution kernel framework. Let us consider $x \in X$ as a composite structure such that $x_1, \ldots, x_N$ are its parts, where x_n belongs to the set X_n for each $1 \leq n \leq N$, and N is a positive integer. We consider $X_1, \ldots, X_n$ as countable sets, however, they can be more general separable metric spaces [8]. Let us denote shortly $\widehat{x} = x_1, \ldots, x_N$. Then the relation "$x_1, \ldots, x_N$ are parts of x" can be expressed as a relation R on the set $X_1 \times \ldots \times X_N \times X$ such that $R(\widehat{x}, x)$ is true if $\widehat{x}$ are the parts of x. Then we can define $R^{-1}(x) = \{\widehat{x} : R(\widehat{x}, x)\}$. Now let us suppose that $x, y \in X$ and there exist decompositions such that $\widehat{x} = x_1, \ldots, x_N$ are the parts of x and $\widehat{y} = y_1, \ldots, y_N$ are the parts of y. If we have the following kernel functions

$$k_n(x_n, y_n) = \langle \Phi(x_n), \Phi(y_n) \rangle, 1 \leq n \leq N,$$

to measure similarity between elements of X_n, then the kernel $k(x, y)$ measuring the similarity between x and y is defined to be the following generalized convolution:

$$k(x, y) = \sum_{\widehat{x} \in R^{-1}(x)} \sum_{\widehat{y} \in R^{-1}(y)} \prod_{n=1}^{N} k_n(x_n, y_n). \tag{6}$$

To demonstrate how presented above string kernels could be considered within convolution framework let us consider two strings p, q and let $p = (p_1 \ldots p_{|p|})$ and $q = (q_1 \ldots q_{|q|})$ be the characters these strings consist of. The similarity of p and q is obtained with kernel

$$k(p, q) = \sum_{i=1}^{|p|} \sum_{j=1}^{|q|} \kappa(i, j). \tag{7}$$

By defining κ in the general formulation (7), we obtain different similarity functions between strings. If we set $\kappa(i, j) = \delta(p_i, q_j)$, where

$$\delta(x, y) = \begin{cases} 0, & \text{if } x \neq y \\ 1, & \text{if } x = y \end{cases}$$

then (7) equals to the number of matching characters in two strings. Then the spectrum kernel is obtained by using

$$\kappa(i,j) = \prod_{l=0}^{h-1} \delta(p_{i+l}, q_{j+l}), \tag{8}$$

in (7). The intuition behind mismatch kernel is that similarity between two sequences is large if they share many similar subsequences. By restricting number of mismatches to m between the subsequences of length h, the (h,m)-mismatch kernel is obtained by using

$$\kappa(i,j) = \begin{cases} 0, & \text{if } \sum_{l=0}^{h-1} \delta(p_{i+l}, q_{j+l}) < h - m \\ 1, & \text{otherwise} \end{cases} \tag{9}$$

in (7). The spectrum kernel (8) is a special case of the mismatch kernel where $m = 0$. Note, that we consider characters as mismatched in (8) and (9), when the indices $i + l$ and $j + l$ are not valid, that is, $i + l > |p|$ or $j + l > |q|$.

The above presented example of calculation is simple, but not an efficient way of computing string kernels. It is provided solely for explanation purposes and actual computations should not be conducted in this manner. It can be observed that computing cost of using this approach is $O(k|p||q|)$. A much more efficient approach is to use tree data structure as described in [4, 13] that could reduce the cost to $O(k(|p| + |q|))$. Using this approach each k-gram feature corresponds to the path from the root to the leaf node of the tree and the data structure is used to organize a traversal of all the inexact matching instance patterns in the data that contribute to each k-gram feature count.

2.6 Graph Kernels

Another type of kernels widely used with textual data are graph kernels. Graph kernels represent drift from string type of representation described previously and are usually applied in the situations when graph based annotation of text is naturally present. For example, Figure 1 depicts a sentence parsed with the link grammar parser [14] that adds to the text syntactical annotation describing different relations. Thus, we obtain a structure that naturally can be represented as a graph rather than as a sequence, although sequential representation described in [15] is also possible. Graph kernels can be used to take advantage of these graph based structures and information contained within them.

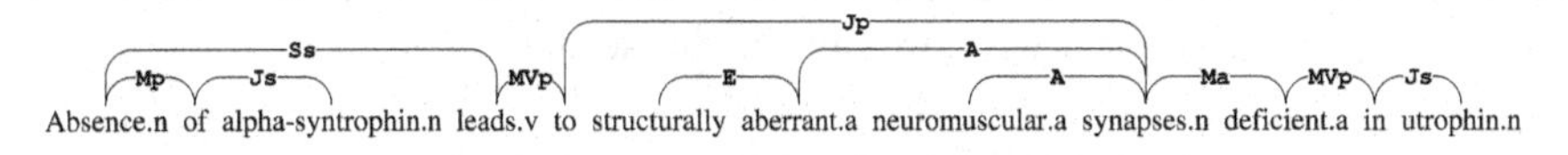

Fig. 1. Example of parsed sentence

We follow [10,16], and define a labeled graph representation for the data. We denote the set of real valued matrices of dimension $i \times j$ by $\mathcal{M}_{i\times j}(\mathbb{R})$ and $[M]_{i,j}$ defines the element of matrix M that is located in the i-th row and j-th column. To enumerate all possible labels that could occur in the graph we use the set $\mathcal{L} = \{l\}_r, r \in \mathbb{N}$. Then let $G = (V, E, h)$ be a graph that consists of the set of vertices V, the set of edges $E \subseteq V \times V$, and a function $h : V \to \mathcal{L}$ that assigns a label to each vertex of a graph. We represent the edge set of G as an adjacency matrix $A \in \mathcal{M}_{|V|\times|V|}(\mathbb{R})$ whose rows and columns are indexed by the vertices V, and

$$[A]_{i,j} = \begin{cases} 1, & \text{if } (v_i, v_j) \in E \\ 0, & \text{otherwise.} \end{cases}$$

We also assume that the function h is represented as a label allocation matrix $L \in \mathcal{M}_{|\mathcal{L}|\times|V|}(\mathbb{R})$ so that

$$[L]_{i,j} = \begin{cases} 1, & \text{if } v_j \in V \wedge l_i \in \mathcal{L} \\ 0, & \text{otherwise.} \end{cases}$$

In following we define a class of kernel functions on labeled graphs that are closely related to the ones described in [16]. Let us consider the nth power A^n of the adjacency matrix of the graph G. It is easy to show that $[A^n]_{i,j}$ is the number of walks of length n from vertex v_i to vertex v_j. By taking into account the labels of the vertices, we observe that $[LA^nL^{\mathrm{T}}]_{i,j}$ is the number of walks of length n between vertices labeled l_i and l_j. Let G and G' be labeled undirected graphs and let $\langle M, M'\rangle_F$ denote the Frobenius product of matrices M and M', that is, $\langle M, M'\rangle_F = \sum_{i,j}[M]_{i,j}[M']_{i,j}$. Let further $\gamma \in \mathcal{M}_{n\times n}(\mathbb{R})$ be a positive semidefinite matrix. We define the kernels k_n between the graphs G and G' as follows:

$$k_n(G, G') = \textstyle\sum_{i,j=0}^{n}[\gamma]_{i,j}\langle LA^iL^{\mathrm{T}}, L'A'^jL'^{\mathrm{T}}\rangle_F. \tag{10}$$

By specializing $[\gamma]_{i,j}$ in (10), we obtain several kernel functions with different interpretations. For example, if we set $[\gamma]_{i,j} = \theta^i\theta^j$, where $\theta \in \mathbb{R}^+$ is a parameter, we obtain the kernel

$$\widehat{k_n}(G, G') = \langle L\Big(\sum_{i=0}^{n}\theta^iA^i\Big)L^{\mathrm{T}}, L'\Big(\sum_{i=0}^{n}\theta^iA'^i\Big)L'^{\mathrm{T}}\rangle_F. \tag{11}$$

This kernel can be interpreted as an inner product in a feature space in which there is a feature $\phi_{k,l}$ per each label pair (k, l) so that its value $\phi_{k,l}(G)$ for a graph G is a weighted count of walks of length up to n from the vertices labeled l to the vertices labeled k. On the other hand, by setting $[\gamma]_{i,j} = \theta^i$ when $i = j$ and zero otherwise, we obtain the kernel

$$\widetilde{k_n}(G, G') = \sum_{i=0}^{n}\theta^{2i}\langle LA^iL^{\mathrm{T}}, L'A'^iL'^{\mathrm{T}}\rangle_F, \tag{12}$$

which can be interpreted as an inner product in a feature space in which there is a feature $\phi_{i,k,l}$ per each tuple (i, k, l), where l and k are labels and i is a length of a walk. Its value $\phi_{i,k,l}(G)$ for a graph G is θ^i times the count of walks of length i from the vertices labeled l to the vertices labeled k. Finally, with certain conditions on the coefficients $[\gamma]_{i,j}$, we can also define $k_\infty(G, G') = \lim_{n\to\infty} k_n(G, G')$. One such kernel function is, for example, the exponential graph kernel

$$k_{exp}(G, G') = \langle Le^{\beta A}L^{\mathrm{T}}, L'e^{\beta A'}L'^{\mathrm{T}}\rangle_F, \tag{13}$$

where β is a parameter and $e^{\beta A}$ can be written as

$$e^{\beta A} = \lim_{n\to\infty}\sum_{i=0}^{n}\frac{\beta^i}{i!}A^i. \tag{14}$$

In this case, the coefficients $[\gamma]_{i,j}$ are determined by the parameter β as follows:

$$[\gamma]_{i,j} = \frac{\beta^i}{i!}\frac{\beta^j}{j!}. \tag{15}$$

3 Application

In Sects. 2.3 and 2.6, we described string and graph kernels that could be in particular useful when applied to textual data in classification, ranking or regression problems. In this section, we give an example of an application and demonstrate how these kernels could be adapted for it.

3.1 Bag of Features

The features used by a learning algorithm are essential to its performance, and in the problem that deals with textual data, particular attention to the extracted features is required due to the sparseness. Moreover, when learning from textual data, going beyond simple bag of words approach can be beneficial. One possibility for this is to include syntactical information of the text into the feature space by using a parser. A typical example of parsed sentence is presented in Fig. 1. Below we provide possible features that can be extracted from a parsed sentence as described in [17]. These features are grammatically relevant and applicable even when relatively few training examples are available.

The output of the Link Grammar (LG) parser [14] contains the following information for each input sentence: the linkage consisting of pairwise dependencies between pairs of words termed links, the link types (the grammatical roles assigned to the links), and the part-of-speech (POS) tags of the words. As LG does not perform any morphological analysis, the POS tagset used by LG is limited, consisting mostly of generic verb, noun and adjective categories. Different parses of a single sentence have a different combination of these elements. Each of the features we use are described below.

Grammatical bigram. This feature is defined as a pair of words connected by a link. In the example linkage of Fig. 1, the extracted grammatical bigrams are *absence—of*, *of—alpha-syntrophin*, *absence—leads*, etc. These grammatical bigrams can be considered a lower-order model related to the grammatical trigrams proposed as the basis of a probabilistic model of LG in [18]. Grammatical bigram features allow the learning machine to identify words that are commonly linked, such as *leads—to* and *binds—to*. Further, as erroneous parses are provided in training, the learning machine also has the opportunity to learn to avoid links between words that should not be linked.

Word & POS tag. This feature contains the word with the POS tag assigned to the word by LG. In the example, the extracted word & POS features are *absence.n*, *alpha-syntrophin.n*, *leads.v*, etc. Note that if LG does not assign POS to a word, no word & POS feature is extracted for that word. These features allow the ranker to learn preferences for word classes; for example, that "binds" occurs much more frequently as a verb than as a noun in the domain.

Link type. In addition to the linkage structure and POS tags, the parses contain information about the link types used to connect word pairs. The link types present in the example are *Mp*, *Js*, *Ss*, etc. The link types carry information about the grammatical structures used in the sentence and allow the ranker to learn to favor some structures over others.

Word & Link type. This feature combines each word in the sentence with the type of each link connected to the word, for example, *absence—Mp*, *absence—Ss*, *of—Js*, etc. The word & link type feature can be considered as an intermediate between grammatical unigram and bigram features, and offers a possibility for addressing potential sparseness issues of grammatical bigrams while still allowing a distinction between different linkages, unlike unigrams. This feature can also allow the ranker to learn partial selectional preferences of words, for example, that "binds" prefers to link directly to a preposition.

Link length. This feature represents the number of words that a link in the sentence spans. In Fig. 1, the extracted features of this type are *1*, *1*, *3*, etc. This feature allows the ranker to learn the distinction between parses, which have different link length. The total summed link length is also used as a part of LG ordering heuristics, on the intuition that linkages with shorter link lengths are preferred [14].

Link length & link Type. This feature combines the type of the link in the sentence with the number of words it spans. In Fig. 1, the extracted features of this type are *1—Mp*, *1—Js*, *3—Ss*, *1—MVp*, etc. The feature is also related to the total length property applied by the LG parser heuristics, which always favor linkages with shorter total link length. However, the link length & link type feature allows finer distinctions to be made by the ranker, for example, favoring short links overall but not penalizing long links to prepositions as much as other long links.

Link bigram. The link bigram features extracted from the parse are combinations of two links connected to the word, ordered leftmost link first. In the example, the link bigrams are *Mp—Ss*, *Mp—Js,Ss—MVp*, etc. When there are more than two links associated with a word, we extract all link pairs. For example, all link bigrams that have word 'synapses' in common are *Jp—A,Jp—A,Jp—Ma, A—A, A—Ma, A—Ma.*

Presented above features can be joined into a feature vector and, therefore, every parse can be treated as a bag of features (BOF). This would provide us with additional information about the parse, that extends beyond simple bag of words representation. On the other hand we can design a feature space using graph kernels by utilizing natural structure of the parse as follows.

3.2 Graph Representation

Let p be a parse generated from a sentence s, and let $G = (V, E, h)$ denote the graph representation of p. The graph representation of the example parse of Fig. 1 is presented in Fig. 2. Let us first define the vertices of p. For each word in the sentence s, there is a corresponding vertex $v \in V$ and the vertex is labeled with the word (the word vertices and their labels do not depend from the parse). Thus, if a word occurs several times in the sentence s, all of the occurrences have their own vertices in the graph but the labels of the vertices are equal. For each link in the parse p, there is a corresponding vertex in V that is labeled with its link type. Similarly to the word vertices, a parse may have several occurrences of the same link type. In p, each word is assigned a part-of-speech, for which there is a corresponding vertex in V labeled with the part-of-speech. Further, each link in p has a length (the number of words that a link in the sentence spans) for which there is a corresponding vertex in V labeled with the length. In Fig. 2, they are the vertices labeled with integers, for example, *1*, *1*, *3*, etc.

The edges of G are defined as follows. A word vertex and its corresponding part-of-speech vertex are connected with an edge. A link vertex and its corresponding length vertex are connected with an edge. This type of edge allows the ranker to learn the distinction between parses, which have different link length. If two words are connected with a link in the parse p, the corresponding link vertex is connected with an edge to both of the corresponding word vertices. The connection of a word vertex in the graph with a link vertex (for example in Fig. 2: *absence—Mp*, *absence—Ss*, *of—Js*, etc.) can be considered

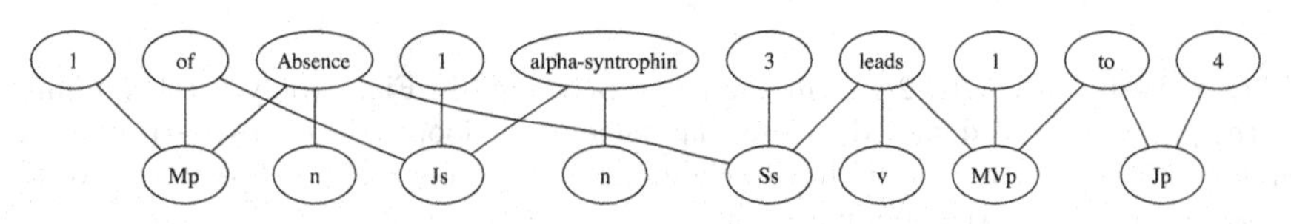

Fig. 2. Graph representation of the parse presented in Fig. 1 (only the beginning of the parse graph is depicted)

as the word & link type feature. Below, we demonstrate how these connections are used to create word bigram and link bigram described previously [17].

Random Walk features. From the graph representation of the parses, we observe that the graphs have the bipartite property which is defined as follows. A graph G is called bipartite, if its vertex set V has a partition to two subsets $X \subseteq V$ and $Y \subseteq V$ such that each edge $(u, v) \in E$ connects a vertex of X and a vertex of Y. In our case, the vertices that correspond to the words or to the link lengths form the first subset, and the other vertices belong to the second subset. Let $G = (V, E, h)$ be a graph representation of a parse. Let us consider the second power of the adjacency matrix A of G, that is, the walks of length 2 in the graph representation of the corresponding parse. Those walks in the graph representation of the example parse of Fig. 1 are illustrated in Fig. 3. Due to the bipartite property of G, all of the walks of length 2 have both the start and the end vertices in the same subset. Among the walks of length 2 there is, for example, a walk between two word vertices if they are connected with a link in the parse, and between two link vertices if the links are connected to the same word in the parse. Such connections are called grammatical bigrams and link bigrams in [17]. The grammatical bigrams can be considered a lower-order model related to the grammatical trigrams proposed as the basis of a probabilistic model of LG in [18]. Grammatical bigram features allow the learning machine to identify words that are commonly linked, such as *leads—to* and *binds—to*. Further, as erroneous parses are provided in training, the learning machine also has the opportunity to learn to avoid links between words that should not be linked. We also obtain walks between part-of-speech vertices and link vertices, and between word vertices and link length vertices. Finally, a vertex has as many cycles (cycle is a walk that starts and ends at the same vertex) as there are edges connected to it. If we consider the higher powers of the adjacency matrices, we obtain new features, for example, link length pairs in the fourth power. In the higher powers, we also obtain word and link bigrams, where the words and links are not connected to each other in the parse.

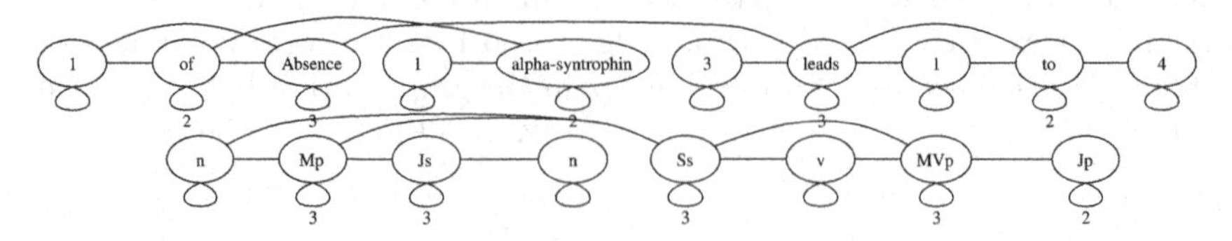

Fig. 3. Walks of length 2 within the parse presented in Fig. 1 (only the beginning of the parse graph is depicted). Note that some of the loops (the edges starting from and ending to the same node) have weights larger than one, that is, there are several repetitions of the corresponding walks

3.3 Evaluation Using Bag of Features

To evaluate the proposed features described in Sect. 3.1 on the problem of parse ranking, we conduct experiment on 1,100 annotated sentences from BioInfer corpus [19]. For detailed description of the experimental setup, we refer to [17].

We divide the set of 1,100 sentences in two parts so that the parses originating from the first 500 sentences form a training set and the parses from the latter 600 sentences are used to form a validation set. The instances in the training and validation sets consist of the graph representations of the candidate parses and their F-scores are used as the corresponding target output values. The task of the learning machine is, for a sentence, to rank its candidate parses in the order of their score values. The learning algorithm is, in fact, trained to regress the score values of the parses, but here, we are only interested of the ranking of the candidate parses for each sentence. Therefore, we measure the ranking performance by calculating Kendall's τ_b correlation coefficient [20] for the parse candidate set of each sentence. Note that the coefficient is calculated for each sentence separately, since we are not interested of the mutual order of the parses originating from different sentences. The overall performance for the whole data set is obtained by taking the average of the correlation coefficients of the sentences in the data set.

We use a regularized least-squares (RLS) as a learning algorithm and LG ranking heuristics as a baseline. The performance of RLS ranker with the seven features on validation set is 0.42 and the improvement is statistically significant when compared to 0.18 obtained by the LG parser. To address the issue of applicability of the proposed method to very sparse datasets, we measure performance of the RLS ranker with respect to two main criteria: the number of sentences and the number of parses per sentence used for training. In these experiments all grammatical features are used.

Number of Sentences. The training dataset of 500 sentences is divided into several parts and for testing a separate set of 500 sentences is used. The validation procedure is applied for each of the parts, representing sets of sizes 50, 100,..., 500 sentences. The number of parses used per sentence for training is 5 and for testing 20. We observe that even with a very sparse dataset our method gives a relatively good performance of 0.37 while the learning set size remains as small as 100 sentences. The learning procedure reflects expected tendency of the increased ranker performance with increased number of sentences, reaching 0.42 with 500 sentences.

Number of Parses. We measure performance of the RLS ranker based on the number of parses per sentence used for training with dataset size fixed to 150 sentences. Number of parses per sentence in training is selected to be 5, 10,..., 50 for each validation run. Test dataset consists of 500 sentences each containing 20 parses. We observe that major improvement in ranker performance occurs while using only 10 or 20 parses per sentence for training corresponding to 0.41 and 0.43 performance respectively. When using 50

Table 1. Ranking performance with different number of sentences and parses

# Sentences	# Parses	Correlation	Difference in correlation
20	50	0.3303	0.0841
30	30	0.3529	0.0615
50	20	0.3788	0.0357
100	*10*	*0.4145*	*0.0000*
200	5	0.3798	0.0347
300	3	0.3809	0.0335
500	2	0.3659	0.0485

parses per sentence, performance is 0.45, indicating a small positive difference compared to results obtained with less number of parses.

Parse-Sentence Tradeoff. In this experiment, we fix the number of training examples, representing number of sentences multiplied by number of available parses per sentence, to be approximately one thousand. Datasets of 20, 30, 50, 100, 200, 300, 500 sentences with number of parses per sentence 50, 30, 20, 10, 5, 3, 2, respectively, are validated with 500 sentences each containing 20 parses. The results of these experiments are presented in Table 1. The best performance of ranker is achieved using 100 sentences and 10 parses per each sentence for training, corresponding to 0.41 correlation. The decrease in performance is observed when either having large number of parses with small amount of sentences or vice versa.

3.4 Evaluation Using Graph Feature Representation

In our experiments (see [11]), we evaluate the kernels $\widehat{k_n}$ and $\widetilde{k_n}$ up to the third power with different parameters. In addition, we evaluate the following version of the exponential graph kernel

$$\bar{k}_{exp}(G, G') = \langle LAe^{\beta A}L^{\mathrm{T}}, L'A'e^{\beta A'}L'^{\mathrm{T}}\rangle_F. \tag{16}$$

We multiply the exponentiated adjacency matrix $e^{\beta A}$ by A, because then by setting $\beta = 0$, we obtain the original adjacency matrix A as a special case, and we can set the preferred weight of the higher powers by a grid search on β.

We start by evaluating the graph kernel k_0 with $\gamma_{0,0} = 1$, that is, the zeroth power of the adjacency matrix. The obtained performance of the ranker with k_0 is 0.377. This corresponds to a kernel that counts the elementary features that are present in both parses, that is, the words, link types, part-of-speech tags, and link lengths. We continue by evaluating k_1 with $\gamma_{i,j} = 1$ when $i = j = 1$ and zero otherwise. The result of this experiment is 0.406 correlation points. The performance differences in these two experiments are in correspondence to [17], in which it is observed that the ranking performances with the four elementary features are low compared to their combinations that are present in the first power of the adjacency matrices.

The second powers of the adjacency matrices contain the elementary features present in the zeroth power, the rest of the combination features proposed in [17] are present in the first and the second powers, and also some new combination features in the second power. To get a weighted combination of the first and the second power features, we evaluate the following kernel function

$$k(G, G') = \langle L(A + \theta A^2)L^{\mathrm{T}}, L'(A' + \theta A'^2)L'^{\mathrm{T}} \rangle_F, \tag{17}$$

where θ is a parameter for which we perform a grid search in range 2^{-5}, $2^{-4}, \ldots, 2^5$. The above kernel is similar to $\widehat{k}_2$ in (11) except that we exclude the zeroth power. Note that due to the bipartite property, the kernel is also equal to $\widetilde{k}_2$ in (12) with the zeroth power excluded. The performance with the best θ parameter is 0.429 correlation points.

To analyse the usefulness of the walk features of length longer than 2, we evaluate the following two kernels

$$k(G, G') = \langle L(A + \theta A^2 + \theta^2 A^3)L^{\mathrm{T}}, L'(A' + \theta A'^2 + \theta^2 A'^3)L'^{\mathrm{T}} \rangle_F, \tag{18}$$

and

$$\begin{aligned} k(G, G') = \langle LAL^{\mathrm{T}}, L'A'L'^{\mathrm{T}} \rangle_F + \theta \langle LA^2L^{\mathrm{T}}, L'A'^2L'^{\mathrm{T}} \rangle_F \\ + \theta^2 \langle LA^3L^{\mathrm{T}}, L'A'^3L'^{\mathrm{T}} \rangle_F, \end{aligned} \tag{19}$$

where the best θ parameters are found with a grid search. The first kernel is similar to $\widehat{k}_3$ in (11) with the zeroth power excluded, and the second kernel is similar to $\widetilde{k}_3$ in (12) with the zeroth power excluded. The performance using the first and the second kernels with the best parameters are 0.422 and 0.429 correlation points, respectively. The performance with the exponential graph kernel is 0.425 correlation points. According to the results, is seems that the walks of length larger than 2 are not useful features when this kind of graph representation is used. In fact, they seem to be even harmful when they are mixed with the lower order features, for example, in the kernel (18).

We conduct a final validation experiment on 600 sentences reserved for this purpose. We select the single labeled graph representation and the kernels (17) and (19) with the best performing parameter combinations, that is, the settings that give the highest ranking performance in the parameter estimation experiments. The ranker is trained with the parameter estimation data and the results with the kernels (17) and (19) are 0.444 and 0.447 correlation points, respectively.

3.5 Summary of the Experiments

To convey information the learning algorithm, we use BOF representation containing features such as grammatical bigrams, link types, a combination of link length and link type, part-of-speech information, and others. When evaluating the ranker with respect to each feature separately and all features combined, we observe that most of them let the ranker to outperform Link Grammar parser built-in heuristics. For example, grammatical bigram (pair of words connected by a link) and link bigram (pair of links related by words) underline importance of link dependency structure for ranking. Another feature yielding good performance is link type & word, representing an alternative grammatical structure and providing additional information in case of similar parses. We observe that link length feature, which is related to LG heuristics, leads to poor, below the baseline, performance, whereas other features appear to have more positive effect. We perform several experiments to estimate learning abilities of the ranker, and demonstrate that the method is applicable for sparse datasets.

Furthermore, we demonstrate how the graph kernels can be used to generate features that are based on the start and end labels of random walks between vertices in the graphs. The feature vector corresponding to a data point is determined by its graph representation and the kernel function. Both of them have to be selected carefully in order to ensure a good performance of the machine learning method. Several kernel functions have already been proposed for data that consists of graphs. The results underline the importance of the design of a good graph representation for the data points. For example, random walk features that connect word labels to part-of-speech labels can be generated from both of the proposed graph representations. However, the features generated from the vector labeled representation are too noisy, because walks that connect words to the part-of-speech of their neighboring words have too large weight compared to the walks that connect words to their own part-of-speech. The achieved performance can be even further improved by designing representations capturing additional prior knowledge about the problem to be solved.

References

1. Aronszajn, N.: Theory of reproducing kernels. Transactions of the American Mathematical Society **68** (1950)
2. Scholkopf, B., Smola, A.J.: Learning with Kernels: Support Vector Machines, Regularization, Optimization, and Beyond. MIT, Cambridge, MA (2001)
3. Herbrich, R.: Learning Kernel Classifiers: Theory and Algorithms. MIT, Cambridge, MA (2002)
4. Shawe-Taylor, J., Cristianini, N.: Kernel Methods for Pattern Analysis. Cambridge University Press, New York, NY (2004)
5. Joachims, T.: Text categorization with support vector machines: Learning with many relevant features. In: European Conference on Machine Learning (ECML), Berlin, Springer (1998) 137–142
6. Lodhi, H., Saunders, C., Shawe-Taylor, J., Cristianini, N., Watkins, C.J.C.H.: Text classification using string kernels. J. Mach. Learn. Res. **2** (2002) 419–444

7. Cancedda, N., Gaussier, E., Goutte, C., Renders, J.M.: Word sequence kernels. J. Mach. Learn. Res. **3** (2003) 1059–1082
8. Haussler, D.: Convolution kernels on discrete structures. Technical Report UCSC-CRL-99-10, UC Santa Cruz (1999)
9. Collins, M., Duffy, N.: Convolution kernels for natural language. In Dietterich, T.G., Becker, S., Ghahramani, Z., eds.: NIPS, MIT, Cambridge, MA (2001) 625–632
10. Gärtner, T., Flach, P.A., Wrobel, S.: On graph kernels: Hardness results and efficient alternatives. In Schölkopf, B., Warmuth, M.K., eds.: Sixteenth Annual Conference on Computational Learning Theory and Seventh Kernel Workshop (COLT-2003). Volume 2777 of Lecture Notes in Computer Science., Springer (2003) 129–143
11. Pahikkala, T., Tsivtsivadze, E., Boberg, J., Salakoski, T.: Graph kernels versus graph representations: a case study in parse ranking. In Gärtner, T., Garriga, G.C., Meinl, T., eds.: Proceedings of the ECML/PKDD'06 workshop on Mining and Learning with Graphs (MLG'06). (2006)
12. Cristianini, N., Shawe-Taylor, J., Lodhi, H.: Latent semantic kernels. J. Intell. Inf. Syst. **18** (2002) 127–152
13. Leslie, C., Kuang, R.: Fast string kernels using inexact matching for protein sequences. J. Mach. Learn. Res. **5** (2004) 1435–1455
14. Sleator, D.D., Temperley, D.: Parsing english with a link grammar. Technical Report CMU-CS-91-196, Department of Computer Science, Carnegie Mellon University, Pittsburgh, PA (1991)
15. Tsivtsivadze, E., Pahikkala, T., Boberg, J., Salakoski, T.: Locality-convolution kernel and its application to dependency parse ranking. In Ali, M., Dapoigny, R., eds.: IEA/AIE. Volume 4031 of Lecture Notes in Computer Science., Springer (2006) 610–618
16. Gärtner, T.: Exponential and geometric kernels for graphs. In: NIPS Workshop on Unreal Data: Principles of Modeling Nonvectorial Data. (2002)
17. Tsivtsivadze, E., Pahikkala, T., Pyysalo, S., Boberg, J., Mylläri, A., Salakoski, T.: Regularized least-squares for parse ranking. In: Proceedings of the 6th International Symposium on Intelligent Data Analysis, Springer-Verlag (2005) 464–474 Copyright Springer-Verlag Berlin Heidelberg 2005
18. Lafferty, J., Sleator, D., Temperley, D.: Grammatical trigrams: A probabilistic model of link grammar. In: Proceedings of the AAAI Conference on Probabilistic Approaches to Natural Language, Menlo Park, CA, AAAI Press (1992) 89–97
19. Pyysalo, S., Ginter, F., Heimonen, J., Björne, J., Boberg, J., Järvinen, J., Salakoski, T.: BioInfer: A corpus for information extraction in the biomedical domain. BMC Bioinformatics (2007) Available at http://www.it.utu.fi/BioInfer.
20. Kendall, M.G.: Rank Correlation Methods. 4 edn. Griffin, London (1970)

Discovering Time-Constrained Patterns from Long Sequences

Changzhou Wang, Anne Kao, Jai Choi, and Rod Tjoelker

Boeing Phantom Works, Seattle, WA, USA, changzhou.wang@boeing.com, anne.kao@boeing.com, jai.j.choi@boeing.com, rod.tjoelker@boeing.com

Summary. In recent years, many techniques have been proposed to discover sequential patterns with temporal constraints from long sequences of categorical events. Central to these techniques is the concept of pattern occurrence. A rich set of measures based on pattern occurrence has been developed for sequential pattern discovery, filtering and ranking. Often, recurrences of patterns within the same sequence are ignored. However, the total number of pattern occurrences in each individual sequence can provide valuable insights, especially for applications with long sequences each containing many events.

In this chapter, we propose to use the maximum cardinality of all disjoint occurrence sets as the number of pattern occurrences in an individual sequence. In contrast to previous proposals, our definition (1) depends only on the sequence and the pattern, without requiring additional parameters such as a sliding window size; (2) ensures that patterns occur more often if their temporal constraints are relaxed; and (3) enables us to easily estimate the expected number of pattern occurrences, so that patterns whose counted occurrences deviate significantly from their expectations can be regarded as surprising ones.

In addition, we (1) describe a greedy algorithm which efficiently identifies a single disjoint occurrence set that has the maximum cardinality; (2) develop a formula to calculate the expected value under a uniform distribution assumption; and (3) design an approximation process to efficiently estimate the expected value. Our experiments show that the greedy algorithm is efficient and the approximation is accurate.

1 Introduction

Large volumes of event data have been collected over time in many applications, such as aircraft maintenance records and web click streams. An important analysis task is to discover sequential patterns among events, i.e., co-occurrences of multiple events and some ordinal or temporal relationship among them. Discovered patterns may be then interpreted as rules. An example pattern is *an Engine Oil event followed by an Automatic Flight event in one to three days.* This pattern can be interpreted as a sequential rule as

C. Wang et al.: *Discovering Time-Constrained Patterns from Long Sequences*, Studies in Computational Intelligence (SCI) **116**, 99–116 (2008)
www.springerlink.com

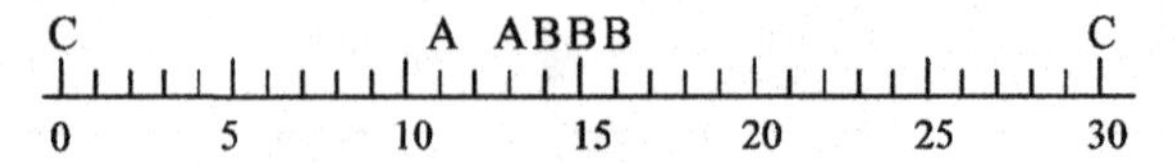

Fig. 1. An example event sequence

follows: *If an Engine Oil event occurs, within one to three days, an Automatic Flight event will occur.*

Many data mining techniques have been proposed to discover, filter and rank sequential patterns or rules using a variety of measures. Most of these measures are based on the concept of pattern occurrence. For example, the support [8] of a sequential pattern is defined as the ratio of the number of pattern occurrences to some support base value; and the confidence [8] of a sequential rule is defined as the ratio of the number of occurrences of the rule antecedent to the number of occurrences of the whole rule.

Many sequential pattern mining techniques deal with multiple data sequences, and only count the number of sequences containing at least one occurrence of a given pattern. Recurrences of a pattern in the same sequence are simply ignored. However, the total number of pattern occurrences in individual sequences can provide valuable insights, especially for applications with long sequences each containing many events. For example, airplane maintenance records are usually kept for the life time of each airplane, and many patterns naturally repeat in the maintenance history of the same airplane. The number of occurrences within sequences might indicate problems of a particular airplane (while the number of occurrences across sequences might indicate problems of a group of airplanes).

While it is easy to check whether a pattern occurs in a sequence, it is not trivial to decide the exact number of occurrences. For example, in Fig. 1, how many times does the pattern $\{A\}1-3\{B\}$ (which means an event A is followed by an event B in one to three time units) occur? Intuitively, one can first identify occurrences of A, and check whether an occurrence of B follows within the constrained temporal region. However, extra care needs to be taken when multiple B events satisfy the constraint. In general, one may make different choices to count the exact number, e.g., whether to count the same event instances multiple times, or whether to ignore some *legitimate occurrences*, i.e., sets of event instances matching the pattern structure along with the temporal constraints.

Mannila et al. [8] proposed[1] to use a sliding window on the input sequence to obtain a set of (overlapping) subsequences, and to report the number of subsequences in which the pattern occurs. Recurrences within a subsequence are ignored. For instance, Table 1 shows the different reported numbers of occurrences for pattern $\{A\}1-3\{B\}$ in the sequence given in Fig. 1. When the window size is large enough, all legitimate occurrences will be

[1] Although their original definition does not support temporal constraints, it is straight forward to extend it.

Table 1. Pattern occurrences using the sliding window approach

Sliding window size	2	3	4
Number of occurrences	1	2	3
Windows containing $\{A\}1\!-\!3\{B\}$	13–14	12–14,13–15	11–14,12–15, 13–16

considered. However, the same event instances or even pattern occurrences may be counted multiple times in multiple sliding windows. For example, there could be five or more occurrences of the pattern, even though there are only two instances of event A. Indeed, the number of pattern occurrences increases as the window size increases. This makes the choice of window size critical and difficult for the user to determine. In addition, this approach is not robust even for ranking patterns: increasing the window size might introduce different numbers of new occurrences for different patterns, and thus change the order of patterns in terms of the number of occurrences or other derived measures.

Alternatively, Mannila and Toivonen [7] proposed to only count the minimal pattern occurrences. An occurrence is minimum if no other occurrence can be found in any proper sub-interval of its time span. Again, consider pattern $\{A\}1 - 3\{B\}$. Now it "occurs" only once in the sequence given in Fig. 1, with the event instances $\{A@13, B@14\}$ (here, we denote the event instance of category C occur at time unit T as $C@T$). Other legitimate occurrences, such as $\{A@11, B@14\}$ or $\{A@13, B@16\}$, are not minimal and thus ignored. However, the more constrained pattern $\{A\}2 - 3\{B\}$ "occurs" twice (the two minimal occurrences are $\{A@11, @B14\}$ and $\{A@13, B@15\}$)! This counter-intuitive phenomenon is a consequence of the "minimal" principle which excludes some legitimate occurrences.

In this chapter, we consider all legitimate occurrences and put them into sets of "disjoint occurrences". Different occurrences in the same set do not share event instances. A sequence may contain many different sets of disjoint occurrences. We define the *number of occurrences* to be the maximum cardinality of these sets, and call it **occurrence-based frequency** or simply **o-frequency.** For the example sequence shown in Fig. 1, pattern $\{A\}1 - 3\{B\}$ has the following six (non-empty) sets of disjoint occurrences (separated by semicolons):

$$\{\{A@11, B@14\}\}; \{\{A@13, B@14\}\}; \{\{A@13, B@15\}\}; \{\{A@13, B@16\}\};$$
$$\{\{A@11, B@14\}, \{A@13, B@15\}\}; \{\{A@11, B@14\}, \{A@13, B@16\}\};$$

Under our definition, this pattern occurs twice in this sequence. Similarly, the more constrained pattern $\{A\}2 - 3\{B\}$ also occurs twice. In general, our definition ensures that (1) each event instance is only used at most once in counting pattern occurrences; (2) patterns never occur more frequently when additional constraints are added.

It is very expensive to enumerate all sets of disjoint occurrences for each pattern against each sequence. Instead, we describe a simple greedy algorithm which only identifies a single disjoint occurrence set. We prove that this set indeed has the maximum cardinality. Hence, the algorithm correctly calculates the o-frequency.

In practice, users often want to look for surprising patterns. Measures such as support or confidence often cannot reveal whether a pattern is expected or not. For example, if one knows that event category B occurs in every time unit, she will not be surprised at all to find many temporal rules with B as their right hand sides with 100% confidence.

We propose to calculate the expected value of the o-frequency directly using knowledge of event distributions. When the counted o-frequency deviates significantly from the expected values, the pattern can be regarded as surprising if the estimation is accurate. As an example, we develop a formula to calculate the expected value of the o-frequency under the independent uniform distribution assumption, and verify it through experiment. The calculation is often expensive. We design a process to approximate the expected value by efficiently deriving tight lower and upper bounds, as validated in our experiments.

The rest of the chapter is organized as follows: Sect. 2 discusses some related work; Sect. 3 describes the concept of disjoint occurrence and occurrence frequency; Sect. 4 introduces the greedy algorithm with its correctness proof; Sect. 5 presents the formula to calculate the expected occurrence frequency and provides an efficient algorithm to estimate it; Sect. 5 concludes the chapter.

2 Related Work

Sequential pattern mining has attracted much attention recently. See [3, 12] for recent surveys. In this section, we can only review a few items which are most relevant to our work.

Based on the work on association rule mining, [1] introduced the sequential pattern mining problem, and [13] generalized the problem to support temporal constraints (and taxonomy of event types). Later, [4,10,16] proposed alternative efficient mining methods. All these studies assumed multiple sequences and ignored pattern recurrences within single sequences.

Heikki Mannila and co-workers [7–9] focused on discovering episodes (which are more general than sequential patterns) from a single sequence. However, their counting approaches may have some drawbacks as we discussed in the introduction.

Mahesh Joshi [5] discussed a few different ways to count sequential pattern occurrences. Their CDIST method counts the number of disjoint occurrences ("distinct occurrence with no event timestamp overlap allowed" in [5]) using an unspecified algorithm.[2] Instead, our definition is independent from any specific counting algorithm.

Many other works contributed to sequential pattern mining research from different aspects. For example, [14] introduced methods to integrate sequential pattern mining into database systems, [6] proposed a general approach to speed up frequency counting, [11] developed a framework to handle constraint-based sequential pattern mining, and [15] extended sequential pattern mining techniques to handle noisy data. These techniques may be directly applied or extended using our definition of pattern occurrence.

Subsequence counting problems have also been studied extensively in the context of string matching [2]. However, matching criteria are usually simpler (e.g., without temporal constraints), and the matching algorithms cannot be directly applied in sequential mining. Nevertheless, some ideas may be borrowed to improve, for example, the greedy algorithm in Sect. 4.

3 Disjoint Occurrences

In this chapter, we assume that an *event instance* has a unique ID, a category name and a timestamp. In addition, we consider a *sequence* as a list of event instances ordered by their timestamps. Multiple event instances with the same category and/or the same timestamp are allowed. For simplicity, we assume that the same time unit and time origin are used for all sequences. Hence we use integer numbers to denote the event timestamps. We use capital letters A, B, C, D, E to denote categories, and lower case letters e to denote event instances. We also use $\mathcal{C}(e)$ and $\mathcal{T}(e)$ to denote the category and timestamp of the event instance e, respectively. Table 2 summarizes the symbols used in this chapter.

We extend the work in [13] and define a *sequential pattern*, or simply a *pattern*, as a collection of event categories with structural and temporal constraints, specified in Fig. 2. Specifically, a group is a bag[3] of event categories with a group window size constraint. A pattern is an ordered list of groups with minimum and maximum time gap constraints between any two consecutive groups. The gaps and window sizes are integers indicating time differences with the implicit time unit. Different gaps and window sizes may be used in the same pattern. If a group contains only a single category, its window size is 0 and is omitted with the colon separator.

For example, $\{A\}1-3\{B\}$ is a simple pattern with two groups each having a single category, while the minimum and maximum gaps between these two groups are 1 and 3 time units, respectively. In pattern $\{2 : A, B\}6-8$

[2] The only description in [5] is "the timeline is scanned in forward direction". We suspect that the algorithm is similar in spirit to our greedy algorithm.

[3] A bag is similar to a set except that a bag may contain multiple copies of the same item.

Table 2. Symbols

Symbols	Comments
A, B, C, D, E	Event category
e	Event instance
$\mathcal{C}(e)$	Category of event e
$\mathcal{T}(e)$	Timestamp of event e
P	Pattern
O, K	Pattern occurrence
$\mathcal{O}$	Set of disjoint pattern occurrences
n	Number of event categories in a pattern
g	Number of groups in a pattern
n_i	Number of event categories in the i-th group in a pattern
α_i	Minimum time gap between the i-th group and the $(i+1)$-th group
β_i	Maximum time gap between the i-th group and the $(i+1)$-th group
ω_i	Window size of the i-th group

⟨ pattern ⟩ := ⟨ group ⟩ (⟨ min gap ⟩ "-" ⟨ max gap ⟩ ⟨ group ⟩)*
⟨ group ⟩ := "{" ⟨ category ⟩ "}" |
"{" ⟨ window size ⟩ ":" ⟨ category ⟩ ⟨ category ⟩+ "}"
⟨ min gap ⟩ := ⟨ integer ⟩
⟨ max gap ⟩ := ⟨ integer ⟩
⟨ window size ⟩ := ⟨ integer ⟩

Fig. 2. Definition of sequential pattern

$\{3 : B, A, B\}13 - 15\{C\}$, there are three groups. The second group has two copies of category B and the third group only has a single category.

In general, we use α_i and β_i to denote the minimum and maximum gaps between the i-th group and the $(i+1)$-th group, and use ω_i to denote the window size of the i-th group. Hence, a general pattern form is $\{\omega_1 : E_{1,1}, \ldots, E_{1,n_1}\}\alpha_1 - \beta_1\{\omega_2{:}E_{2,1}, \ldots, E_{2,n_2}\}\alpha_2 - \cdots - \beta_{g-1}\{\omega_g{:}E_{g,1}, \cdots, E_{g,n_g}\}$. An *occurrence* of this pattern in a given sequence is a subset of (event instances in) the sequence, denoted $\{e_{1,1}, \ldots, e_{1,n_1}, e_{2,1}, \ldots, e_{2,n_2}, \ldots, e_{g,1}, \ldots, e_{g,n_g}\}$, such that (1) $\mathcal{C}(e_{i,j}) = E_{i,j}$ for $1 \le i \le g, 1 \le j \le n_g$; (2) $\max_j \mathcal{T}(e_{i,j}) - \min_j \mathcal{T}(e_{i,j}) \le \omega_i$ for $1 \le i \le g$; (3a) $\max_j \mathcal{T}(e_{i+1,j}) - \min_j \mathcal{T}(e_{i,j}) \le \beta_i$ for $1 \le i \le (g-1)$; and (3b) $\min_j \mathcal{T}(e_{i+1,j}) - \max_j \mathcal{T}(e_{i,j}) \ge \alpha_i$ for $1 \le i \le (g-1)$.

Clearly, a pattern occurrence is simply a set of event instances satisfying the structural and temporal constraints. First, the event categories in the pattern and the categories of these event instances are in one-to-one correspondence. Second, event instances mapped to the same group must occur in the corresponding window size. Third, the time gap between any two event instances mapped to two consecutive groups shall be no greater than the specified maximum gap and no less than the specified minimum gap. For each event instance in the occurrence, we say that it *matches* an event category in the pattern if the instance is mapped to the category in one-to-one correspondence.

Two different occurrences (of the same pattern) are said to be *disjoint* if the intersection of the two sets is empty. A set of occurrences (of a pattern) is said to be a *disjoint occurrence set* if any two occurrences in the set are disjoint. An event sequence often contains multiple disjoint occurrence sets for a given pattern. The *occurrence-based frequency*, or simply *o-frequency*, of a pattern in a data sequence is the maximum cardinality of all its disjoint occurrence sets in that sequence.

Clearly, o-frequency is well defined. If a sequence does not contain any occurrence of a pattern, the maximum cardinality is 0. If it does contain at least one occurrence, there must be a disjoint occurrence set with the maximum cardinality. Indeed, o-frequency is lower-bounded by 0 and upper-bounded by the number of occurrences of any event category specified in the pattern.

A pattern may be relaxed by (1a) dropping some categories without dropping all categories from any group, (1b) dropping the first or the last group, (2) increasing the window size for some groups, (3a) increasing some maximum group gaps, (3b) decreasing some minimum group gaps. Clearly, all occurrences of a pattern will be legitimate occurrences of its relaxed patterns. Hence, the maximum cardinality of all disjoint occurrence sets, i.e., the o-frequency, of the relaxed pattern will be no less than the o-frequency of the original one.

4 Counting Algorithm

A naive method to determine the o-frequency of a pattern in a given sequence is to first test all possible subsets of event instances in the sequence to decide whether it is a legitimate occurrence, then test any subset of the discovered occurrences to decide whether it is a disjoint occurrence set, and finally find the maximum cardinality. This clearly incurs an exponential computation cost.

In this section, we describe a simple greedy algorithm adapted from [13] to identify only a single disjoint occurrence set and prove that this set has the maximum cardinality. Figure 3 outlines the algorithm.

The algorithm works by forward scanning the data sequence along the time line. To find a pattern occurrence, the algorithm tries to match group by group (2.1). Once all categories in a group are matched to some event instances, the window size constraint on that group is checked (2.1.3). If the constraint is violated for a group j, the algorithm moves backward ω_j time units and tries to match the same group again. If a group is matched without violating the

Input	A sequence $S: e_1, \ldots, e_m$ and a pattern P
Output	O-frequency of P in S
1	initialization
1.1	c=0 // number of matched occurrences
1.2	j=0 // number of matched groups
1.3	i=0 // index of event (instance) in S
2	while ($1 \leq i \leq m$) // within the range of S
2.1	while (group j not fully matched AND $i \leq m$)
2.1.1	match event i against group j
2.1.2	increase i by 1
2.1.3	if (group j fully matched BUT ω_j violated)
2.1.3.1	let h be the smallest index such that $\mathcal{T}(e_h) \geq \mathcal{T}(e_i) - \omega_j$
2.1.3.2	let i be h
2.1.3.3	remove all matches in group j
2.2	if (group j cannot be matched) // but $i > m$
2.2.1	return c
2.3	if (β_{j-1} is violated (group j is too far))
2.3.1	let i be the smallest index such that $\mathcal{T}(e_h) \geq$(latest time in group j) $-\beta_j$
2.3.2	decrease j by 1 // rework previous group
2.4	else // group j is matched
2.4.1	increase i so that $\mathcal{T}(e_i) > \alpha_j +$ (latest time in group j)
2.4.2	if ($c > 0$) // with some matched occurrence
2.4.2.1	increase i so that $\mathcal{T}(e_i) >$ (earliest time in group j in last matched occurrence)
2.4.3	increase j by 1
2.5	if (current occurrence is fully matched)
2.5.1	increase c by 1
2.5.2	remove event instances in current matched occurrence from S
2.5.3	reset $j = 0$
2.5.4	let i point to event right after earliest event in current matched occurrence
3	return c // all the event instances are checked

Fig. 3. Counting distinct occurrences

window size constraint, but it is too far away from its preceding group (2.3), the algorithm discards the match for this group and its preceding group, moves backward with the step of the maximum gap, and tries to match its preceding group. Otherwise, the group is successfully matched, the algorithm moves forward to skip the minimum gap (2.4.1), and to skip the first item matched in this group in the previously matched pattern occurrence (2.4.2), to match the next group. Once a pattern occurrence is identified (2.5), all matched instances are removed from the sequence (2.5.2), event instances that occur before the first one in the matched pattern occurrence are also ignored (2.5.4), and the rest are used to find the next occurrence.

Many optimization details are omitted in Fig. 3. For example, if there are some event categories in the pattern with very few occurrences in the sequence, we can search such occurrences first and find pattern occurrences around them without wasting time on matching other parts of those patterns. Also, in step 2.1.3.3, instead of fully resetting the matching process for the j-th group, we can reuse the part of the matched occurrences satisfying the ω_j constraints.

The computational complexity of the algorithm is clearly polynomial instead of exponential in terms of sequence size and pattern length. Consider the group match step (2.1). The algorithm never tries to start at the same time (i) to match the same group (due to 2.4.2, and other backward steps in 2.3.2 and 2.1.3.1) more than once. Hence there will be at most $O(mg)$ loops in step 2, where m is the number of events in the sequence and g is the number of groups in the pattern. For each loop, the group match step (2.1) costs $O(mn)$ where n is the number of categories in the pattern. Other steps (2.2–2.5) cost $O(m)$. Hence the overall cost is bounded by $O(m^2gn)$.

In practice, the total cost can be much lower. To study the scalability of the algorithm regarding temporal constraints, pattern length and sequence size, we conduct some experiments using synthesized data. We first generate a data set of 25 sequences each containing 10^5 events. Each event is independently, uniformly and randomly assigned to one of 10 given categories and distributed in the time range $[1, 10^7]$. Therefore any time range $[t+1, t+100]$ is expected to contain a single event on average.

We then apply the algorithm to count different patterns in the data set. The algorithm is implemented purely in Java and the experiment is run within Sun JRE 1.4.1 on a Windows 2,000 laptop with A Pentium 1.2 GHz CPU and 512 MB main memory. All 25 sequences are first loaded in main memory to apply the greedy algorithm. This is feasible as one million events can be encoded using only 8MB main memory (time and type can be each encoded by a four byte integer), yet many real life data sets contain far less than one million events in each sequence. All numbers reported in the following are the averaged results using the 25 sequences. Only the wall clock time spent by the counting algorithm is recorded and reported in milliseconds.

Table 3 reports the averaged runtime and count returned by the algorithm for patterns with different temporal constraints. We consider two types of patterns, $\{C_1\}1-\beta\{C_2\}1-\beta\{C_3\}$ and $\{\omega : C_1, C_2, C_3\}1-\beta\{C_4\}$, where C_1, C_2, C_3, C_4 are the first four event categories, and $\omega = \beta$ varies from 10 to 10^6. In other words, the expected number of events in the search region (to match the next event in the same group or the next group) varies from

Table 3. Average runtime vs. temporal constraints (search region size)

	(a) $\{C_1\}1-\beta\{C_2\}1-\beta\{C_3\}$						(b) $\{\omega : C_1, C_2, C_3\}1-\beta\{C_4\}$					
$\omega=\beta$	10	10^2	10^3	10^4	10^5	10^6	10	10^2	10^3	10^4	10^5	10^6
time	6.4	7.2	11.6	19.2	19.6	18.8	9.6	10.4	16.8	49.3	298.4	709.4
count	1.1	89.0	2980.3	8683.3	9778.4	9793.0	0.04	8.3	1858.4	8442.0	9800.5	9857.0

Table 4. Average runtime vs. pattern length (categories in pattern)

	(a) $\{C_1\}1-\beta\{\ldots\}1-\beta\{C_k\}$				(b) $\{\omega : C_1, \ldots, C_{k-1}\}1-\beta\{C_k\}$			
k	2	3	4	5	2	3	4	5
time	9.2	11.6	12.4	12.8	9.2	13.2	16.0	18.8
count	4993.5	2980.3	1909.8	1274.9	4993.5	2974.8	1858.4	1187.8

Table 5. Average runtime vs. sequence size (events in sequence)

	(a) $\{C_1\}1-\beta\{C_2\}1-\beta\{C_3\}$				(b) $\{\omega : C_1, C_2, C_3\}1-\beta\{C_4\}$			
Sequence size	10^3	10^4	10^5	10^6	10^3	10^4	10^5	10^6
time	0	1.2	11.6	114.2	0	1.6	16.8	167.8
count	28.6	301.4	2980.3	29817.8	18.4	186.2	1858.4	18696.0

0.1 to 10,000. From Table 3, patterns with multiple categories in a group are usually more difficult to count and more sensitive to the increase in the size of the search region. Nevertheless, the algorithm scales sub-linearly with the search region size, probably because a matching event instance can often be found without exhausting the whole search region (this is indicated by the small increase of the counted number). The absolute runtime is also reasonable as the time is in milliseconds.

Table 4 reports the averaged runtimes and counts for patterns with different lengths. Again, we consider two types of patterns, $\{C_1\}1-\beta\ldots1-\beta\{C_k\}$ (each group contains only a single category) and $\{\omega : C_1, \ldots, C_{k-1}\}1-\beta\{C_k\}$, where $C_1, \ldots, C_k$ are the first k event categories. $\beta = 1000$, i.e., the search region (for a category) contains approximately 10 events on average in the sequence. Clearly, the algorithm scales at most linearly with the length of patterns.

To see how the algorithms scale with the sequence size, we also generate three additional data sets using similar settings except that the total number of events ($10^3, 10^4$ and 10^6 respectively) and time range ($[1, 10^5], [1, 10^6]$ and $[1, 10^8]$ respectively) in each sequence are different for different data sets. Again, any time period $[t+1, t+100]$ is expected to contain a single event on average for all sequences. Table 5 reports the averaged runtimes and counts for pattern $\{C_1\}1-\beta\{C_2\}1-\beta\{C_3\}$ and $\{\omega : C_1, C_2, C_3\}1-\beta\{C_4\}$ in all four data sets, where $C_1, \ldots, C_4$ are the first four event categories and $\omega = \beta = 1{,}000$. The total runtime for counting a single pattern in 25 sequences in the smallest data set is less than 1 millisecond and hence 0 is used in the table. Clearly, both the count and the runtime increase linearly with the sequence size (instead of quadratically as suggested by the preliminary analysis given earlier).

The performance of the greedy algorithm on real life data may vary as the events are often correlated and may occur in bursts while patterns are usually more complex. However, the results obtained using the synthesized data can be a guideline. For example, if the search region for a group contains at most 1,000 events and the sequence contains 100,000 events in total, Table 3 and 5 suggest that the runtime is likely to be around $24.4 \times 10 = 244$ ms. When the sequence contains large sparse regions, the runtime can be much less.

4.1 Correctness of Algorithm

To prove the correctness of the algorithm, we first introduce a partial order and two operators on pattern occurrences.

Definition 1. *Given a pattern P with the form $\{\omega_1 : E_{1,1}, \ldots, E_{1,n_1}\}\alpha_1 - \cdots - \beta_{g-1}\{\omega_g : E_{g,1}, \cdots, E_{g,n_g}\}$. For any two occurrences of P, $O_1 : \{e^1_{i,j} | i = 1, \ldots, g; j = 1, \ldots, n_i\}$ and $O_2 : \{e^2_{i,j} | i = 1, \ldots, g; j = 1, \ldots, n_i\}$ where $e^1_{i,j}$ and $e^2_{i,j}$ match $E_{i,j}$ in the pattern P, $O_1 \preceq O_2$ if $\mathcal{T}(e^1_{i,j}) \leq \mathcal{T}(e^2_{i,j})$ for $j = 1, \ldots, n_i$ and $i = 1, \ldots, g$.*

Intuitively, for any event category in P, $O_1 \preceq O_2$ if and only if the matched event instance in O_1 occurs no later than the matched one in O_2.

Definition 2. *Given a pattern P with the form $\{\omega_1 : E_{1,1}, \ldots, E_{1,n_1}\}\alpha_1 - \cdots - \beta_{g-1}\{\omega_g : E_{g,1}, \cdots, E_{g,n_g}\}$. For any two occurrences of P, $O_1 : \{e^1_{i,j} | i = 1, \ldots, g; j = 1, \ldots, n_i\}$ and $O_2 : \{e^2_{i,j} | i = 1, \ldots, g; j = 1, \ldots, n_i\}$ where $e^1_{i,j}$ and $e^2_{i,j}$ match $E_{i,j}$ in P. $O_1 \curlyvee O_2$ is defined as $\{e^3_{i,j} | i = 1, \ldots, g; j = 1, \ldots, n_i\}$ and $O_1 \curlywedge O_2$ is defined as $\{e^4_{i,j} | i = 1, \ldots, g; j = 1, \ldots, n_i\}$, respectively, where for $i = 1, \ldots, g$ and $j = 1, \ldots, n_i$, if $\mathcal{T}(e^1_{i,j}) \leq \mathcal{T}(e^2_{i,j})$, let $e^3_{i,j} = e^1_{i,j}$ and $e^4_{i,j} = e^2_{i,j}$; otherwise, let $e^3_{i,j} = e^2_{i,j}$ and $e^4_{i,j} = e^1_{i,j}$.*

Intuitively, for any category in P, among the two matched event instances in O_1 and O_2, we put the one that occurred earlier into $O_1 \curlyvee O_2$, and the other into $O_1 \curlywedge O_2$. Note that $O_1 \curlyvee O_2 = O_2 \curlyvee O_1$ and $O_1 \curlywedge O_2 = O_2 \curlywedge O_1$.

Lemma 1. *For any two disjoint occurrences O_1 and O_2 of P, $O_3 = O_1 \curlyvee O_2$ and $O_4 = O_1 \curlywedge O_2$ are also two disjoint occurrences of P such that $O_3 \preceq O_4$.*

Proof. First, since $\mathcal{C}(e^1_{i,j}) = \mathcal{C}(e^2_{i,j}) = E_{i,j}$, it is clear that $\mathcal{C}(e^3_{i,j}) = \mathcal{C}(e^4_{i,j}) = E_{i,j}$, for $i = 1, \ldots, g; j = 1, \ldots, n_i$.

To check the time constraints, note that, for $i = 1, \ldots, g$, $\min_j \mathcal{T}(e^3_{i,j}) = \min(\min_j \mathcal{T}(e^1_{i,j}), \min_j \mathcal{T}(e^2_{i,j}))$ and $\max_j \mathcal{T}(e^4_{i,j}) = \max(\max_j \mathcal{T}(e^1_{i,j}), \max_j \mathcal{T}(e^2_{i,j}))$. In addition, for $1 \leq k \leq n_i$, $\mathcal{T}(e^3_{i,k}) \leq \mathcal{T}(e^1_{i,k}) \leq \max_j \mathcal{T}(e^1_{i,j})$ and $\mathcal{T}(e^3_{i,k}) \leq \mathcal{T}(e^2_{i,k}) \leq \max_j \mathcal{T}(e^2_{i,j})$. Hence, $\max_j \mathcal{T}(e^3_{i,j}) \leq \min(\max_j \mathcal{T}(e^1_{i,j}), \max_j \mathcal{T}(e^2_{i,j}))$. Similarly, $\min_j \mathcal{T}(e^4_{i,j}) \geq \max(\min_j \mathcal{T}(e^1_{i,j}), \min_j \mathcal{T}(e^2_{i,j}))$.

Without loss of generality, assume[4] $\min_j \mathcal{T}(e^1_{i,j}) \leq \min_j \mathcal{T}(e^2_{i,j})$, it is clear that

[4] If the assumption does not hold, the same conclusion can be drawn using $e^2_{i,j}$'s.

$$\begin{array}{lll} \max_j \mathcal{T}(e^3_{i,j}) - \min_j \mathcal{T}(e^3_{i,j}) & \leq \max_j \mathcal{T}(e^1_{i,j}) - \min_j \mathcal{T}(e^1_{i,j}) & \leq \omega_i \\ \max_j \mathcal{T}(e^3_{i+1,j}) - \min_j \mathcal{T}(e^3_{i,j}) & \leq \max_j \mathcal{T}(e^1_{i+1,j}) - \min_j \mathcal{T}(e^1_{i,j}) & \leq \beta_i \\ \min_j \mathcal{T}(e^3_{i,j}) - \max_j \mathcal{T}(e^3_{i-1,j}) & \geq \min_j \mathcal{T}(e^1_{i,j}) - \max_j \mathcal{T}(e^1_{i-1,j}) & \geq \alpha_{i-1}. \end{array}$$

Similarly, without loss of generality, assume $\max_j \mathcal{T}(e^1_{i,j}) \leq \max_j \mathcal{T}(e^2_{i,j})$, then

$$\begin{array}{lll} \max_j \mathcal{T}(e^4_{i,j}) - \min_j \mathcal{T}(e^4_{i,j}) & \leq \max_j \mathcal{T}(e^2_{i,j}) - \min_j \mathcal{T}(e^2_{i,j}) & \leq \omega_i \\ \max_j \mathcal{T}(e^4_{i+1,j}) - \min_j \mathcal{T}(e^4_{i,j}) & \leq \max_j \mathcal{T}(e^2_{i+1,j}) - \min_j \mathcal{T}(e^2_{i,j}) & \leq \beta_i \\ \min_j \mathcal{T}(e^4_{i,j}) - \max_j \mathcal{T}(e^4_{i-1,j}) & \geq \min_j \mathcal{T}(e^2_{i,j}) - \max_j \mathcal{T}(e^2_{i-1,j}) & \geq \alpha_{i-1}. \end{array}$$

Hence O_3 and O_4 are both legitimate occurrences of P satisfying both the structural and the temporal constraints specified in Sect. 3.

Since O_1 and O_2 are disjoint, no duplicate event instances exists in $\{e^1_{i,j}, e^2_{i,j} | i = 1, \ldots, g; j = 1, \ldots, n_i\}$, i.e., $\{e^3_{i,j}, e^4_{i,j} | i = 1, \ldots, g; j = 1, \ldots, n_i\}$. Hence O_3 and O_4 are also disjoint. Finally, the way we assign event instances to O_3 and O_4 clearly ensures $O_3 \preceq O_4$. □

Lemma 2. *For any disjoint occurrence set $\mathcal{O}$ of a pattern P, we can use all the event instances in $\mathcal{O}$ to obtain a new disjoint occurrence set $\mathcal{O}'$ of P, such that all occurrences in $\mathcal{O}'$ are totally ordered under $\preceq$.*

Proof. Let $\mathcal{O}$ contains k disjoint occurrences. We can obtain $\mathcal{O}'$ in an iterative way: Initially, $\mathcal{O}'$ is empty. In the first iteration, remove an arbitrary occurrence, denoted O^1_1, from $\mathcal{O}$ and add it to $\mathcal{O}'$. In the $(h+1)$-th iteration $(1 \leq h < k)$, let $\mathcal{O}'$ be $\{O^h_1, \ldots, O^h_h\}$ at the beginning, remove any of the remaining occurrences, denoted K^h_1, from $\mathcal{O}$, let $O^{h+1}_1 = K^h_1 \curlyvee O^h_1$, and $K^h_2 = K^h_1 \curlywedge O^h_1$. Replace O^h_1 in $\mathcal{O}'$ with O^{h+1}_1. Continue this exchange and replace process for K^h_i and O^h_i for $1 < i \leq h$ to get the new set $\{O^{h+1}_1, \ldots, O^{h+1}_{h+1}\}$ where $O^{h+1}_{h+1} = K^h_{h+1}$. Finally, in the k-th iteration, the set $\mathcal{O}'$ contains k disjoint occurrences.

By LEMMA 1, all O^i_j and K^i_j are valid occurrences. In addition, by the definitions of $\curlyvee$ and $\curlywedge$, it is easy to verify that only and all event instances from $\mathcal{O}$ are used to obtain $\mathcal{O}'$, and $O^i_1 \preceq O^i_2, \ldots, O^i_i$ for $i = 2, \ldots, k$. □

Clearly, this leads to the following conclusion:

Theorem 1. *For any pattern and a sequence, either the pattern does not occur in the sequence, or there is a disjoint occurrence set with the maximum cardinality such that all occurrences in the set can be totally ordered under $\preceq$.*

Now we are ready to prove the correctness of the algorithm.

Theorem 2. *The counting algorithm presented in Fig. 3 is correct, i.e., it reports the maximum cardinality of the disjoint occurrence sets of P in S.*

Proof. If the pattern never occurs in the sequence, the algorithm will stop at step 2.2 and correctly report 0.

When the pattern does occur in the sequence, the algorithm will identify a set of disjoint occurrences, since all matched events are removed in step 2.5.2.

By Theorem 1, let $\{O_1, \ldots, O_k\}$ be a disjoint occurrence set with maximum cardinality such that $O_1 \preceq O_2 \preceq \cdots \preceq O_k$. Let the occurrence set enumerated by the algorithm in Fig. 3 be $\{O'_1, \ldots, O'_c\}$. By the greedy nature of the algorithm, $O'_1 \preceq \ldots \preceq O'_c$.

We now show that $O'_i \preceq O_i$ for $i = 1, \ldots, \min(k, c)$. First, $O'_1 \preceq O_1$ is clear by the greedy nature of the algorithm. Now assume $O'_j \preceq O_j$ for $1 \leq j < i$ but $O'_i \not\preceq O_i$. In other words, O'_i includes some event instance e' occurring later than the corresponding instance e in O_i, but e' and e match the same category in the pattern. This contradicts the greedy nature of the algorithm.

Hence, the algorithm will not stop unless $c \geq k$ (if $c < k$, there is at least one legitimate occurrence after the last discovered occurrence, and this can be located by the algorithm easily). On the other hand, since k is the maximum cardinality, $c \leq k$. Hence, the algorithm reports the maximum cardinality of disjoint occurrence sets. □

Indeed, a greedy algorithm working in the opposite direction by scanning sequence backward along the time line will also work. The proof is similar to that which we present here. Other algorithms and optimization techniques may be designed using the properties of disjoint occurrence sets presented in this section.

5 Calculating and Estimating O-Frequency

As discussed in the introduction, o-frequency can be used to help identify surprising patterns. In this section, we calculate the expected value of o-frequency under some simple assumptions and estimate the value using tight lower and upper bounds.

We focus on the pattern $\{A\}\alpha - \beta\{B\}$. For simplicity, assume (1) the data sequence spans the time range $[1, l]$; (2) in total there are i event instances of category A uniformly distributed over $[1, l]$; (3) in total there are j event instances of category B uniformly distributed over $[1, l]$; (4) $\alpha < \beta \ll l$; and (5) all event instances are independently distributed.

Now we calculate the expected o-frequency of $\{A\}\alpha - \beta\{B\}$, denoted $g(i, j)$. First, $g(0, j) = g(i, 0) = 0$. For $i = 1$, $g(1, j)$ equals to the probability that at least one of the j instances of B follows the (only) uniformly distributed instance of A within $\alpha - \beta$ time period, i.e.,

$$g(1, j) \approx 1 - (1 - \rho)^j, \text{ where } \rho = \frac{\beta - \alpha + 1}{l}.$$

Here ρ is the probability that a specific instance of B follows the instance of A within $\alpha - \beta$ time period. This estimation is not accurate due to the fact that the probability decreases when the instance of A is located at $[l-\beta+1,l]$. However, the error can be ignored since $\beta \ll l$. Similarly,

$$g(i,1) \approx 1-(1-\rho)^i.$$

Next, consider the general case $g(i,j)$. Choose any instance of A, with $g(1,j)$ probability, it is followed by at least one instance of B within $\alpha-\beta$ time period. If so, we already have one occurrence of $\{A\}\alpha-\beta\{B\}$, remove both event instances and we are now in the situation where $g(i-1,j-1)$ can be calculated. If not, just discard the instance of A, and we are in the situation to calculate $g(i-1,j)$. To summarize,

$$g(i,j) = g(1,j)(1+g(i-1,j-1)) + (1-g(1,j))g(i-1,j).$$

To verify the calculation of $g(i,j)$, we generate two groups of sequences under the uniform distribution assumptions (1–4). Each of the 100 short sequences in the first group consists of $2i$ event instances of category A_i for $i=1,\ldots,50$ and $20j$ instances of category B_j for $j=1,\ldots,5$. All event instances are independently, randomly and uniformly distributed over the time range $[1,10^9]$. Then we count the o-frequencies of pattern $\{A_i\}1-\beta\{B_j\}$ for any combination of $i=1,\ldots,50; j=1,\ldots,5; \beta=10^8,10^7,10^6,10^5$. The second group contains 100 long sequences. Each sequence consists of $2i$ event instances of category A_i for $i=1,\ldots,500$, and $200i$ instances of category B_j for $j=1,\ldots,5$. We count the o-frequencies of pattern $\{A_i\}1-\beta\{B_j\}$ for any combination of $i=1,\ldots,500; j=1,\ldots,5; \beta=10^8,10^7,10^6,10^5$.

Our experimental results show that the calculated expected o-frequencies are very close to the counted o-frequencies using these generated data. Due to space limitations, we only show the average value of the counted o-frequencies for $\beta=10^8$ and $\beta=10^5$ in Fig. 4. The horizontal axis is the number of event instances of category A_i's, and the vertical axis is o-frequency. The five groups of curves correspond to $B_1,\ldots,B_5$, from bottom to top. In general, the estimate is very close to the counted average. In addition, the standard deviations (not shown) of the counted o-frequencies are usually less than 1% of the average values.

Unfortunately, $g(i,j)$ is defined in a recursive form. The computation time is quadratic in terms of i and j. In practice, the sequence can be long and the number i and j can be very large. To avoid the expensive calculation, we can derive tight lower and upper bounds of the expected value very efficiently, using the following formulas:

$$\begin{aligned}
c_{1,0} &= 1 \\
c_{1,1} &= -1 \\
c_{k,h} &= \frac{c_{k-1,h-1}}{(1-\rho)^h - 1} \quad (\mathit{for}\ k \geq h > 0)
\end{aligned}$$

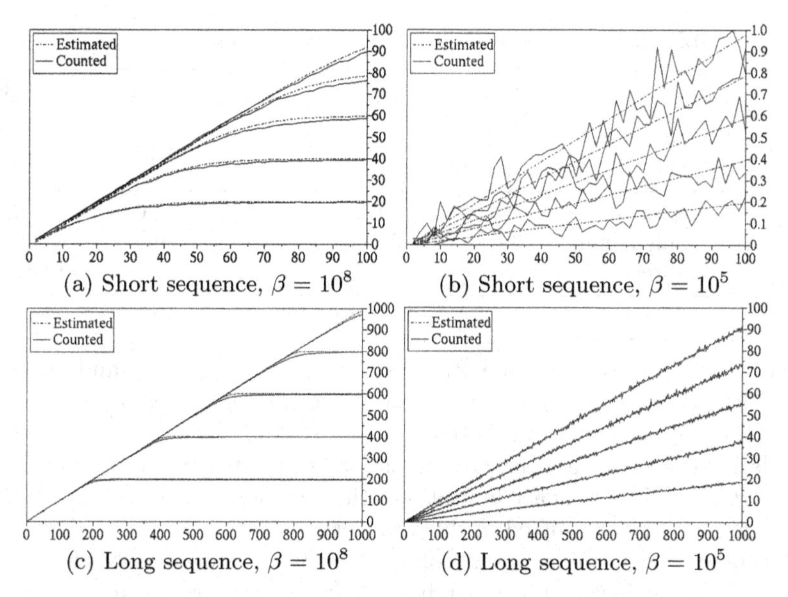

Fig. 4. Average o-frequency and its estimated expected value for $\{A_i\}1 - \beta\{B_j\}$

$$
\begin{aligned}
c_{k,0} &= -\sum_{h=1\ldots k} c_{k,h}(1-\rho)^h \quad (\textit{for } k > 1) \\
\lambda_1 &= (1-\rho)^{2(j-i+1)} \\
\lambda_k &= -\lambda_{k-1}(1-(1-\rho)^k)(1-\rho)^{j-i+1} \quad (\textit{for } k > 1) \\
d_{k,l} &= \lambda_l \sum_{h=0\ldots l} c_{l,h}(1-\rho)^{k(h+1)} \quad (\textit{for } k \geq 0, l \geq 1) \\
\alpha_0 &= 0 \\
\alpha_1 &= \rho(1-\rho)^{2j-2i+3} \\
\alpha_{2m} &= \alpha_{2m-2} + \sum_{l=1\ldots 2m-2} (d_{2m-1,l} + d_{2m,l}) \quad (\textit{for } m > 0) \\
\alpha_{2m+1} &= \alpha_{2m-1} + \sum_{l=1\ldots 2m-1} (d_{2m,l} + d_{2m+1,l}) \quad (\textit{for } m > 0) \\
\beta &= i - \frac{(1-\rho)^{j-i+1}(1-(1-\rho)^i)}{\rho} \\
\gamma_k &= \sum_{l=1\ldots k} \lambda_l \sum_{h=0\ldots l} \frac{c_{l,h}((1-\rho)^{(h+1)(k+1)} - (1-\rho)^{i(h+1)})}{1-(1-\rho)^{h+1}} \quad (\textit{for } k > 0) \\
g_m^+ &= \beta + \alpha_{2m+1} + \gamma_{2m+1} \quad (\textit{for } m > 0) \\
g_m^- &= \beta + \alpha_{2m} + \gamma_{2m} \quad (\textit{for } m > 0)
\end{aligned}
$$

Here g_m^+ and g_m^- are the upper and lower bounds of the expected value $g(i,j)$. The larger the integer m, the tighter the bounds are. A key step

Table 6. Cost to derive tight bounds on expected o-frequency

k		1,024	2,048	4,096	8,192	16,384
Step (m)	average	2.83	2.19	2.04	2.01	2.00
	maximum	12	12	12	12	12
Time (ms)	average	0.025	0.023	0.023	0.023	0.023
	maximum	30	60	70	100	80
	direct	20	81	281	1,022	5,687

in the derivation is to define $\theta_{0,i,j} \equiv g(i,j) - g(i,j-1)$ and $\theta_{k,i,j} \equiv (1-\rho)^k\theta k - 1, i, j - \theta_{k-1,i,j-1}$. It can be shown that $\theta_{k,i,j} = \theta_{k,1,j} + \theta_{k,i-1,j} + (1-\rho)^{j-k-1}\theta_{k+1,i-1,j}$. In addition, $\theta_{2m,i,j} > 0$ and $\theta_{2m+1,i,j} < 0$. One can then derive the upper and lower bounds for $\theta_{k,i,j}$ using the upper and lower bounds $\theta_{k+1,i,j}$. Due to space limitations, the complete derivation process and correctness proof are omitted in this chapter.

Although the formulas seem complicated, it is indeed very efficient to derive very tight lower and upper bounds in practice, where m is typically less than 3. Table 6 reports the average and maximum integer m and time (in milliseconds) used by a Java-based implementation to derive the bounds such that $0.997g_m^+ \geq g(i,j) \geq 1.003g_m^-$ for $512 \leq i \leq j \leq k$. The last row in the table shows the time spent to calculate the true $g(i,j)$ using the original recursive formula. Clearly, the direct calculation is quadratic. On the other hand, the bounds can be derived in almost constant time regardless the value of i or j. The anomaly of the maximum time for $k = 8,192$ vs. $k = 16,384$ is due to the variations in computation environment of the experiment's computer (different background processes run in different experiments and the Java virtual machine is not always given the same time slots). We also conduct experiments to derive even tighter bounds (e.g., $0.9996g_m^+ \geq g(i,j) \geq 1.0004g_m^-$) and with different ρ values. The average runtime and maximum runtime are still in the same range.

Notice that assumptions (1–4) are only used for deriving $g(i,1)$ and $g(1,i)$, not for deriving $g(i,j)$ for $i > 1$ and $j > 1$. Sometimes, events might be distributed under other distributions such as Poisson. In these cases, only $g(i,1)$ and $g(1,i)$ need to be changed to appropriate formulas.

For complex patterns, estimation following this approach can be very complicated. Instead of estimating the o-frequency of a complex pattern using the distribution of individual event instances, we suggest determining the estimate using knowledge of its sub-patterns. For example, we can split a complex pattern into two parts where the second part has the last group and the first part has the rest of the pattern. Then we can assume that these two parts are independently distributed, and use the counted o-frequencies of both parts as the i and j values used in the previous calculation.

6 Conclusion

In this chapter, we proposed a novel definition for the number of sequential pattern occurrences in a single sequence. Our intuitive definition has many beneficial properties: (1) it does not require extra parameters such as a sliding window size, or depend on particular algorithms; (2) it is monotonic in the sense that patterns do not occur less frequently when their constraints are relaxed; (3) the number can be calculated using an efficient greedy algorithm; and (4) the expected value of this number can be estimated and compared to the counted value to evaluate the pattern's significance.

The accurate estimation of the o-frequency under different assumptions of data sequences is essential for identifying surprising patterns according to the o-frequency measure. In this chapter, we estimate the expected value under the independent uniform distribution assumption, and conduct experiments to illustrate the accuracy of the estimation. Synthesized data are used intentionally to observe the independent uniform assumption, since the assumption is made to reflect users' prior knowledge, instead of the posterior knowledge (e.g., dependency between events or the counted o-frequencies) in the data. In practice, other assumptions (e.g., Poisson distribution) may be adopted based on domain knowledge, and the estimation needs to be adapted. In addition, further study is needed to provide estimation of the confidence interval and confidence level.

Pattern occurrence is a fundamental concept in sequential pattern mining problems. We believe our work will benefit the research community in this area. Existing techniques can be adapted, and new techniques may be developed, to discover, filter and rank sequential patterns according to this new basic concept.

For example, using the properties of disjoint occurrence sets described in Sect. 4, more efficient algorithms or optimization techniques can be developed to count the o-frequency. In addition, the monotonic property can be exploited to discover frequent patterns efficiently.

Although the concept of o-frequency is defined for a single sequence, it is possible to extend it to support multiple sequences. It may also be possible to design measures on top of o-frequency and use them to guide the pattern discovery, filtering and ranking in order to assist the analyst in finding patterns of interest.

References

1. Rakesh Agrawal and Ramakrishnan Srikant. Mining sequential patterns. In *The Eleventh International Conference on Data Engineering*, pages 3–14, Taipei, Taiwan, 1995.
2. Dan Gusfield. *Algorithms on Strings, Trees, and Sequences.* Cambridge University Press, 1997.

3. Jiawei Han, Laks V. S. Lakshmanan, and Jian Pei. Scalable frequent-pattern mining methods: An overview. In *The Seventh ACM SIGKDD International Conference on Knowledge Discovery and Data Mining Tutorial Notes*, San Francisco, California, 2001.
4. Jiawei Han, Jian Pei, and Y. Yin. Mining frequent patterns without candidate generation. In *2000 ACM-SIGMOD International Conference on Management of Data*, Dallas, Texas, 2000.
5. Mahesh Joshi, George Karypis, and Vipin Kumar. A universal formulation of sequential patterns. In *The Seventh ACM SIGKDD International Conference on Knowledge Discovery and Data Mining Workshop on Temporal Data Mining*, San Francisco, California, 2001.
6. Kai-Sang Leung, Raymond T. Ng, and Heikki Mannila. OSSM: A segmentation approach to optimize frequency counting. In *18th International Conference on Data Engineering*, San Jose, California, 2002.
7. Heikki Mannila and Hannu Toivonen. Discovering generalized episodes using minimal occurrences. In *Proceedings of the Second International Conference on Knowledge Discovery and Data Mining*, pages 146–151, Portland, Oregon, 1996.
8. Heikki Mannila, Hannu Toivonen, and A. Inkeri Verkamo. Discovering frequent episodes in sequences. In *Proceedings of the First International Conference on Knowledge Discovery and Data Mining*, pages 210–215, Montreal, Canada, 1995.
9. Heikki Mannila, Hannu Toivonen, and A. Inkeri Verkamo. Discovery of frequent episodes in event sequences. Technical Report C-1997-15, Department of Computer Science, University of Helsinki, Finland, 1997. Series of Publications C.
10. Jian Pei, Jiawei Han, Behzad Mortazavi-Asl, Helen Pinto, Qiming Chen, Umeshwar Dayal, and Meichun Hsu. PrefixSpan: Mining sequential patterns by prefix-projected growth. In *17th International Conference on Data Engineering*, pages 215–224, Heidelberg, Germany, 2001.
11. Jian Pei, Jiawei Han, and Wei Wang. Mining sequential patterns with constraints in large databases. In *Eleventh International Conference on Information and Knowledge Management*, pages 18–25, McLean, Virginia, 2002.
12. John F. Roddick and Myra Spiliopoulou. A survey of temporal knowledge discovery paradigms and methods. *IEEE Transactions on Knowledge and Data Engineering*, 14(4):750–767, 2002.
13. Ramakrishnan Srikant and Rakesh Agrawal. Mining sequential patterns: Generalizations and performance improvements. In *Fifth International Conference on Extending Database Technology*, Avignon, France, 1996.
14. Shiby Thomas and Sunita Sarawagi. Mining generalized association rules and sequential patterns using SQL queries. In *Fourth International Conference on Knowledge Discovery and Data Mining*, pages 344–348, New York, 1998.
15. Jiong Yang, Wei Wang, Philip S. Yu, and Jiawei Han. Mining long sequential patterns in a noisy environment. In *21st ACM SIGMOD International Conference on Management of Data*, pages 406–417, Madison, Wisconsin, 2002.
16. Mohammed Javeed Zaki. SPADE: An efficient algorithm for mining frequent sequences. *Machine Learning*, 42(1/2):31–60, 2001.

Gauging Image and Video Quality in Industrial Applications

Weisi Lin

School of Computer Engineering, Nanyang Technological University, Singapore, wslin@ntu.edu.sg

As a result of market, technology and standardization drives, products and services based upon image and video, as well as their delivery, have grown at an explosive rate, with either wired or wireless terminals. Visual signal can be acquired, synthesized, enhanced, watermarked, compressed, transmitted, stored, reconstructed, retrieved, authenticated and presented (e.g., displayed or printed) for various applications. Obviously there is need for accurate quality assessment for images and video, because: (a) visual quality has to be evaluated at different stages of a system or process; (b) how quality is gauged plays a central role in various decision making processes and therefore shaping many (if not all) visual signal manipulating algorithms. Traditionally, simple and mathematically defined measures (e.g., MSE (mean square error), SNR (signal to noise ratio), PSNR (peak signal to noise ratio), or their relatives) have been widely accepted to gauge the distortion of processed visual signal.

The human visual system (HVS) is the ultimate receiver and appreciator for the majority of processed images and video. As a result of many million years' evolution of the mankind, the HVS develops its unique characteristics. It has been well acknowledged that MSE/SNR/PSNR does not align with the HVS' perception [1]. There is therefore an obvious gap in most visual related products and services: a non-perception based criterion is used in the engineering design, while the device or service is for the HVS consumption. Beyond doubt, it is better to use a perceptual criterion in the system design and optimization, in order to make the resultant system more customer-oriented and also bring about benefits for the system; the possible benefits include performance improvement (e.g., in perceived visual quality, traffic congestion reduction, new functionalities, size of device, price of service) and/or resource saving (e.g., for bandwidth allocation, computing requirements or power dissipation in handheld devices).

Perceptual visual quality (or distortion) can be evaluated by subjective viewing tests with the standard procedures [2]. However, this is time-consuming, expensive, and not feasible for on-line manipulations (like encoding, transmission, relaying, etc.), since the resultant mean opinion scores (MOSs) need to

W. Lin: *Gauging Image and Video Quality in Industrial Applications*, Studies in Computational Intelligence (SCI) **116**, 117–137 (2008)
www.springerlink.com

be obtained by many observers through the repeated viewing sessions. Even in the situations where human examiners are allowed (e.g., for visual inspection in a factory environment) and manpower cost is not a problem, the assessment depends upon the viewers' physical conditions, emotional states, personal experience, and the context of preceding display. Hence, it is mandatory to build intelligent, computational models to predict an average observer's evaluation towards the picture under consideration.

The odyssey for perceptual visual quality evaluation proves to be a difficult one [3–5], due to the problem's complex and multi-disciplinary (computer engineering and psychophysics) nature and the mankind's limited knowledge about the functioning of our own eyes and brain. Although a general HVS model is still elusive (now and even in the near future), considerable research and development effort [6–10] has been directed to emulate the HVS' behavior regarding visual quality assessment in specific industrial applications, since a reasonable perceptual visual quality metric can be a differentiating factor to gain the competitive advantage.

In Sect. 1 of this chapter, the basic requirements and classification will be given toward the perceptual quality metrics for industrial usage. Section 2 to Sect. 5 are to present the four computational modules that are oft-used in many metrics. Sect. 6 will present three cases of applications in different practical situations, while the last section highlights the major points of the chapter and discusses some possible future research and development directions. Throughout this chapter, the presentation is with a unified formulation system. The most practical solution for a task is selected and the characteristics of different approaches are discussed whenever this is possible. The chapter serves as a systematic introduction in the field to date and a practical user's guide to the relevant methods.

1 Overview of Practical Quality Metrics

Perceptual visual quality metrics are designed to provide a close approximation to the MoSs, which agree well with the human perception when the number of subjects is sufficiently large. In comparison with the subjective viewing tests, they have additional advantage in repeatability due to their nature of objective measurement. In this section, different practical requirements and approaches of the metrics are discussed.

1.1 Basic Requirements

There are different requirements for the metrics in industrial applications [8, 9, 11]. First of all, we discuss the metrics that carry out direct evaluation of the actual picture under consideration, rather than some predefined signal patterns that have been undergone the same processing [12]. This is because picture quality is a function of visual contents, and the changes of the test signal through a system is usually not a reliable source of visual quality measurement for the actual signal.

Most quality evaluation tasks have to be performed in *in-service* environment. That is, quality is evaluated during the course of the process (like production lines, telecommunication, etc.), and therefore *real time* evaluation is needed. This imposes computational-complexity constraints on algorithm development and selection. To further reduce the work load, evaluation can be done with analysis of *samples* (instead of a continuous stream) of data.

1.2 Metric Classification

A quality metric evaluates the overall quality of an image or a video segment, while a distortion metric is to determine either a particular type of distortion (e.g., blurring) or overall distortion at the presence of multiple impairments. However, a quality metric is closely related to a distortion one in many situations (i.e., higher quality usually means lower distortion, and vice versa), so they are not to be distinguished hereinafter in this chapter.

If we classify the metrics according to the source requirements, there are two major types [11,13,14]: double-ended and single-ended. The former needs both the reference (usually the original) signal and the processed (usually the distorted) signal, and can be further divided into two sub-types: full-reference (FR) and reduced-reference (RR) (only part of the referenced data is needed). The majority of the existing perceptual metrics belong to the FR sub-type, and can be adopted in situations where reference images are available, such as image/video encoders, quality monitoring prior to transmission, and many other visual processing tasks (e.g., enhancement, watermarking). The single-ended (also called blind or no-reference (NR) evaluation) type uses only the processed signal and can be adopted in the situations where reference data are not available or too expensive to transmit/process.

From the viewpoint of methodology, there are two major classes of metrics: *top–down* and *bottom–up* [14]. In the first class, the perceptual models are built with the anatomic and physiological knowledge on the HVS and/or the results of psycho-visual experiments. These metrics are usually based upon signal decomposition into temporal/spatial/color channels, and the other HVS characteristics being modeled typically include spatial/temporal contrast sensitivity function (CSF), luminance adaptation, and various masking effects. In the second class, image data are analyzed for statistical features, luminance/color distortion, and the common visual artifacts (e.g., blockiness and blurring).

A visual quality evaluation module can serve as a *stand-alone* metric or an *embedded* one in a system. *Stand-alone* metrics can be used for quality measurement or monitoring. Quality measurement refers to more accurate assessment required for system design (e.g., video encoder optimization/implementation, equipment selection, acceptance/conformance tests, trouble-shooting, system installation/calibration, or repeatable measurements required for equipment specifications; FR metrics are often used since more information is available for assessment. On the other hand, quality monitoring requests fast speed (necessary for operational monitoring, production line testing, etc.), massive processing (e.g., multi-channel monitoring), applicability to remote sites, and low cost (to enable extensive deployment); NR or RR metrics are more suitable for these scenarios. An *embedded* metric is used inside a visual processing algorithm/system to facilitate better decision making process. Examples include almost every type of visual signal manipulation (be it enhancement, compression, coding, synthesis, transmission or presentation).

The computational modules related to many metrics are those for just-noticeable difference (JND), visual attention, signal decomposition and common artifact detection. These will be introduced in the following four sections. Hereinafter, we use $I_\theta(x, y)$ to denote the image intensity of a pixel at (x, y) for a color component θ; $0 \leq x < X$ and $0 \leq y < Y$; in YC_bC_r representation, $\theta = Y, C_b, C_r$. Accordingly, $\mathbf{I}_\theta$ $(= \{I_\theta(x, y)\})$ and $\{\mathbf{I}_\theta\}$ represent a color component of an image and an image sequence (i.e., video), respectively.

2 Just-Noticeable Difference (JND)

It is well known that the HVS is unable to detect every change in an image. Just-noticeable Difference (JND) refers to the visibility threshold below which any change cannot be detected by the HVS. The simplest JND is described by the well-known Weber–Fechner law, which states that the minimum contrast necessary for an observer to detect change in intensity remains constant over the most meaningful range of luminance intensity. In this section, we will discuss JNDs in different situations.

2.1 JND with Sine-Wave Gratings

The contrast sensitivity function (CSF)[1] demonstrates the varying visual acuity of the human eye towards signal of different spatial and temporal frequencies. In psychophysical experiments, the just-noticeable threshold contrast (the reciprocal of the contrast sensitivity) is measured for traveling sine wave gratings at various spatial frequencies and velocities[2], for both achromatic and chromatic channels [15, 16]. Figure 1 shows the typical spatiotemporal CSF. As can be seen from the figure, the CSF takes on an approximate paraboloid in the spatio-temporal space; at low temporal frequencies, the contrast sensitivity curve is of a band-pass shape; while at high temporal frequencies, the contrast sensitivity curve is of a low-pass shape. It can also be observed that the HVS sensitivity decreases with the increase of spatial and temporal frequencies.

[1] Also called the modulation transfer function (MTF) of the HVS.

[2] The standing sine waves can be regarded as traveling waves at 0 velocity.

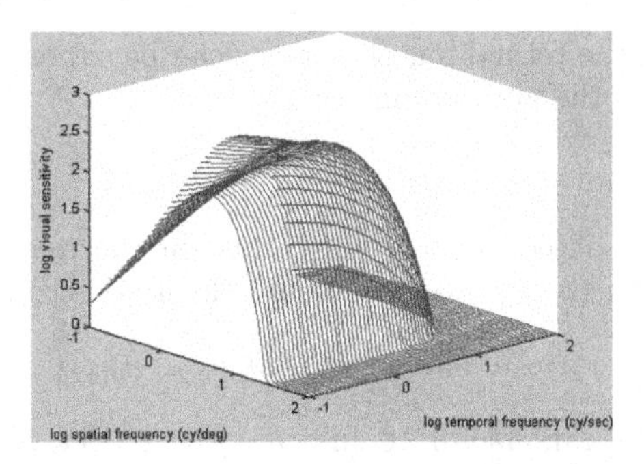

Fig. 1. Spatiotemporal contrast sensitivity surface

Contrast masking refers to the reduction in the visibility of one image component (the target) by the presence of another one (the masker). The relationship of target threshold contrast versus maker contrast has been derived based on the experiments for the sine-wave gratings (e.g., [17, 18]). In general, the threshold contrast increases with increment of the maker contrast. Masking is strong when the interacting stimuli have similar characteristics, i.e., similar frequencies, orientation, colors, etc.

2.2 Formulation of CSF in DCT Domain

In [19] where the work in [15, 20] has been extended, the visibility threshold based upon CSF for the (i, j)-th subband of the n-th DCT block is approximated as:

$$t(n, i, j) = \frac{exp(k_3 \cdot \rho_{i,j}^2(\varepsilon \cdot v(n) + 2))}{\rho_{i,j}^2 \cdot v(n) \cdot (k_1 + k_2|log(\varepsilon \cdot v(n)/3)|^3)} \tag{1}$$

where i, j = 0,1,...,$N - 1$, and N is the dimension of the DCT block; $\rho_{i,j}$ is the corresponding spatial subband frequency while $v(n)$ depicts the retinal image velocity of the block; for the model parameters, $\varepsilon = 1.7$, $k_1 = 175\pi^2$, $k_2 = 208\pi^2$, and $k_3 = 0.05\pi$ [19]. The spatial subband frequency is:

$$\rho_{i,j} = \frac{1}{2N}\sqrt{\frac{i^2}{\omega_x^2} + \frac{j^2}{\omega_y^2}} \tag{2}$$

where $\omega_\hbar$ $(\hbar = x, y)$ is the horizontal and vertical visual angles of a pixel, which can be calculated based on viewer distance ξ and the display width of a pixel, Λ_x or Λ_y, on the monitor:

$$\omega_\hbar = 2 \cdot arctan(\frac{\Lambda_\hbar}{2\xi}) \tag{3}$$

The spontaneous eye movement tends to track the moving object, reduces the retinal velocity of the image, and thus compensates the loss of sensitivity due to motion. In [20], the retinal image velocity can be expressed as below, with the consideration of the eye movement:

$$v(n) = v_o(n) - v_e(n) \tag{4}$$

where $v_o(n)$ is the object velocity in retina on the assumption of no eye movement, and $v_e(n)$ is the eye movement velocity determined as:

$$v_e(n) = \min\lfloor g \cdot v_o(n) + v_{min}, v_{max} \rfloor \tag{5}$$

where $g = 0.92$ (a gain factor to indicate the object tracking efficiency), $v_{min} = 0.25$ deg/sec (the minimum eye-velocity due to the drift movement for static-image perception [21]), $v_{max} = 80.0$ deg/sec (the maximum eye-velocity before the saccadic movement for rapidly moving objects [22]), and $\lfloor \cdot \rfloor$ represents the operator to get the biggest integer.

Let $f_m(n)$ be the motion vector for the block, and it can be detected via a motion estimation algorithm. Then,

$$v_o(n) = \delta \cdot f_m(n) \tag{6}$$

where δ is the frame rate of video (in frames per second). The image plane velocity $v_o(n)$ is converted to the retinal image velocity $v(n)$ via (4) and (5).

When $v(n) \equiv 0.15$ deg/sec (i.e., only natural drift movement occurs), the spatiotemporal CSF becomes the spatial (static) CSF, and the formulae derived above are applicable for still images.

2.3 JND for Real-World Video

In real-world visual signal, various factors affect the actual JND. In DCT domain, the visibility threshold $t(n, i, j, t)$ can be combined with the influence from luminance adaptation [24] and contrast masking [23]. We will present the JND in pixel domain because of its operating efficiency in practice.

The spatial JND of pixel (x,y) in a YC_bC_r image can be calculated as [25] (an extension of [26, 27]):

$$T_\theta^S(x, y) = T^l(x, y) + T_\theta^t(x, y) - \psi_\theta^{lt} \cdot \min\{T^l(x, y), T_\theta^t(x, y)\} \tag{7}$$

where $T_\theta^t(x, y)$ are the visibility thresholds for luminance adaptation and texture masking, respectively; ψ_θ^{lt} accounts for the overlapping effect in masking, and $0 < \psi_\theta^{lt} \leq 1$. A reference set of ψ_θ^{lt} values is: $\psi_Y^{lt} = 0.3$, $\psi_{C_b}^{lt} = 0.25$,and $\psi_{C_r}^{lt} = 0.2$ [25].

$T^l(x, y)$ accounts for the visibility threshold affected by the background luminance in the image neighborhood, and can be determined as follows for 8-bit luminance representation [26]:

$$T^l(x, y) = \begin{cases} 17(1 - \sqrt{\frac{\bar{I}_Y(x,y)}{127}}) + 3, & \text{if } \bar{I}_Y(x, y) \leq 127 \\ \frac{3}{128}(\bar{I}_Y(x, y) - 127) + 3 & \text{otherwise} \end{cases} \tag{8}$$

where $\bar{I}_Y(x,y)$ is the average background luminance at (x,y) within a small (e.g., NxN) neighborhood.

Texture masking can be estimated as [25]:

$$T^t_\theta(x,y) = \eta_\theta \cdot G_\theta(x,y) \cdot e_\theta(x,y) \quad (9)$$

where η_θ denotes control parameters with $\eta_Y < \min\,(\eta_{C_b}, \eta_{C_r})^3$, and η_Y, η_{C_b} and η_{C_r} are set as 0.12, 0.65 and 0.45 [25], respectively; $G_\theta(x,y)$ denotes the maximum weighted average of gradients around (x,y); $e_\theta(x,y)$ is an edge-related weight of the pixel at (x,y), and its corresponding matrix $\mathbf{e}_\theta$ is computed as [25]:

$$\mathbf{e}_\theta = \mathbf{l}_\theta * \mathbf{h} \quad (10)$$

where $\mathbf{l}_\theta$ is a matrix whose elements are assigned to 0.1, 0.3 and 1.0 for smooth, edge and texture pixels[4], respectively; $\mathbf{h}$ is a $b \times b$ Gaussian low pass filter with standard deviation σ, to smooth $\mathbf{l}_\theta$ for avoidance of dramatic changes in a neighborhood. An appropriate parameter selection is: $b = 7$ and $\sigma = 0.8$.

$G_\theta(x,y)$ is determined as [26]:

$$G_\theta(x,y) = \max_{k=1,2,3,4}\{grad_{\theta,k}(x,y)\} \quad (11)$$

with

$$grad_{\theta,k}(x,y) = \frac{1}{16}\sum_{i=1}^{5}\sum_{j=1}^{5} I_\theta(x-3+i, y-3+j) \cdot g_k(i,j) \quad (12)$$

where $\{g_k(i,j), k = 1,2,3,4\}$ are four directional high-pass filters as shown in Fig. 2.

Object movement or change affects the visibility threshold, and bigger inter-frame difference usually leads to larger temporal masking. As an efficient implementation, the overall JND for video can be therefore expressed as:

$$T_\theta(x,y) = f(d(x,y)) \cdot T^S_\theta(x,y) \quad (13)$$

g_1:

0	0	0	0	0
1	3	8	3	1
0	0	0	0	0
-1	-3	-8	-3	-1
0	0	0	0	0

g_2:

0	0	1	0	0
0	8	3	0	0
1	3	0	-3	-1
0	0	-3	-8	0
0	0	-1	0	0

g_3:

0	0	1	0	0
0	0	3	8	0
-1	-3	0	3	1
0	-8	-3	0	0
0	0	-1	0	0

g_4:

0	1	0	-1	0
0	3	0	-3	0
0	8	0	-8	0
0	3	0	-3	0
0	1	0	-1	0

Fig. 2. Directional high-pass filters for texture detection

[3] The HVS is more sensitive for a difference in Y space than in C_b or C_r space.

[4] The masking effect in smooth regions is not obvious, and the effect in edge regions is modest since the distortion around edge is more noticeable than a same amount of distortion in texture regions.

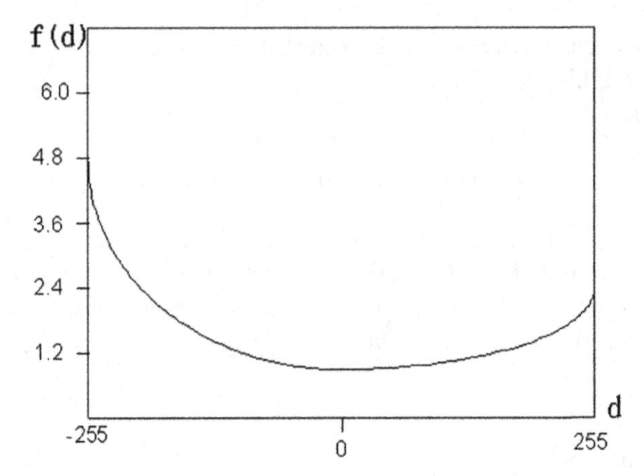

Fig. 3. Temporal masking vs. inter-frame difference (from [27], ©1996 IEEE)

where $d(x,y)$ represents the inter-frame luminance difference between the current frame $\mathbf{I}$ and the previous frame $\mathbf{I}^p$; $f(d(x,y))$ increases with the increase in $d(x,y)$, as denoted in Fig. 3 [27], and

$$d(x,y) = \frac{I_Y(x,y) - I_Y^p(x,y)}{2} + \frac{\bar{I}_Y - \bar{I^p}_Y}{2} \tag{14}$$

where $\bar{I}_Y$ and $\bar{I^p}_Y$ are the average luminance of $\mathbf{I}$ and $\mathbf{I}^p$, respectively.

3 Visual Attention

The HVS does not process and perceive all the information equally, as described as the phenomenon of visual attention [28]. It is therefore necessary to treat signal changes in a picture unequally in an image, in order to emulate the human vision. Not all the changes in an image lead to distortion, nor does same amount of signal change results in same perceptual distortion. In addition, excessive perceptually-insignificant data may disturb the measurement and hence hinder the effectiveness of metrics.

Visual attention is determined by many visual stimuli. First of all, it is the contrasts (instead of the absolute values) of sensory data (e.g., for color, luminance, texture and motion) against the background attract the human visual attention. Other factors include cognitive features (such as skin color, human faces, and prominent objects). For application-specific models, we may need to consider the influence from task-oriented features (like objects with certain characteristics to be located/recognized) and from other media (speech, audio, text, etc.). In this section, a basic model of visual attention is to be presented, following the methodology described in [29], in which color, texture and motion can be detected as sensory features, while skin color and human face are examples of cognitive features.

3.1 Feature Extraction

A pixel (and a region) usually draws a bigger level of attention if its difference from the background is bigger. Motion vector $f_m(x, y)$ at a pixel can be detected against the background (e.g., via [30]). For color and texture, the *k-means* clustering algorithm can be used to group all NxN image blocks. If the size of the largest cluster is greater than a threshold (e.g., a fraction of the image size), this cluster is regarded as the background area for the image. Otherwise, the corresponding feature (color or texture) would be excluded from the computation for visual attention, because the contrast is only meaningful if obvious reference (background) exists. When the background cluster is available, the mean of the feature for the cluster is calculated; let $f_c(x, y)$ and $f_t(x, y)$ denote the color and texture contrasts at pixel (x, y), respectively. Then, $f_c(x, y)$ and $f_t(x, y)$ are calculated as their Euclidean distances from the means of corresponding background cluster.

Each feature, $f_p(x, y)$, $p = c$, t, and m, reflects the contrast in comparison with its background. Furthermore, we need to consider the occurrence of different contrast levels in the same frame of image; if more pixels in the image are with big contrast, the corresponding impact would not be as significant as the situation otherwise. We use $n_p(d)$ to represent the number of pixels in the image with $f_p(x, y) = d$. Then $f_p(x, y)$ is modified as:

$$F_p(x, y) = \frac{\sum_d n_p(d)}{\sum_d n_p(d) \cdot d} f_p(x, y) \tag{15}$$

The appearance of human skin and face attract additional attention of the HVS besides the effect of the sensory data that has been accounted for. The occurrence of skin color and face is detected as $F_s(x, y)$ [31], and $F_f(x, y)$ (from the luminance image) [32], respectively, and their correlation is to be considered in the next subsection.

3.2 Integration

The integration for the overall effect of all features can be fulfilled as follows [29] (as an extension of the nonlinear additivity model in [33]):

$$V(x, y) = \sum_p F_p(x, y) - \sum_p \min(\beta_{p,q} \cdot F_q(x, y), \beta_{q,p} \cdot F_p(x, y)) \tag{16}$$

where $p = c, t, m, s$ and f; the main feature $q = arg\ \max_p(F_p(x, y))$; the first term in (16) accumulates all features; $\beta_{p,q}$ (in the range of [0,1]) represents the cross-dimensional coupling factor, and $\beta_{p,p}$=0; the second term in (16) accounts for the coupling effect of $F_q(x, y)$ with each $F_p(x, y)$.

If a feature correlates with another, there exists a non-zero $\beta_{.,.}$. The higher the correlation is, the bigger value $\beta_{.,.}$ takes. As can be seen from (16), a bigger

$\beta_{\cdot,\cdot}$ tends to yield smaller compound effect. The accurate determination of $\beta_{\cdot,\cdot}$ calls for more psychological evidence and also depends upon the specific application. A reference set of valuation is given and demonstrated in [29], $\beta_{c,t}$ and $\beta_{t,c}$ are set to zero since color and texture attract attention independently [34]. A bigger value is assigned to $\beta_{s,a}(=\beta_{a,s}=0.75)$ because of the high correlation between skin color and human face. Due to the modest correlation between motion and skin color (as well as human face), $\beta_{m,s}=\beta_{s,m}=\beta_{m,f}=\beta_{f,m}=0.5$. All other $\beta_{\cdot,\cdot}$ values are set to 0.25.

Formula (16) gives a relative indication of visual significance in an image for a pixel or region[5], and reflects the global characteristics of the HVS's spatial selectivity. The working of the HVS in this regard resembles that of a personal computer: the HVS has limited computing power and cache memory, so the most eye-catching portion of the image is first fetched into the cache memory for processing.

3.3 Modulation for JND

In essence, visual attention captures the HVS' global selectivity profile within an image, while JND describes the local sensitivity level if the pixel falls in the foveal area. Therefore, the resultant visibility threshold at a pixel is the JND value modulated by visual attention:

$$T_\theta^M(x,y)=T_\theta(x,y)\cdot M(V(x,y)) \tag{17}$$

where $M(V(x,y))$ denotes the modulatory function determined by $V(x,y)$, and takes a small value with a larger $V(x,y)$ [29].

4 Signal Decomposition

Since the HVS has different sensitivities towards different frequencies and orientations, a proper image decomposition regarding these aspects facilitates the HVS-compliant evaluation of a signal change, for both achromatic and chromatic parts. The decomposition allows unequal treatment of each component to emulate the HVS response.

4.1 Spatiotemporal Filtering

Firstly, $\mathbf{I}_\theta$ can be separated as a sustained (low-pass) channel $\mathbf{I}_{\theta,su}$ and a transient (band-pass) channel $\mathbf{I}_{\theta,tr}$, with two temporal filters whose impulse responses are expressed as [35]:

$$h_{su}(t)=e^{-(\frac{ln(t/\tau)}{\iota})^2} \tag{18}$$

[5] The methodology in this section can be applied to a region (i.e., a group of pixels).

Table 1. Coefficients for the sustained and transient 9-tap FIR filters

tap	0	1	2	3	4	5	6	7	8
$h_{su}(t)$	0.0004	0.0121	0.0672	0.2364	0.3677	0.2364	0.0672	0.0121	0.0004
$h_{tr}(t)$	−0.0260	−0.0567	−0.2242	0.0337	0.5289	0.0337	−0.2242	−0.0567	−0.0260

and its second derivative:

$$h_{tr}(t) = h''_{su}(t) = \frac{2}{\iota^2 t^2} e^{-(\frac{ln(t/\tau)}{\iota})^2} \{ \frac{2}{\iota^2} ln^2(t/\tau) + ln(t/\tau) - 1 \} \tag{19}$$

respectively, with $\tau = 160$ msec and $\iota = 0.2$. This can be materialized via two temporal FIR (or IIR) filters [13]. Table 1 shows the 9-tap FIR filter coefficients for $h_{su}(t)$ and $h_{tr}(t)$.

Based upon the fact that receptive fields in the primary visual cortex resemble Gabor patterns [36] which can be characterized by a particular spatial frequency and orientation, a filter band (e.g., Gabor filters, Cortex filters [37], wavelets, Gaussian/sterrable pyramid filters [38]) can be used to spatially decompose each temporal channel ($\mathbf{I}_{\theta,\varrho}$, $\varrho = su, tr$).

For instance, the steerable pyramid filters [38, 39] decomposes a temporal channel with various spatial frequencies and orientations. Each $\mathbf{I}_{\theta,\varrho}$ may be decomposed into three frequency bands (band-pass ones) and then four orientations each, as well as one isotropic low-pass band. This results in 13 spatial channels: $\{\mathbf{I}_{\theta,\varrho,\mu,\nu}, \mu = 1,2,3, \nu = 1,2,3,4\}$, and $\mathbf{I}_{\theta,\varrho,4,0}$ (the isotropic low-pass channel); the four possibilities of non-zero ν (1–4) indicate the orientations of 0°, 45°, 90° and 135°, respectively; μ being 1 corresponds to the highest frequency band and resolution, and an increment by 1 with μ specifies a lower frequency band and a 2:1 down-sampling in both dimensions.

4.2 Contrast Gain Control

The contrast of $I_{\theta,\varrho,\mu,\nu}(x,y)$ is denoted as $c_{\theta,\varrho,\mu,\nu}(x,y)$ and can be obtained via division by a local sum in the neighborhood [6, 14]. The contrast gain control stage that follows is to set appropriate gain for each contrast by factoring in the inter-channel masking effect [39, 40]:

$$\tilde{c}_{\theta,\varrho,\mu,\nu}(x,y) = \chi \frac{c^2_{\theta,\varrho,\mu,\nu}(x,y)}{r + \phi * c^2_{\theta,\varrho,\mu,\nu}(x,y)} \tag{20}$$

where χ and r are the gain control parameters, and ϕ is a pooling function over different channels by the convolution operator "*". In practice, only those channels with similar characteristics are pooled, because masking is most prominent when interacting stimuli are close in frequency, orientation and color. If only channels with same frequency range and equal level of contribution among the channels are considered [41], Eq. (20) becomes:

$$\tilde{c}_{\theta,\varrho,\mu,\nu}(x,y) = \chi \frac{c^2_{\theta,\varrho,\mu,\nu}(x,y)}{r + \sum_{\nu_0=1(\nu_0 \neq \nu)}^{4} c^2_{\theta,\varrho,\mu,\nu_0}(x,y)} \tag{21}$$

Signal in each temporal-spatial channel $I_{\theta,\varrho,\mu,\nu}(x,y)$ can be weighted before the use of Eq. (20) or Eq. (21). The weighting parameters and all model parameters (like χ and r) can be determined via the fitting the resultant model to CSF and contrast masking curves [39], or subjective viewing test scores [41,42].

5 Common Artifact Detection

Certain artifacts occur in the signal compression and delivery process [51], and cause annoying effects to the HVS. The common ones that need to be detected in practice are blockiness [7–9], blurring [8, 9] and picture freeze [7–9]. The traditional measures, like MSE, SNR and PSNR, fail to reflect the perceptual effect of such structural artifacts. Effective detection of these artifacts with the available domain knowledge greatly facilitates visual quality evaluation in industrial applications. This is a typical *bottom–up* process, usually performs without the reference signal (i.e., in the NR case), and only uses the luminance image for operational efficiency.

5.1 Blockiness

Blocking artifact is a prevailing degradation caused by the block-based Discrete Cosine Transform (BDCT) coding technique, especially under low bit-rate conditions. It is unavoidable due to the different quantization steps used in the neighboring blocks and the lack of consideration for inter-block correlation. Because of the occurrence at a NxN block boundaries, the most straightforward measure is [43]:

$$M_h = \left[\sum_{k=1}^{Y/N-1} \sum_{x=0}^{X-1} (I_Y(x, k \cdot N - 1) - I_Y(x, k \cdot N))^2 \right]^{1/2} \tag{22}$$

for horizontal blockiness, and

$$M_v = \left[\sum_{l=1}^{X/N-1} \sum_{y=0}^{Y-1} (I_Y(l \cdot N - 1, y) - I_Y(l \cdot N, y))^2 \right]^{1/2} \tag{23}$$

for vertical blockiness.

The overall measure can be obtained by an appropriate weighted sum of M_h and M_v [43]. Alternatively, blockiness can be located via harmonic analysis [44], and this approach may be used when block boundary positions are unknown beforehand (e.g., with video being cropped or re-taken by a camera).

5.2 Blurring

Blurring can be effectively measured around edges in an image, because it is most noticeable there and such detection is efficient (since edges only constitute a small fraction of all image pixels). A method to measure the blur in an image has been proposed in [45] with edge spread detection. For each edge pixel (e.g., extracted by Canny edge detector [46]), the gradient's direction is calculated as:

$$\varpi(x,y) = tan^{-1}\frac{G_y(x,y)}{G_x(x,y)} \tag{24}$$

where

$$G_x(x,y) = \frac{\partial I_Y(x,y)}{\partial x}, G_y(x,y) = \frac{\partial I_Y(x,y)}{\partial y} \tag{25}$$

For the convenience of implementation, $\varpi(x,y)$ is divided into eight quadrants. The edge spread is obtained by counting the number of the pixels along $\varpi(x,y)$ and the opposite direction, until the pixel's greylevel value stops significant changes. The average edge spread in the image can be then used as an indication of image blur.

5.3 Frame Freeze

Overloaded networks cause the bitrate controller at a video encoder to skip frames, and transmission loss (e.g., packet loss) forces a decoder to repeat the display of the previous frame if the loss is not recoverable. Both of these situations result in the so-called frame freeze phenomenon to viewers. Frame freeze is a major source of perceived jerkiness.

The preliminary inter-frame difference can be evaluated via the change detection [47,48] of the current luminance frame and the previous one:

$$\triangle I_Y(x,y) = |I_Y(x,y) - I_Y^p(x,y)| \tag{26}$$

Apart from the motion in video, the other cause to a non-zero $\triangle I_Y(x,y)$ includes random noise (e.g., the camera noise if the video is re-taken). A proper morphological or simply an averaging operation in the local neighborhood can alleviate greatly the influence of such noise. A non-zero $\triangle I_Y(x,y)$ represents the movement in the video at (x,y).

The total inter-frame motion can be obtained by

$$D = \sum_{x,y} \triangle I_Y(x,y) \tag{27}$$

If D is insignificant, there may exist frame freeze. In industrial use, D is usually checked for a multiple of $1/\delta$ second, where δ is the frame rate of video, before frame freeze is declared, since the HVS cannot sense single frame freeze in most situations.

6 Case Studies

A visual quality/distortion metric can be used in industries as both *stand-alone* test equipment and an embedded module in a system. As introduced in Sect. 1.2, a quality measurement system provides more accurate assessment required for system design, implementation and optimization; quality monitoring systems are often designed for speedy, massive, on-line and low-cost processing. A metric in line with the HVS perception ensures the related process effective, efficient and customer-oriented. Three cases of application to be discussed in this section are based upon the computational modules introduced earlier.

6.1 JNDmetrix^{TM} as Quality Measurement

Sarnoff Corporation developed the HVS-based model called JNDmetrix^{TM} [6] for visual quality measurement, using FR and top–down strategies, to measure perceptual difference for full-color video. This model has become the basis of Tektronix's PQA200 [7, 11], which is a real-time, in-service picture quality measurement system. In JNDmetrix^{TM}, the processing is just a variation of what is described in Sect. 4. Figure 4 illustrates the flow chart of the processing for luminance in the JNDmetrix^{TM} system. Firstly, each luminance field of video is decomposed into four levels (Levels 0 to 3) of Gaussian pyramid [49]. Similar to the notations in Sect. 4.1, an increment by 1 with levels specifies a lower frequency band and a 2:1 down-sampling in both dimensions, and Level 3 represents the lowest-resolution pyramid image. Afterward, Level 3 is subjected to temporal filtering with four consecutive fields, while the other three levels are subjected to spatial filtering with different orientations. The contrast here is calculated as a local difference of the reference signal and the processed one, and then scaled by a local sum. The contrast gain masking stage is to divide each oriented contrast by a function of all the other local stimuli from different levels and orientations. As a result, the sensitivity to distortion decreases in activity-rich image regions, temporal structure masks spatial differences, and spatial structure also masks temporal differences.

Processing of chrominance signal (as shown in Fig. 5) is similar to that for the luminance counter part. The pyramid decomposition here gives seven levels in favor of higher color sensitivity in the resultant system. Due to the inherent chrominance insensitivity to flicker, temporal averaging is carried out over four consecutive image fields. Contrast gain masking is simpler than in luminance processing, since the division is done by a function of only a luminance stimulus.

JNDs with sine-wave gratings were used to calibrate various model parameters. For determination of the parameters related to pyramid channels, the system was made to output 1 (i.e., 1 unit of JND), when the inputs were a test image of sine wave with the amplitude specified by the spatial CSF

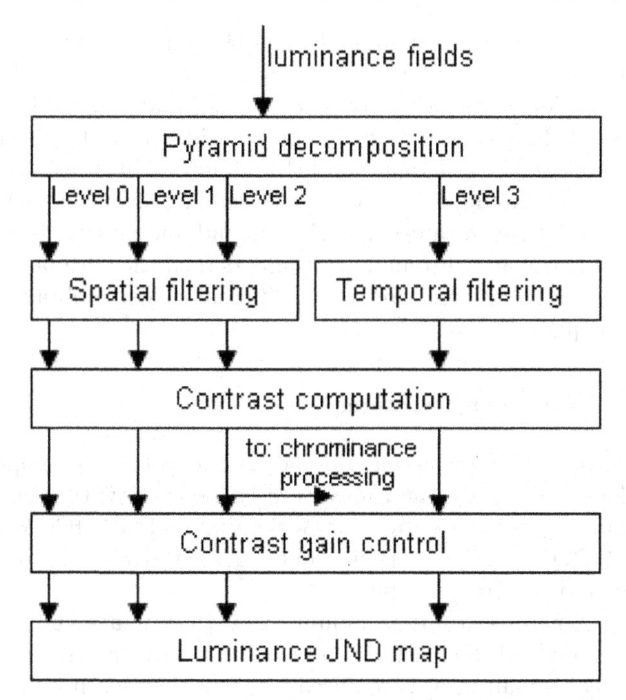

Fig. 4. Processing of luminance signal in JNDmetrix^{TM}

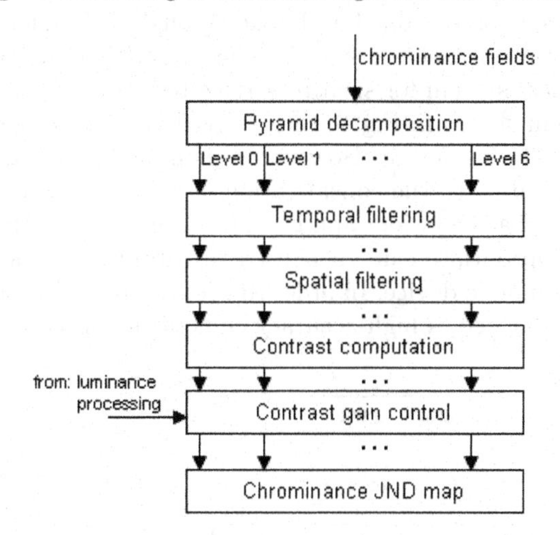

Fig. 5. Processing of chrominance signal in JNDmetrix^{TM}

at a frequency, and a reference image of a uniform field with the luminance specified by the average luminance of the test image. Similar fitting was performed for deciding the temporal filter coefficients and the contrast gain control parameters with the temporal CSF, and contrast masking data, respectively. As a result, the luminance and chrominance differences yielded by the processing denoted in Figs. 4 and 5 are measured in the units of JNDs.

There are different ways to arrive at a single number to represent the visual quality of the processed video under consideration. In JNDmetrix^{TM}, for each image field, the luminance and chrominance difference maps are first reduced to a single number each for summary. This is accomplished by histogramming all difference values in each map and finding the 90th percentile. Then, the luminance and chrominance numbers (i.e., the 90th percentiles) are combined. The same process repeats for different fields to obtain the overall single-number quality score.

6.2 Quality Monitoring Systems

Quality monitoring is often performed via NR and bottom–up approaches for a fast and low-cost process, and may be used together with a speech/audio metric. Examples include products of Genista [8] and OPTICOM [9], as well as Tektronix's PQA300 [7]. The typical outputs are the measures of blockiness, frame freeze, and blurring/sharpness.

There are numerous situations in industries where image and video quality needs to be monitored. For instance, in cable TV broadcasting and video on demand services, monitoring is required for each channel in a relay point or at a user's site; in manufacturing lines for mobile devices (like PDAs, handphones), test video can be coded and gone through a channel simulator, and the received picture artifacts are to be checked with the devices. Figure 6 illustrates a typical system for such *in-service* testing.

For the system shown in Fig. 6, the received video can be extracted from the frame store of the device. However, a non-intrusive acquisition means may be more useful, with the consideration of speedy, massive, and low-cost monitoring, as well as the wider application scope. To acquire the video non-intrusively, an appropriate camera is used to capture the visual signal from the screen of the mobile device. In order to avoid introducing significant additional noise to the video, a high-accuracy camera should be used, and proper

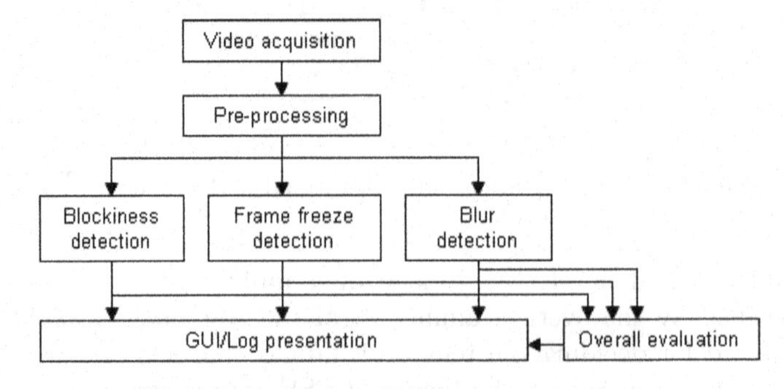

Fig. 6. A visual quality monitoring system

setting needs to be exploited (e.g., structured lighting or use of blinds). The captured video is pre-processed first: compensation of camera noise (an adaptive method is described in [48]), extraction of the portion of signal belonging to the device under monitoring, etc. The detection of the common artifacts has been discussed in Sect. 5, and real-time processing is possible even with high-level language (e.g., C/C++) programming. The joint effect of all artifacts can be also evaluated. For most existing systems, the results are presented in a graphic form against the timing parameter via GUI (graphical user interface), and recorded in log files for off-line diagnosis and statistical analysis.

6.3 Modulated JNDs in Visual Communication

The JND map modulated by the visual attention (Sect. 2) gives the relative visibility of distortion within an image or a frame of video. Therefore the generator of a such map can be embedded in a visual communication system to guide the allocation of the scarce bits to the most perceptually significant part of the signal. The implementation has to be sufficiently efficient in order not to impede the performance of the overall system.

All existing standard video codecs (i.e., MPEG-1/2/4, and H.261/3/4) tend to code foreground objects with lower quality than background ones. This is because foreground objects usually are with more motion and deformation, and therefore more difficult to be coded [50]. In addition, the foreground attracts more attention from the viewer, so the distortion appearing there is definitely more noticeable. A solution is to incorporate the aforementioned modulated JND model to the bitrate controller in a video encoder to reverse this tendency.

A simplified modulated JND model in spatial domain (for operating efficiency) can be used in both wired and wireless videoconferencing (or videophony), where the interested foreground objects are the talking heads. In [50], skin color is used as the cognitive feature to modulate the JNDs as derived in Sect. 2.3, and the resultant visibility thresholds control the determination of quantization steps for a given bitrate: the higher the local visibility threshold (i.e., within a less perceptually significant region) is, the higher the quantization step (hence with less bits required) is chosen. The system is shown in Fig. 7. Subjective viewing tests have proven the effectiveness of the approach.

7 Concluding Remarks

We addressed the increasing need of gauging visual quality in line with the perception of the human visual system (HVS) in industrial applications, due to the calls from both the technology (for further algorithm improvement) and the marketplace (to be more customer-oriented). The recent advance of computer engineering in the related areas has enabled the deployment of some

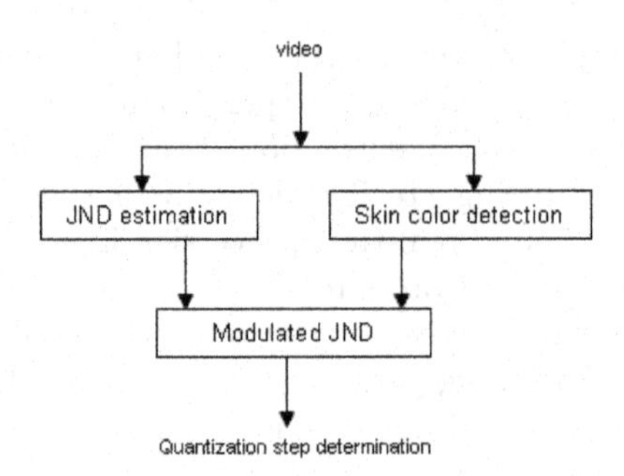

Fig. 7. Modulated JNDs in visual communication

pro-HVS metrics in commercial products, although the related research is still in its infancy, and also demonstrated good prospects for further research and development.

This chapter firstly presented an overview of perceptual visual quality metrics, with the discussion on the basic requirements and classification. Under a unified formulation system, we then introduced several computational modules that are often found in different visual-quality related processing. The computational modules introduced in sufficient details include those for just-noticeable difference (JND), visual attention, signal decomposition and common artifact detection. Based upon the knowledge about the aforementioned modules, three examples of application have been briefed, namely, two *stand-alone* metrics (one with *full-reference* and *top–down* approach, and one for *no-reference* and *bottom–up* approach) and an embedded JND estimator modulated by visual attention.

It is anticipated that more and more applications of perceptual models will be found in industrial environment, as a result of the aforementioned technological and market drives, as well as the wider acceptance of an objective means to predict the subjective visual evaluation in test equipment (more companies buying the idea than five years' ago). For future research effort, more emphasis should be placed in *no-reference* metrics, because: a) they align well with the HVS perception (the human being does not need an explicit reference when judging a picture's quality); b) they are computationally efficient (no need to process the reference image); c) they are applicable to wider scope of use (in situations with and without the reference). Furthermore, metrics in transform domain (DCT and wavelet) are less investigated, but there is clear demand from industries because of the need of multi-channel monitoring (for IPTV, cable broadcasting, video on demand, and so on) where quality has be assessed directly in the compressed form to avoid the full decoding complexity for multiple channels. Besides, more convincing method is needed to evaluate multiple artifacts that co-exist in images or video, for gauging overall quality to match the subjective viewing result; this requires good understanding and modeling of the latest findings in the relevant physiological and psychological research. In addition, joint evaluation with other media (like audio, speech, text, etc.) and task-oriented information is crucial for many products and services to come.

References

1. B. Girod (1993) What's wrong with mean-squared error? In: A.B. Watson (ed) Digital Images and Human Vision. MIT
2. ITU-R Recommendation 500-10 (2000) Methodology for the Subjective Assessment of the Quality of Television Pictures. ITU, Geneva, Switzerland
3. Z. Wang, A. C. Bovik, L. Lu (2002) Why is image quality assessment so difficult? In: Proc. of IEEE International Conference on Acoustics, Speech, and Signal Processing
4. Video Quality Expert Group (2000) Final Report from the Video Quality Expert Group on the validation of Objective Models of Video Quality Assessment. http://www.vqeg.org
5. Video Quality Expert Group (2003) Final Report from the Video Quality Experts Group on the Validation of Objective Models of Video Quality Assessment, Phase II. http://www.vqeg.org
6. Sarnoff Corporation, J. Lubin (ed.) (1997) Sarnoff JND vision model. Contribution to IEEE G-2.1.6 Compression and Processing Subcommittee
7. Tektronix, Inc. (2007) Picture Quality Analysis System. http://www.tektronix.com/video_audio
8. Genista Corporation (2006) Quality Assurance Solutions for IPTV and Mobile Media. http://www.genista.com
9. OPTICOM GmbH (2007) Perceptual Voice, Audio and Video Quality. http://www.opticom.de
10. Symmetricoms QoE Assurance Division (2007) QoS Metrics. http://qoe.symmetricom.com
11. Tektronix, Inc. (2001) A Guide to Maintaining Video Quality of Service for Digital Television Programs. http://www.tek.com/Measurement/App_Notes/Technical_Briefs/digital_QoS/25W_14000_1.pdf
12. ITU-T Recommendation J.147 (2002) Objective Picture Quality Measurement Method by Use of In-service Test Signals. ITU, Geneva, Switzerland
13. S. Winkler (2005) Perceptual Video Quality Metrics–A Review. In: H. R. Wu, K. R. Rao (eds) Digital Video Image Quality and Perceptual Coding. CRC
14. W. Lin, L. Dong, P. Xue (2005) IEEE Trans. Circuits Syst. Video Technol. 15(7):900–909
15. D. H. Kelly (1979) J. Opt. Soc. Amer. 69(10):1340–1349
16. P. Barten (1999) Contrast Sensitivity of the Human Eye and Its Effects on Image Quality. SPIE Press
17. G. E. Legge (1981) Vision Res. 21:457–467
18. C. Carlson, R. Cohen (1980) Proc. Soc. Inf. Disp. 21:229–245
19. Y. Jia, W. Lin, A. A. Kassim (2006) IEEE Trans. Circuits Syst. Video Technol. 16(7):820–829
20. S. Daly (2001) Engineering observations from spatiovelocity and spatiotemporal visual models. In: Vision Models and Applications to Image and Video Processing. Kluwer, Norwell, MA

21. R. W. Ditchburn (1973) Eye Movements and Perception. Clarendon, Oxford, UK
22. P. E. Hallett (1986) Chapter 10. In: Handbook of Perception and Human Performance. Wiley, New York
23. H. Y. Tong, A. N. Venetsanopoulos (1998) In: Proc. of IEEE International Conference on Image Processing 3:428-432
24. X. Zhang, W. Lin, P. Xue (2006) Signal Processing 85(4):795–808
25. X. Yang, W. Lin, Z. Lu, E. Ong, S. Yao (2005) IEEE Trans. Circuits Syst. Video Technol. 15(6): 742–750
26. C. H. Chou, Y. C. Li (1995) IEEE Trans. Circuits Syst. Video Technol. 5(6):467–476
27. C. H. Chou, C. W. Chen (1996) IEEE Trans. Circuits Syst. Video Technol. 6(2):143–156
28. M. M. Chun, J. M. Wolf (2001) Visual attention. In: B. Goldstein (ed) Blackwell Handbook of Perception. Blackwell, Oxford (UK) p. 272310
29. Z. Lu, W. Lin, X. Yang, E. Ong, S. Yao (2005) IEEE Trans. Image Process 14(11):1928–1942
30. K. Zhang, J. Kittler (1998) Global motion estimation and robust regression for video coding. In: Proc. of IEEE International Conference on Acoustics, Speech, and Signal Processing
31. D. Chai, K. N. Ngan (1999) IEEE Trans. Circuit Syst. Video Technol. 9(6):551–564
32. H. A. Rowley, S. Baluja, T. Kanade (1998) IEEE Trans. Pattern Anal. Mach. Intell. 20(1):23–38
33. H.-C. Nothdurft (2000) Vision Res. 40(10–12):1183–1201
34. M. C. Morrone, V. Denti, D. Spinelli (2002) Curr. Biol. 12(13):1134–1137
35. R. E. Frederickson, R. F. Hess (1998) Vision Res. 38(7):1023–1040
36. J. G. Daugman (1980) Vision Res. 20(10):847–856
37. A. B. Watson (1987) Comp. Vis. Graphics Ima. Proc. 39(3):311–327
38. E. P. Simoncelli, W. T. Freeman, E. H. Adelson, D. J. Heeger (1992) IEEE Trans. Info. Theory 38(2):587–607
39. S. Winkler (2000) Vision models and quality metrics for image processing applications. Ecole Polytecnique Federale De Lausanne (EPFL), Swiss Federal Institute of Technology, Thesis No. 2313
40. A. B. Watson, J. A. Solomon (1997) J. Opt. Soc. Am. A 14(9):2379–2391
41. E. Ong, W. Lin, Z. Lu, S. Yao, M. Etoh (2004) IEEE Trans. Circuits Syst. Video Technol. 14(4):559–566
42. Z. Yu, E. R. Wu, S. Winkler, T. Chen (2002) Proc. IEEE. 90(1):154–169
43. H. R. Wu, M. Yuen (1997) IEEE Sig. Proc. Lett. 4(11):317–320
44. K. T. Tan, M. Ghanbari (2000) IEEE Sig. Proc. Lett. 7(8):213–215
45. E. Ong, W. Lin, Z. Lu, S. Yao, X. Yang, L. Jiang (2003) No Reference JPEG-2000 Image Quality Metric. In: Proc. of IEEE International Conference on Multimedia and Expo (ICME). 545–548
46. J. Canny (1986) IEEE Trans. Pattern Anal. Mach. Intell. 8(11):679–698
47. T. Aach, A. Kaup (1993) Signal Processing 31:165–180
48. E. Ong, W. Lin, B. Tye M. Etoh (2006) Video Object Segmentation for Content-based Applications. In: Y. J. Zhang (ed.) Advances in Image and Video Segmentation. Idea Group, Inc.

49. P. J. Burt, E. H. Adelson (1983) IEEE Trans. Comm. 31(4): 532–540
50. X. Yang, W. Lin, Z. Lu, X. Lin, S. Rahardja, E. Ong, S. Yao (2005) IEEE Trans. Circuits Syst Video Technol. 15(4):496–507
51. M. Yuen, H. R. Wu (1998) Signal Processing 70(3):247–278

Model Construction for Knowledge-Intensive Engineering Tasks

Benno Stein

Faculty of Media, Media Systems, Bauhaus University Weimar, 99421 Weimar, Germany, benno.stein@medien.uni-weimar.de

Summary. The construction of adequate models to solve engineering tasks is a field of paramount interest. The starting point for an engineering task is a single system, S, or a set of systems, $\mathcal{S}$, along with a shortcoming of information, often formulated as a question:

- Which component is broken in S? (diagnosis $\sim$ analysis)
- How does S react on the input u? (simulation $\sim$ analysis)
- Does a system with the desired functionality exist in $\mathcal{S}$? (design $\sim$ synthesis)

If such an analysis or synthesis question shall be answered automatically, both adequate algorithmic models along with the problem solving expertise of a human problem solver must be operationalized on a computer. Often, the construction of an adequate model turns out to be the key challenge when tackling the engineering task. Model construction – also known as model creation, model formation, model finding, or model building – is an artistic discipline that highly depends on the reasoning job in question.

Model construction can be supported by means of a computer, and in this chapter we present a comprehensive view on model construction, characterize both existing and new paradigms, and give examples for the state of the art of the realization technology. Our contributions are as follows:

- In Sect. 2 we classify existing model construction approaches with respect to their position in the model hierarchy. Nearly all of the existing methods support a top-down procedure of the human modeler; they can be characterized as being either structure-defining (top), structure-filling (middle), or structure propagating (down).
- Domain experts and knowledge engineers rarely start from scratch when constructing a new model; instead, they develop an appropriate model by modifying an existing one. Following this observation we analyzed various projects and classified the found model construction principles as model simplification, model compilation, and model reformulation. In Sect. 3 we introduce these principles as *horizontal modeling construction* and provide a generic characterization of each.

B. Stein: *Model Construction for Knowledge-Intensive Engineering Tasks*, Studies in Computational Intelligence (SCI) **116**, 139–167 (2008)
www.springerlink.com

- Section 4 presents real-world case studies to show horizontal model construction principles at work. The underlying technology includes, among others, hybrid knowledge representations, case-based as well as rule-based reasoning, and machine learning.

1 Introduction

To an observer B, an object A^* is a model of an object A to the extent that B can use A^* to answer questions that interest him about A.

[Minsky, 1965, p. 45]

In this chapter, the interesting objects, A, are technical systems. The observer, B, is a domain expert who works on an engineering task, such as a diagnosis or design problem. The questions are embedded in a ternary way: they relate to the technical system, to the engineering task, and to the domain expert.

A system, S, can be considered as a clipping of the real world and has, as its salient property, a boundary. On account of this boundary, it can be stated for each object of the world whether or not it is part of S. The process of developing a model for a system is called model formation, model creation, or modeling.

If we are given a system S along with a question about S, an answer can be found by performing tailored experiments with S. Experiments are purposeful excitations of S while observing its reactions or modifications, which are called behavior. Note that experimenting with a system is not possible if a question is of hypothetical nature and relates to an, up to now, non-existing system. And, even if S exists, various reasons can forbid experimenting with S: the system's reactions may be too slow or too fast to be observed, the experiment is too expensive or too risky to be executed, or the experiment's influence onto the system cannot be accepted. A common a way out is the creation of a model M from the system S and to execute the experiment on M (see Fig. 1). Performing an experiment on a model is called simulation [21].

As shown in Fig. 1, a model does not depend on the system as the only source of information but is formed purposefully, in close relation to the interesting question and/or the experiment to be performed on it. Under the

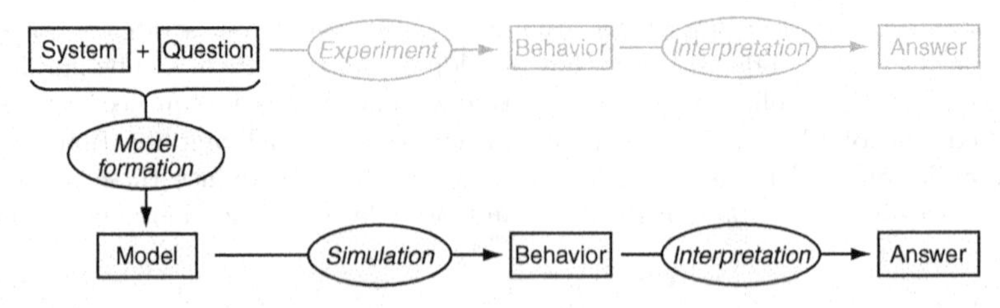

Fig. 1. Simulation as a concept to eliminate information deficits with respect to a system

problem-solving view of Artificial Intelligence this is common practice: the interesting questions (diagnosis or planning problems) prescribe the problem representation (the model) and yet the problem-solving method (the "simulation" algorithm) [31].

2 Top-Down Model Construction

Creating a model of a technical system S means to shape its underlying concept or idea – a process that is first of all mental, and that always involves three major steps:

1. Identification of the system's boundary $\Rightarrow$ black-box model.
   This step relates to the modeling focus and serves two main purposes: complexity reduction and behavior isolation. In the field of engineering, focusing is realized among others by the well-known Method of Sections, typically used to cut out a part of a truss [17].
2. Identification of the subsystems and relations of $S \Rightarrow$ structure model.
   This step relates to the structural granularity and defines all subsystems of S that can be identified in the model as such, and, at which level of detail the interplay between subsystems is represented. For this purpose the system's and subsystems' parameters along with input, output, and state variables are defined.
3. Characterization of constraints between variables $\Rightarrow$ behavior model.
   This step relates to the modeling accuracy or modeling fidelity and determines the complexity of the mathematical relations that define the constraints between a model's state variables and algebraic variables. Both structural granularity and modeling fidelity define the modeling deepness.

The outlined modeling steps happen in our mind, and, a model at this stage is called a *mental model.* To communicate a mental model it must be given a representation. A mental model becomes representational as a physical or material model by craftsmanship. A mental model becomes representational as a symbolic or formal model by developing a prescription for the constraints that are stated in the third model formation step.

We use the term model construction as a collective term for all kinds of processes where a given model is transformed into another model. Model construction takes very different forms. The common situation where one encounters model construction is the transformation of an abstract model, which is close to the human understanding, into a computer model, that is to say, a program. The execution of this program is an experiment at a symbolic model and hence a simulation. We call a model construction process that maps a model M onto a less abstract model *top-down model construction*; it is closely related to system analysis [20].

The transformation of a mental model into a computer model usually happens in several steps wherein intermediate models are constructed: a structure

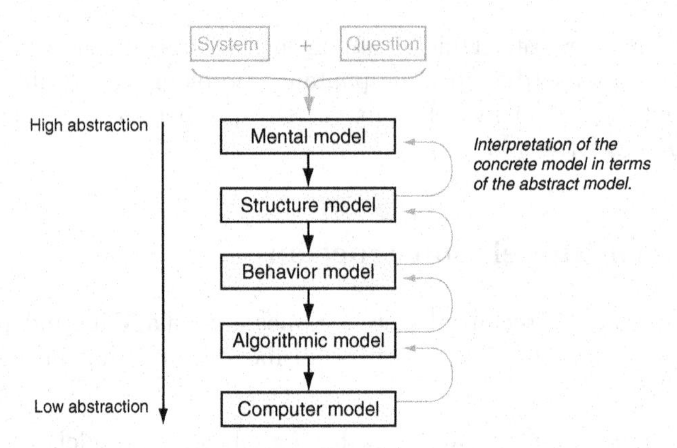

Fig. 2. A hierarchy of models. Top-down model construction means to map abstract models onto less abstract models. Final objective is the operationalization of the mental model in the form of a computer program

model, a behavior model, and an algorithmic model (see Fig. 2, which shows a modified version from [47]). Typically, the human designer materializes his mental model as a textual or graphical structure model, for instance as a drawing or a diagram of the system S. The structure model, which defines subsystems and relations of S, becomes a behavior model of S by specifying a state prescription function for each subsystem. Behavior models are also designated as mathematical models [47]. Usually, the behavior model must be prepared with respect to simulation algorithms. This step includes the introduction of a causal order for the behavior equations, the normalization of behavior equations, or other algebraic operations. Finally, the specification of an algorithmic model within a computer language or a simulation language yields a computer model.

Moving down the model hierarchy means going in structure-to-behavior direction and is a process of increasing concretization. Model construction processes of this type resemble so-called association mappings [51]. As already mentioned, model construction needs not to be a vertical process; in Sect. 3 we consider model construction as a mapping between models at the same level.

2.1 Top-Down Model Construction Support: A Classification Scheme

To support model construction many approaches have been developed. Most of them are top-down; they aim at a reduction of the distance between more abstract models and less abstract models. Ideally, a mental model should be translated automatically into a computer program. We relate top-down approaches for model construction to the three areas shown in Fig. 3, for which

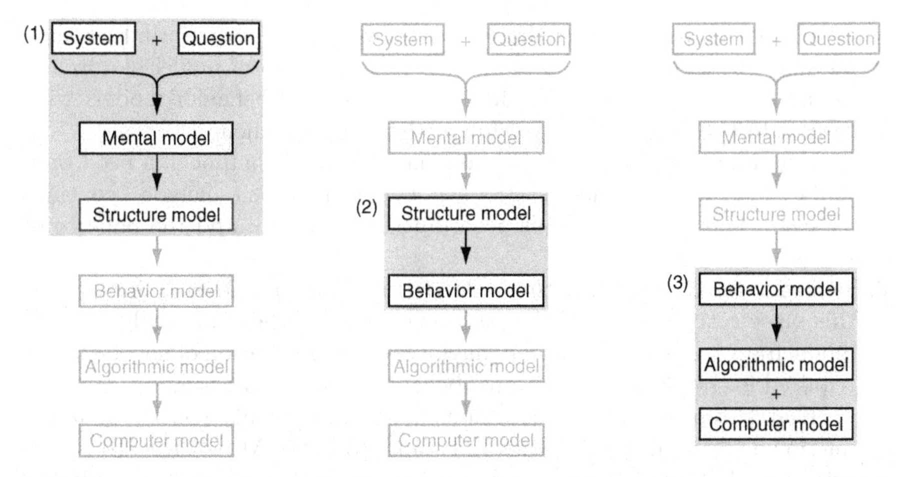

Fig. 3. The *shaded areas* indicate the three major places where top-down model construction is supported (from *left* to *right*): system boundaries and relevant phenomena, model deepness and model coherence, model synthesis and model processing

we state an abstract description, the known representatives, and a running example from the field of fluidic engineering:

1. *System + Question → Mental Model → Structure Model.* Support at this place relates to the model construction steps 1 and 2. Creating a mental model includes the problem of phenomena selection, which in turn is closely connected to the problem of identifying adequate system boundaries. Note that no sharp line can be drawn between mental models and graphical structure models.

   Known representatives: the model fragment idea for the organization of phenomena [29], the reasoning on hypothetical scenarios for the determination of system boundaries by [35], the CFRL paradigm that maps from function to behavior [18], the approaches to ontological reasoning [37], the case-based design approaches that map from demand specifications to technical devices [24].

   Example: For a hydraulic lifting platform the fail-save behavior for overload operation shall be investigated. This requires the analysis of maximum pressures and the functioning of relief valves. Hence, the system structure should represent the basic hydraulic building blocks such as pumps, pipes, cylinders, or valves.
2. *Structure Model → Behavior Model.* Support at this place relates to the model construction steps 2 and 3. Creating a behavior model means to select and compose local behavior models from a given domain theory according to a structural description of a system.

Known representatives: the model composition approach and its variants [5], the graphs of models paradigm [2], the use of model hierarchies in the field of diagnosis [45], the selection of local behavior models with respect to the desired accuracy for predefined phenomena [46].

Example: For the given question the hydraulic components need only to be described by their stationary behavior, which includes continuity conditions, pressure drops according to Bernoulli, and the balance of forces.

3. *Behavior Model → Algorithmic Model + Computer Model.* Support at this place relates to one of the following tasks: synthesis of an algorithmic model from a given set of local behavior models, generation of normal forms required for numerical algorithms, behavior model processing.

   Known representatives: continuity and compatibility constraint introduction as realized in FLUIDSIM and standardized in MODELICA [10], task-oriented selection of efficient numerical procedures, symbolic manipulation of equation systems such as BLT-partitioning or automatic tearing [23], minimum coordinate determination in kinematics computations, coherent synthesis of local behavior models [22], automation of Jordain's principle as realized in AUTOLEV [1].

   Example: The mathematical equations of the hydraulic components are collected, variables are unified according to the circuit structure, and a causal ordering is determined.

Remarks. The operationalization of model construction knowledge at the mental model level does not hit the mark in many modeling problems. Especially when diagnosing or configuring a system, one is rather concerned with aspects of adequacy and efficiency. Phenomena selection is a worthwhile objective though, but it is usually too challenging to be tackled by a computer at present.

Recall the three main places where top-down model construction can be supported, shown in Fig. 3. According to this view we distinguish top-down model construction methods as follows:

1. *Structure-defining* methods derive necessary phenomena from the interesting question, structural information from necessary phenomena, or structural information from the desired function.
2. *Structure-filling* methods derive behavioral information from necessary phenomena, behavioral information from functional requirements, or behavioral information from structural information.
3. *Structure-propagating* methods derive algorithmic information from behavioral information.

This classification tells us a lot about how these methods work. Structure-propagating methods is a collective term for mathematical procedures that form a global algorithmic model by "propagating" the implications of the local behavior models along a given structure. Since structure-propagating methods operate on the mathematical model level, the largest part of them is both domain-independent and task-independent. Structure-defining methods, as well as structure-filling methods, are based on libraries of device models, component models, or model fragments. The models in the libraries describe systems, building blocks, or phenomena from a particular domain.

Example. A model of an analog low-pass filter is a device model from the electrical engineering domain. A resistor of this filter is a component, and one of its model fragments is given by Ohm's law, $v(t) = R \cdot i(t)$; another model fragment may take the resistivity depending on the temperature, T, into account. In this case the model fragment consists of two additional equations, $R = \rho \cdot l/A$ and $\rho = \rho_0(1 + \alpha(T - T_0))$, where l, A, ρ_0, and α designate the resistor length, the cross-sectional area, the resistivity at T_0, and the coefficient of the resistivity respectively.

The models in the libraries are tagged by qualifiers that encode knowledge about their applicability. Generally speaking, the job of structure-defining and structure-filling methods is the selection of models from libraries by processing the qualifier constraints of the chosen models. The qualifiers take different forms:

- In case-based design the qualifiers encode an abstract functional description of devices. The job of a case-based reasoning system is to find for a given functional demand the best fitting case from the device model library [32]. Qualifier processing here relates to the computation of similarity measures and to case retrieval [34].
- Model fragment qualifiers as used by [29] encode relations among phenomena, typically formulated within propositional logics. Given two model fragments, m_1, m_2, the contradictory relation states whether m_1 and m_2 go along with each other; the approximation relation, on the other hand, states whether m_1 is a more approximate description of some phenomena compared to m_2. Additionally, coherence constraints and precondition constraints are used to define domain-dependent modeling restrictions. Qualifier processing here aims at the synthesis of device models that are causal, coherent, and minimum. This synthesis problem is NP-complete.
- Model fragment qualifiers are also employed to encode mathematical properties or hints respecting behavior processing. This includes knowledge about signal ranges, model linearity, model stiffness, system orders, numerical conditioning, steady states, oscillation behavior, damping, processing difficulty, and processing performance. Qualifier processing here means assumption management.

Remarks. (1) The largest group of commercial tools that support model construction concentrate on structure filling; moreover, they are restricted to situations where the model deepness has been predefined by the human designer. (2) An important role for the effectiveness of such tools comes up to the user interface. It can provide powerful graphical support and be close to the mental model of the human designer, for example as realized within FLUIDSIM: based on CAD drawings of even large and complex electro-fluidic systems, FLUIDSIM generates the related algorithmic models without human support [44]. AUTOLEV [38] and the IMECH toolbox [3] are tools for textually describing mechanical multibody systems. Models in AUTOLEV are formulated within a proprietary mathematical language, models in the IMECH toolbox are formulated by instantiating C++ objects. A comparison of these tools can be found in [16]. (3) Because of its popularity the SIMULINK toolbox is worth to be noted here. Although SIMULINK comes along with a fully fledged graphical interface, it does not provide model construction support at one of the mentioned places. Working with SIMULINK means to specify algorithmic models manually, by drawing block diagrams.

3 Horizontal Model Construction

We now describe in which ways from a given model M, called the *source model*, a new model M' can be constructed, which is particularly suited to solve a given analysis or synthesis problem (see Fig. 4). For instance, structure information extracted from a behavior model can form the base when tackling a related configuration problem, or, by evaluating a complex behavior model at selected points in time a heuristic model for diagnosis or control purposes can be derived. Our text deals with the following horizontal model construction principles:

- *Model Simplification.* Coarsening a model by stripping off unused parts
- *Model Compilation.* Making a model more efficient by introducing shortcuts
- *Model Reformulation.* Switching to another modeling paradigm

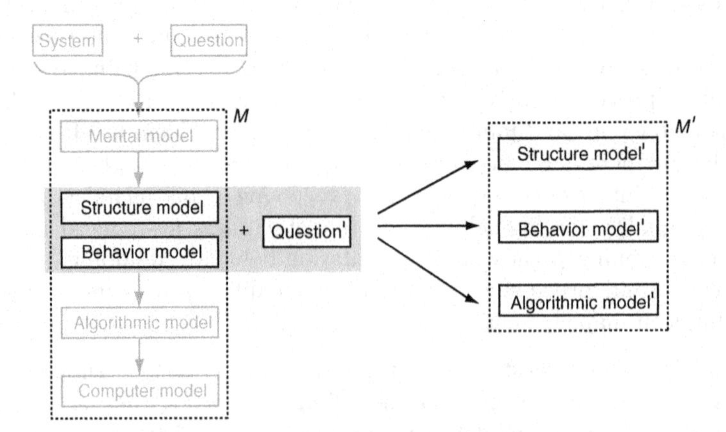

Fig. 4. Horizontal model construction uses a given source model to construct an adequate model regarding some question. Question′ designates an analysis or synthesis task; the *shaded area* indicates the underlying knowledge source

These principles are applied to overcome computational intractability, to improve runtime performance, to simplify model maintenance, and the like. At the end of the next paragraphs, the passage "Formal description" provides details about the pros and cons of each principle, while Table 2 provides for a short characterization. Note that these principles are not complete, but they are – often subconsciously – applied in engineering practice. Moreover, both the compilation principle as well as the reformulation principle have not been described in an abstract or formal way so far. Other horizontal model construction methods include model refinement or model envisioning.

While model construction support as depicted in Fig. 3 means moving down the model abstraction hierarchy, model construction that is based on source models operationalizes a horizontal mapping. A source model, M, which guides the model construction, establishes the primary knowledge source when building the new model M'. However, additional knowledge sources, which give answers to the following questions, are inevitable:

- *Model Simplification.* Which parts shall be stripped off?
- *Model Compilation.* Where is a hidden shortcut?
- *Model Reformulation.* How do the migration rules look like?

The nature of the additional knowledge sources unveils that instances of these model construction approaches are much more specialized with respect to the domain and the interesting question then the top-down approaches of the previous section. This in turn means that we cannot expect recipes for horizontal model construction: model compilation, for instance, designates a construction *principle* rather than a construction *method.* The case studies presented in Sect. 4 contain new ideas for model construction, which pertain either to conceptual aspects or to the realized algorithms.

Methods for model construction based on source models transform a structure model, M_S, into another structure model, M'_S, and a behavior model, M_B, into another behavior model, M'_B:

$$M_S \xrightarrow{\gamma_S} M'_S, \quad M_B \xrightarrow{\gamma_B} M'_B$$

As opposed to structure-defining and structure-filling methods, the functions γ_S and γ_B cannot be classified in a uniform way. This is in the nature of things: the effects of a structure model mapping, γ_S, and a behavior model mapping, γ_B, may range from a superficial variation up to an entire reformulation of the source model. Both functions address deficits of the source model or its utilization – deficits that are quantifiable by means of one or more of the following properties: time, place, handling, maintenance, intricateness, comprehensibility, algorithms, representation. To render horizontal model construction approaches comparable to each other they are characterized according to the properties listed in Table 1.

Remarks. The functions γ_S and γ_B can be compared to the "system morphism" idea [49, 51]. System (homo-, iso-) morphisms are a concept to

Table 1. Properties to characterize the model construction approaches

Property	Semantics
Characteristics	Modification of the structure model and the behavior model
Modeling effects	Effects of a modification from the modeling perspective
Processing effects	Effects of a modification with respect to model processing: 1. Processing efficiency, typically the runtime complexity 2. Processing difficulty, which describes how intricate the employed algorithms are 3. Model handling, which relates to the maintenance effort of the modified model
Area of application	Problem classes, where the model construction approach is used
Techniques	Techniques, algorithms, and strategies to implement the model construction approach; say, the functions γ_S and γ_B

transform one system description into another. The main contribution in [49, 51] is the development of a generic framework for system description, system design, and system analysis. Nevertheless, their work is less concerned with the identification and characterization of special instances of morphisms.

3.1 Model Simplification

Model simplification aims at a reduction of a model's analytic complexity, a model's constraint complexity, or both. A reduction of the *search space complexity* may be an additional consequence but is not top priority. While analytic complexity and constraint complexity are crucial for analytical problem solving tasks, the search space complexity is of paramount importance within synthesis tasks: it is a measure for the number of models that have to be synthesized and analyzed in order to solve a design problem. The concept of model simplification has been discussed by several researches before [12]. Formal description:

1. *Characteristics.* Simplification of behavior models usually happens within two respects. First, the set of functionalities is restricted to a subset, which leads to a simpler structure model (cf. Fig. 5). Second, the complexity of the functions in the state prescription function are reduced. The former establishes an aggregation of function and directly leads to a structure simplification [14]; the latter falls into the class of behavior aggregation.
2. *Modeling effects.* The lessened interaction between submodels results in a disregard of physical effects. The simplification results in a coarsening of physical phenomena or in physically wrong connections. The behavior is rendered inaccurately up to certain degree, but easier to understand [12].

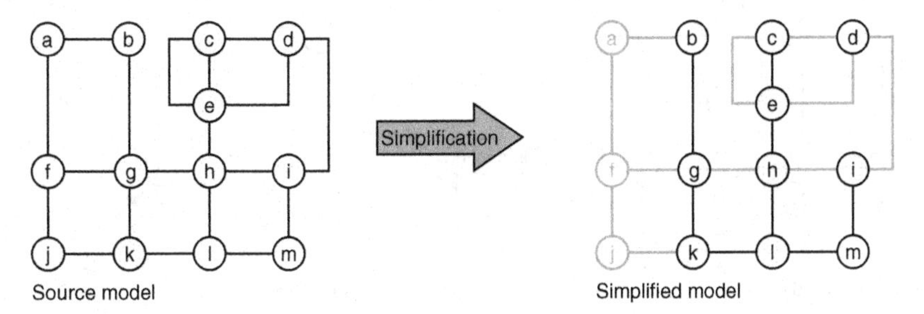

Fig. 5. Illustration of the source structure model (*left*) and a simplified structure model (*right*)

3. *Processing effects.* Both structure model and behavior model can be processed more efficiently; the processing difficulty is reduced; model handling is simplified.
4. *Area of application.* Analysis of large or numerically demanding behavior models; synthesis of behavior models without knowledge about the model structure.
5. *Techniques.* γ_S: elimination of feedback loops; equalization of the node degree in the structure model; elimination of edges to create a causal ordering, i.e., a unique computation sequence when determining unknown functionalities [28].

   γ_B: fuzzification of equations the state prescription; piecewise linear reasoning [36]; linearization of higher order polynomials; balancing of dominant terms in complex equations [50]; state combination by aggregating similar states; numerical coarsening by means of scaling down numerical precision of the functionalities; order of magnitude reasoning [33]; reduction to a structure model by omitting behavior descriptions, which is called "representational abstraction" [12].

Model simplification represents a more or less serious intervention in the physical underpinning of the source model. Hence, model simplification is always bound up with model evaluation: it must be ensured that the simplified model is able to answer the interesting question in connection with the intended experiment.

3.2 Model Compilation

Model compilation is the anticipation of model processing effort: processing effort is shifted from the model utilization phase to the model construction phase. Model compilation is a powerful principle to address a model's analytic complexity, its constraint complexity, or the search space complexity. A compiled model is created by introducing either (1) computational short cuts within long-winded calculations that are caused by a complex behavior model, or (2) exploratory short cuts within a large search space that results

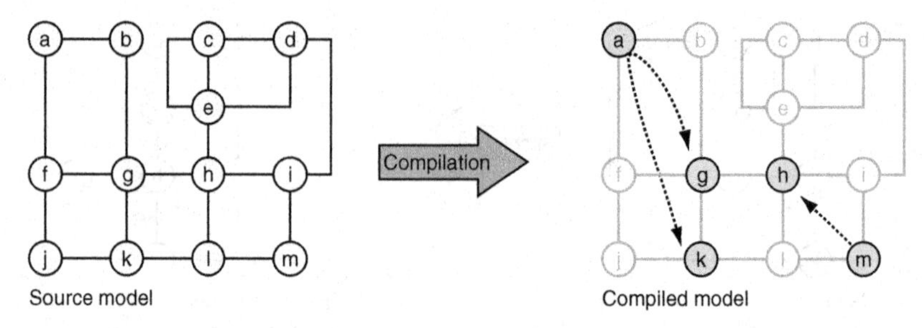

Fig. 6. Computational short cuts: hints in the compiled behavior model short-circuit the computation of constraints between functionalities of different submodels. For example, the input parameters could directly be mapped onto the output parameters

from problem-inherent combinatorics. The idea is to break global connections within the source model down to local connections, which are encoded within the compiled model in the form of special hints. These hints can take one or more of the following forms:

- Hints that use memorization to short-circuit involved state prescription constraints between functionalities of different submodels (see Fig. 6). Cause effect chains are shortened, possibly to simple associations.
- Hints that suppose an order within a sequence of tentative search space decisions. These hints are introduced as tags at the respective choice points in the search space and control the search.
- Hints that restrict the search space by introducing additional state prescription constraints.

Model compilation methods can also be characterized by their *scalability.* Using a scalable method, there is a trade-off between the effort for model *pre*processing and model processing at runtime. The scalable character of a compilation method is bound up with the depth of the analyzed search space, or the precision at which simulations are performed. However, a compilation method that analyzes a model with respect to special structures is usually not scalable. Formal description:

1. *Characteristics.* The set of functionalities remains unchanged. The behavior model is enriched with additional constraints that encode numerical or search-specific hints.
2. *Modeling effects.* The modeling accuracy and the level of detail is not reduced, although the ways of computing model behavior may be completely altered.
3. *Processing efficiency.* The model can be processed much faster. However, the implementation of the necessary inference (simulation) algorithms may be more challenging than before compilation. Moreover, model handling gets more complicated since modifications of the model require a renewed preprocessing.

4. *Area of application.* Whenever processing efficiency is highest bid and processing hints can be computed at all, model compilation is expedient. Given this situation, further prerequisites must be fulfilled: (a) The problem solving task can be split into two phases: a model construction or preprocessing phase, where computing power and/or computing time are given on a large scale, and a model utilization or problem solving phase. (b) The number of problem instances that require a recompilation is small.
5. *Techniques.* γ_S: topological analyses from which computational constraints or search constraints are derived; determination of candidates for a dependency-directed or knowledge-based backtracking; identification of spheres of influence of both revision and trashing [25]; decomposition of equation models into (a) involved subproblems, which must be processed by a global method, and (b) feedback-free subproblems, which can be tackled by a local inference approach; model decomposition by causal analysis [8].

   γ_B (computational short cuts): pre-computation of typical simulation situations and encoding of input/output associations in the form of look-up tables; compilation of rules into a rete-network [13]; utilization of an assumption-based truth maintenance system (ATMS) in order to organize results of computational expensive inference problems [19]; sharing of computation results by identifying and replacing instances of similar submodels, a concept that can be realized algebraically; case-based learning of characteristic features to select suited inference methods for simulation [43].

   γ_B (exploratory short cuts): behavior aggregation by a precedent detection of repeating cycles [48]; coalescing a system by methods from the field of inductive inference [4]; extensive or even exhaustive search in order to analyze the search space or to develop a decision strategy at choice points; ordering of value assignments in constraint-satisfaction problems with finite domains [9].

Remarks. We call a compilation method that is not specially keyed to models of technical systems a *knowledge* compilation method. Knowledge compilation methods presuppose a determined knowledge representation form, but no problem solving task, no domain, and no model. The rete-algorithm mentioned above is such a knowledge compilation method; its prescribed knowledge representation form is the rule form. While the rete-algorithm has been developed for rule languages whose interpretation is defined on the recognize-and-act cycle, the compilation method in [52] exploits the fixed-point convergence of the rule language to be compiled.

The following examples give an idea of the spectrum of knowledge forms where compilation methods can be applied: (1) The syntactic analysis and substitution of algebraic terms. (2) A still more basic knowledge compilation method is based on Horn approximations [40]; its prescribed knowledge representation form are formulas in propositional form. (3) If graphs are the

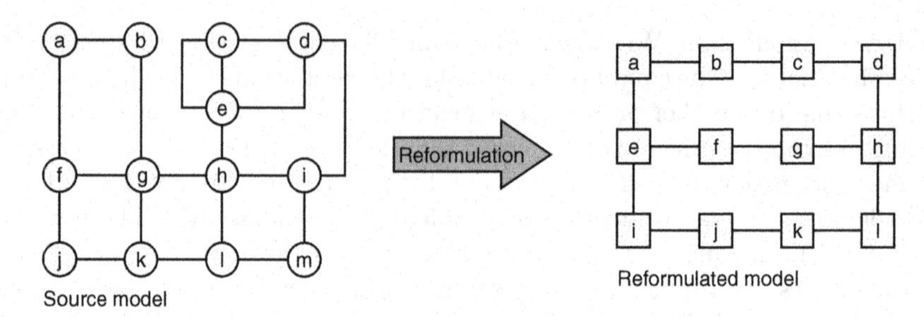

Fig. 7. Model reformulation is usually bound up with a paradigm shift in model processing, entailing both a new structure model and behavior model

interesting knowledge form, knowledge inference means graph matching or isomorphism analysis. For the latter problem a compilation method with scalable preprocessing effort is presented in [26].

3.3 Model Reformulation

In a literal sense, every construction of a new model from a source model could be entitled a reformulation. This not the intention here; rather, the term model reformulation is used as a collective term for model constructions that are indifferent with respect to both modeling effects and processing efficiency (Fig. 7).

Model reformulation aims at issues from one or more of the following fields: knowledge representation, knowledge acquisition, model maintenance, available inference methods, user acceptance. After a model reformulation, the resulting model is in a form ready to become used for the problem solving task in question. Model reformulation does not target on complexity issues. Formal Description:

1. *Characteristics.* The set of functionalities may or may not be altered. Typically, the state prescription function is reformulated, leading to a paradigm shift in model processing.
2. *Modeling effects.* Ideally, there are no effects on the model's accuracy or its level of granularity.
3. *Processing effects.* Ideally, the processing efficiency is not affected. Nothing can be said regarding processing difficulty. The model handling may be simplified.
4. *Area of application.* There is no specific area of application. Model reformulation comes into play if a model that has been developed with respect to a special processing approach shall be transformed for another processing approach.
5. *Techniques.* There is no specific reformulation technique.

As opposed to model simplification or model compilation, there is no pool of techniques by which a model reformulation is to be realized. This is in the nature of things; model reformulation takes a model that has been used successfully with processing paradigm A and tries to transform this model such that it can be used within processing paradigm B.

At first sight model reformulation appears to be a close relative of model compilation. This, however, is not the case. The maxim of model compilation is processing efficiency, and the problem solving task could be done without a compilation – at a lower processing efficiency, of course. We speak about model reformulation, if a model *must* be transformed into another representation in order to be processed within the problem solving task.

3.4 Discussion and Related Work

Table 2 contrasts the presented model construction approaches. The used symbols are interpreted as follows: ↑↑ (↑) means strong (low) positive impact, while ↓↓ (↓) means strong (low) negative impact, the dash stands for no impact. The table shows the dependencies in an oversimplified way and should only be used as a road map.

The three horizontal model construction principles are process-centered: model simplification as well as model compilation relate to the difference between the provided and the required computing power when going to solve a problem with the source model. A model is simplified if the interesting problem solving task cannot be handled with this model – a situation that occurs if, for instance, a design task is addressed with a model conceived for an analysis task. In fact, model compilation can provide a way out in such a situation as well. The essentials for a compilation strategy are twofold: the task in question can be tackled with acceptable computing power, and, the employment of the necessary computing power can be allotted a model construction and a model utilization phase.

Figure 8 shows the model transformation theory from [51]. The vertical arrows in the diagram connect a behavior model at different levels of explicitness. At the lowermost level, behavior is specified by input/output relations; when going up in the hierarchy the models get supplemented: by a global state prescription function, by initial states, by local behavior relations, and,

Table 2. Short characterization of three horizontal model construction principles

Approach	Modeling quality	Processing efficiency	Processing difficulty	Handling difficulty
Simplification	↓↓	↑	↓	↓
Compilation	–	↑↑	↑	↑
Reformulation	–	–	↓	↓↓

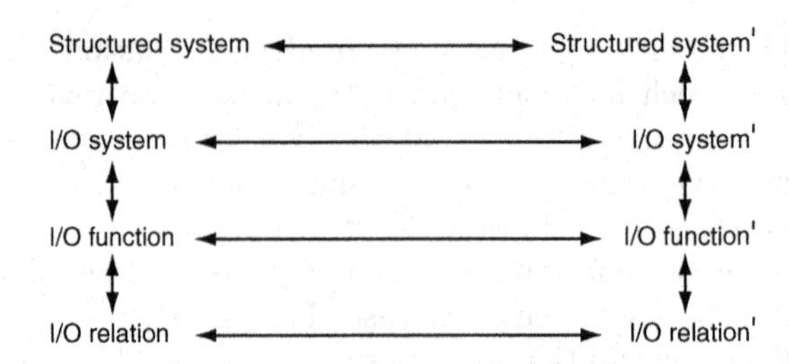

Fig. 8. Hierarchy of system specifications according to [51]. The *horizontal lines* represent mappings for model construction, called morphisms here

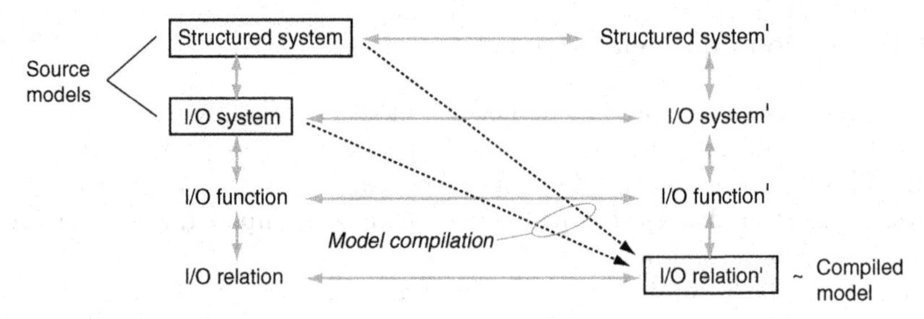

Fig. 9. Integrating the idea of model compilation into Zeigler et al.'s hierarchy of system specifications: a high-level system description, e.g., an equation model, is broken down to plain I/O relations by means of model compilation

finally, on the topmost level, by a component influence structure. The horizontal arrows represent mappings between two models; they are called morphisms here and correspond to our construction functions γ_S and γ_B. We can use this diagram to point up the effects of a model compilation that introduces computational short cuts: it is a mapping from a model on the left-hand side onto the I/O-relation model at the lowermost level on the right-hand side (see Fig. 9).

4 Case Studies

This section presents three case studies where the horizontal model construction paradigm is applied. Note that the section cannot serve with detailed introductions and discussions of algorithmic, technical, or problem-specific background; for this purpose we refer to the respective literature. The case studies have illustrative character and shall sensitize the interested reader to identify the use of horizontal model construction principles in other situations. Nevertheless, they describe non-trivial, knowledge-intensive real-world applications that were tackled in the outlined way.

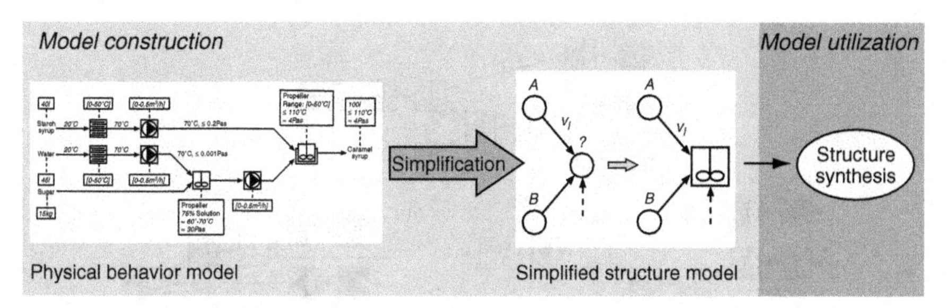

Fig. 10. The behavior model is abstracted to a structure model where physical quantities are represented as linguistic variables. The synthesis space is restricted to a set of labeled graphs

4.1 Case Study 1: Plant Design in Chemical Engineering[1]

A chemical plant can be viewed as a graph whose nodes represent devices or unit-operations, while the edges correspond to pipes responsible for the material flow. Typical unit-operations are mixing (homogenization, emulsification, suspension, aeration), heat transfer, and flow transport. The task of designing a chemical plant is defined by a demand set D, in the form of properties of various input substances along with the desired output substance. The goal is to mix or to transform the input substances in such a way that the resulting output substance meets the requirements.

The design happens by passing through (and possibly repeating) the following five steps: preliminary examination, choice of unit operations, structure definition, component configuration, and optimization. An automation of the steps at a behavioral level would be very expensive – if possible at all. Present systems limit design support to isolated subtasks; they relieve the human designer from special simulation or configuration tasks, and the effort involved there is high enough.

Instead of deriving a concrete solution for supplied demands at the modeling level, the physical model is simplified to an abstract model. On this abstract level, a solution can be efficiently calculated and transferred back to the physical level, although some adjustments may be necessary at this point. See Fig. 10 for an illustration.

Model Simplification

The following model simplification steps include structure model simplifications (S1–S4) and behavior model simplifications (B1–B6), resulting in a tractable design problem. They are oriented at the taxonomy of model abstraction techniques listed in [14].

(S1) *Single task assumption.* It is general practice to combine different chemical processes in order to share by-products or partial process chains.

[1] Supported by DFG grants PA 276/23-1 and KL 529/10-1.

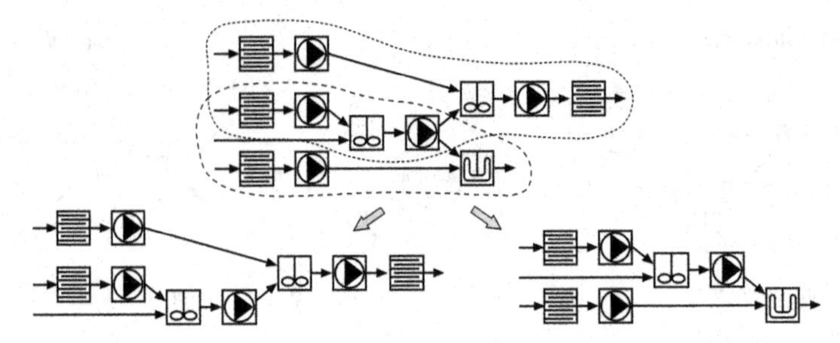

Fig. 11. Single task: overlapping plant structures are split and treated separately

A combined design corresponds to the solution of different tasks simultaneously – a procedure which belongs to optimization. Here, design is restricted to the n:1-case, and overlapping plant structures are split and dealt with separately as shown in Fig. 11.

(S2) *Model context.* The way how models are embedded into a context is clearly defined. Pumps, for example, have a strict structural relationship; they have single input and output, and the predecessor of a pump must be another device.

(S3) *Limited input space.* During the design process, decisions are taken based on the abstract values of a small set of substance properties, such as temperature, viscosity, density, mass and state. Properties such as heat capacity, heat conductivity or critical temperature and pressure are neglected at this point.

(S4) *Approximation.* Instead of using different functions and formulas that apply under different conditions, only one function or formula covering the widest range of restrictions is used. For example, there are more than 50 different formulas to calculate the viscosity of a mixture most of which are very specialized versions and only applicable under very rare circumstances. The formula $ln(\eta) = \sum_i \varphi_i \cdot ln(\eta_i)$, however, is very often applicable and yields a good approximation, even in the complicated cases.[2]

(B1) *Causal decomposition.* To prevent components from exerting influence on themselves, feedback loops are not allowed.

(B2) *Numeric representation.* Although the use of crisp values leads to exact results, fuzzy sets are used to represent essential value ranges. This simplification diminishes the combinatorial impact on our graph grammar approach.

(B3) *State aggregation.* In general, the material being processed within a device is in different states. This behavior is simplified by assuming that, inside a device, a material will be in one single state.

[2] The symbols φ_i and η_i designate the volume portion and the viscosity of input i.

(B4) *Temporal aggregation.* Time is neglected, rendering statements about continuous changes to material properties impossible; changes to material properties are connected to entry and exit points within the plant structure.

(B5) *Entity aggregation by structure.* Devices usually consist of different parts that can be configured separately. For instance, a plate heat transfer device is composed of a vessel and a variable number of plates. The arrangement of the plates within the vessel is a configuration task.

(B6) *Function aggregation.* In contrast to entity aggregation by function, where devices are represented by a special device, we aggregate functions. For instance, mixers are capable of performing different functions, such as homogenization, emulsification, aeration, or suspension.

The DIMOD *Workbench*

Rationale of the aforementioned model simplifications is to automate the generation of adequate structure models for a given set of demands. We developed tools to automate this synthesis step with graph grammars: they allow for knowledge modeling, manipulation, and systematic exploration of the search space, which are essential requirements for a successful synthesis of structure models [39]. Graph grammars generate graphs by applying transformation rules on, initially, some start graph. Here, the start graph represents the unknown chemical plant where the abstracted demands in the form of input and output substance properties are connected to. The successive application of transformation rules corresponds to the application of domain knowledge with the goal to transform the start graph into a design fulfilling all demands.

Figure 12 illustrates the use of graph grammars in chemical engineering design; it shows graphical representations of a small fraction of the transformation rules in the design knowledge base of DIMOD. The rules, which have been developed in close cooperation with domain experts, are used for synthesis, refinement, and optimization purposes when designing plants for food processing.

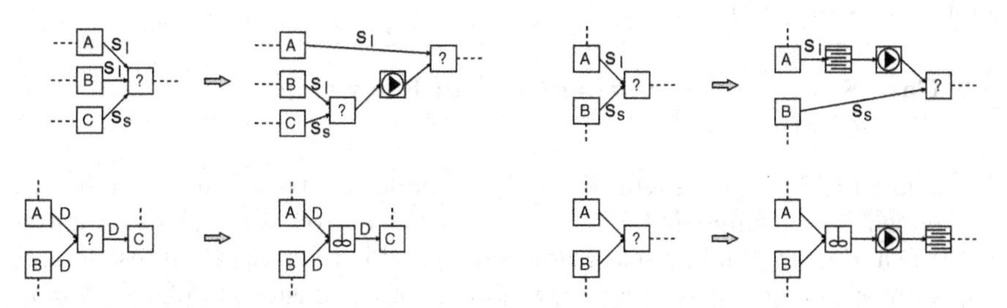

Fig. 12. *Top*: synthesis rules for the splitting of mixing jobs (*left*) and the insertion of a heating chain to improve dissolution (*right*). *Bottom*: refinement of a generic mixer as propeller mixer (*left*) and refinement of a generic mixer as propeller mixer with trailing heating chain (*right*)

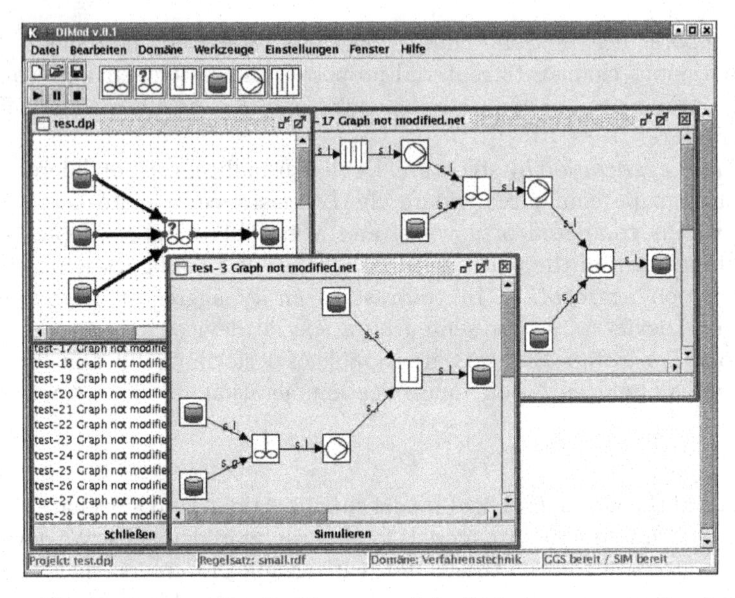

Fig. 13. The DIMOD workbench. The *upper left* window represents the abstracted demands, the windows to the *right* and *center* show two generated structure models

Graph grammars can be seen as a collection of rules that define some search space. Note, however, that they provide no means to *navigate* within this search space; each domain requires a dedicated search method exploiting domain knowledge. Figure 13 shows a snapshot of the DIMOD workbench that has been developed to support human designers when solving synthesis tasks in chemical engineering. The core of the system consists of a generic graph grammar engine, used for modeling and application of knowledge, and a domain-specific module used to guide the search process. A generated structure model is completed towards a tentative behavior model by attaching behavior model descriptions to the components of the structure model. Tentative behavior models are validated by simulation, for which the ASCEND IV simulator is employed [30].

4.2 Case Study 2: Generating Control Knowledge for Configuration Tasks

Configuration is the process of composing a model of a technical system from a predefined set of components. The result of such a process is called configuration too and has to fulfill a set of constraints. Aside from technical restrictions a customer's demands form a larger part of these constraints [6, 15]. A configuration process can rely on a structure model or on a behavior model. If emphasis is on the former the configuration approach is called skeleton configuration; if emphasis is on the latter the configuration task can be solved with a so-called resource-based configuration approach.

Within resource-based configuration the involved components are characterized by simplified functional dependencies, the resources. E.g., when configuring a small technical device such as a computer, one functionality of a power supply unit may be its power output, f_1, and one functionality of a plug-in card may be its power consumption, f_2. Both functionalities are elements of the resource "power": a power supply unit *supplies* some power value $f_1 = a_1$, while each plug-in card *demands* some power value $f_2 = a_2$. In its simplest form, demanded values are represented by negative numbers, which are balanced with the supplied positive numbers. Within a technically sound model the balance of each resource must be zero or positive.

Resource-based modeling provides powerful and user-friendly mechanisms to formulate configuration tasks. However, the solution of resource-based configuration problems is NP-complete, which means that no efficient algorithms exist to solve a generic instance of the problem [41]. When given a real-world configuration task formulated within a resource-based description, the search for an optimum configuration is often realized efficiently with heuristics that are provided by human experts. Put another way, a resource-based system description can be enriched with control knowledge, and model compilation is the idea to generate control knowledge automatically. Consequently, the configuration process is divided into a preprocessing phase, where heuristics are generated, and in a configuration phase, typically at the customer's site, where a concrete configuration problem is solved. Figure 14 illustrates this view.

The Bosch Telecommunication Application

The configuration of telecommunication systems requires technical know-how since the right boxes, plug-in cards, cable adapters, etc. have to be selected and put together according to spatial and technical constraints. Customer demands, which form the starting point of each configuration process, include various telephone extensions, digital and analog services, or software specifications. Typically, there exist a lot of alternative systems that fulfill a customer's demands from which – with respect to some objective – the optimum has to be selected. For this kind of domain and configuration problem the resource-based

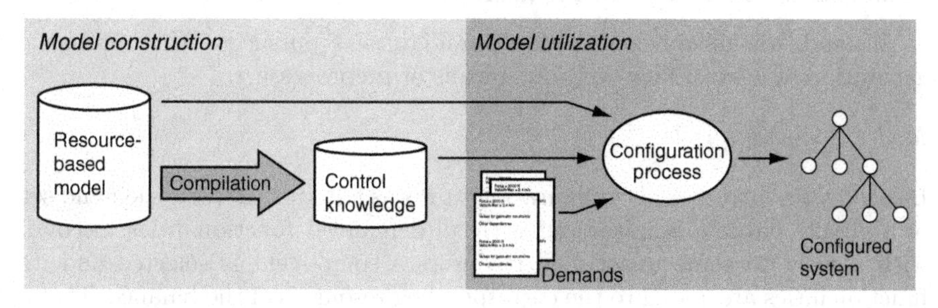

Fig. 14. Partitioning the resource-based configuration process

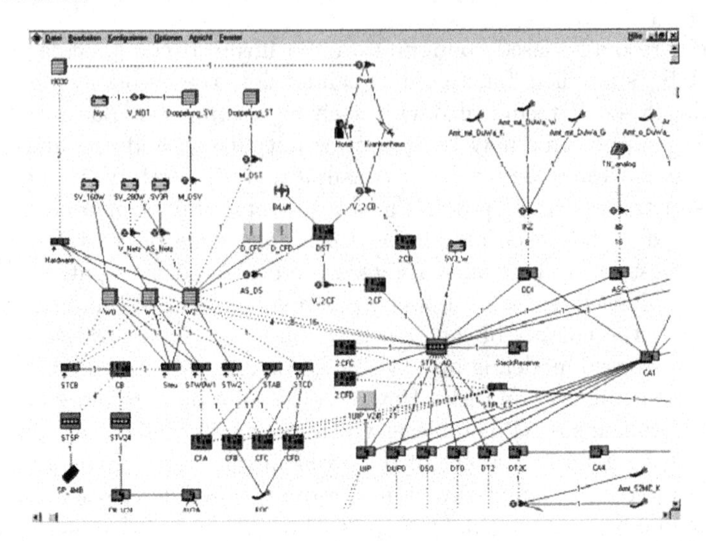

Fig. 15. Resource model representation of a telecommunication system in PREAKON

component model establishes the right level of abstraction: constraints can be considered as a finite set of functionality-value-pairs, which are supplied or demanded from the components.

To cope with their huge and increasing number of telecommunication system variants and to reduce the settling-in period for their sales personnel, Bosch Telenorma, Frankfurt, started the development of the resource-based configuration system PREAKON. Figure 15 shows a part of the knowledge base in PREAKON's acquisition mode. The field tests showed the necessity of a heuristic search space exploration if optimum configurations should be found in an acceptable time. For the following reasons we refrained from a *manual* integration of control knowledge:

1. Control knowledge of domain experts is often contradictory or incomplete.
2. The additional effort bound up with maintenance of heuristic knowledge complicates the configuration system's introduction.
3. Each modification of the knowledge base (e.g., new components) can potentially invalidate existing heuristics.

Instead, the model compilation paradigm was pursued – the automatic generation of control knowledge by means of preprocessing.

Model Compilation

Basically, resource-based configuration works as follows. First, the demand set of a virtual balance is initialized with all demanded functionalities. Second, with respect to some unsatisfied resources a component is selected and its functionalities are added to the corresponding resources of the balance. Third, it is checked whether all resources are satisfied. If so, the selected components form a solution of the configuration problem; otherwise, the configuration process is continued with the second step. Without powerful heuristics that control the functionality and component selection steps, only toy problems can be tackled by resource-based configuration. Resource selection is related to the search space's total depth in first place; component selection affects the effort necessary for backtracking.

Heuristics for resource selection are derived from graph-theoretical considerations which base on the analysis of the strong components in the dependency graph. Heuristics for component selection are derived from a best first search analysis,where for particular demand values the recursive follow-up cost of component alternatives are estimated. These sampling points are used to interpolate for each component and each demanded resource a function. The result of the analysis is a family of functions, which can be evaluated at configuration runtime.

The control knowledge must be recomputed each time the knowledge base is modified; this preprocessing phase takes several minutes on a standard PC. Altogether, the model compilation led to a significant speed-up for realistic instances of the configuration problem: PREAKON was the first configuration system at Bosch Telenorma that provided realistic means for being used at the customer site.

4.3 Case Study 3: Synthesis of Wave Digital Structures[3]

Wave digital structures have their origins in the field of filter design, where they are designated more specifically as wave digital filters [11]. They can be considered as a particular class of signal flow graphs whose signals are linear combinations of the electric current and flow, so-called a/b-waves. The translation of an electrical circuit from the electrical v/i-domain into the a/b-wave-domain establishes a paradigm shift with respect to model processing; it is bound up with considerable numerical advantages. However, since neither the modeling accuracy nor its granularity is affected, the translation into a wave digital structure establishes a model reformulation.

When migrating from a voltage/current description of an electrical circuit S towards a wave digital structure, the model is completely changed: the structure model of S is interpreted as a series-parallel graph with closely connected components and transformed into an adaptor structure (cf. Fig. 16). This reformulation aims at the analysis, say, simulation of S, as illustrated at Gero's design cycle in Fig. 17.

The construction of a wave digital structure is a design task, namely the design of a sophisticated algorithmic model. Since this design task is not trivial and needs experience, its automation is a worthwhile undertaking. In [42], the necessary concepts and algorithms for a WDS design automation are explained in detail. At the same place an algorithm is presented which computes for a

[3] Supported by DFG grant KL 529/10-1.

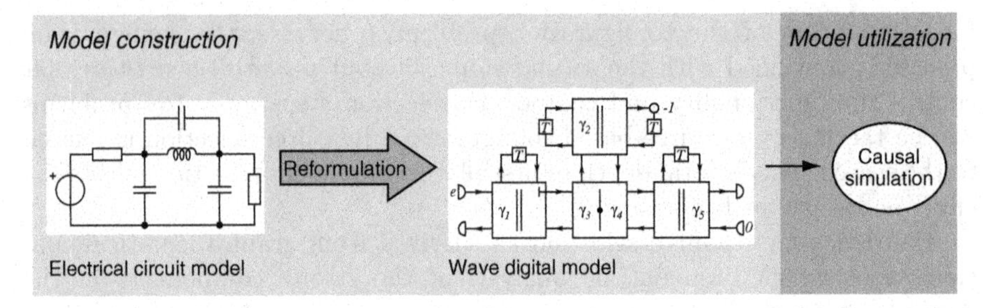

Fig. 16. Reformulation of an electrical circuit model as wave digital structure for model processing reasons

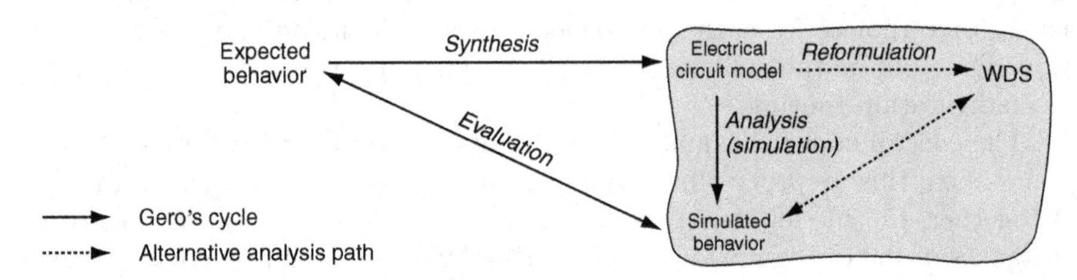

Fig. 17. *Left*: Gero's widely-accepted model of the human design process [15]. The automatic synthesis of WDS provides a numerically attractive alternative to operationalize the analysis step (*right, shown gray*)

given electrical circuit the optimum wave digital structure in linear time – a result which cannot be further improved. In the following we outline the reformulation procedure.

Model Reformulation

As already noted, a wave digital structure is a special kind of signal flow graph. Its topology is realized with series and parallel connectors; the signals that are processed when traveling along the signal flow graph are wave quantities. The reformulation of an electrical circuit as a wave digital structure involves three principal steps:

1. Topology reformulation of the Kirchhoff interconnecting network
2. Description of component behavior in the a/b-wave domain
3. Discretization by numerically approximating the differential equations

These reformulation steps divide into local operations (Step 2 and 3), which pertain to components of the electrical circuit in an isolated manner, and into the global topology reformulation in Step 1. Note that Step 2 and Step 3 are orthogonal to each other, say, their order of application can be interchanged.

Topology Reformulation. Let S be an electrical circuit. The reformulation of the Kirchhoff interconnecting network of S starts with the identification of subsystems in S that are connected in series or in parallel. Both series connections and parallel connections are specializations of a concept called "port", as much as each component with two terminals establishes a (one-) port as well. A port fulfills the port condition, which claims that the currents at terminal 1 and terminal 2, i_1 and i_2, fulfill the constraint $i_1 = -i_2$ at any point in time.

Objective of the topology reformulation is the replacement of series and parallel subgraphs by respective connectors, which guarantee that Kirchhoff's laws are fulfilled and which permit a special network analysis approach. Common network analysis approaches are based on mesh equations, node equations, or state equations [7]. Following a common approach means to set up and transform matrices, such as the mesh-incidence matrix, the branch-impedance matrix, the node-incidence matrix, the branch-admittance matrix, or the state space matrix. Computations on matrices are global computations in the sense that a system of equations must be treated at the same time to find the equations' solutions. By contrast, a computation is local if a single equation at a time is sufficient to compute a solution of that equation, and if this solution is coded explicitly in the equation.

For the part of S that is realized with series and parallel connections model processing effort can be significantly decreased: computational effort can be made up front – during model construction time – resulting in a new behavior model whose equations can be processed locally. Note that a behavior model where all equations can be processed by local propagation, forms a causal behavior model. Such a behavior model is the most efficient algorithmic model possible.

Transfer to the a/b-Wave Domain. The electrical quantities voltage, v, and current, i, can be expressed by so-called wave quantities, a, b, which are linear combinations of v and i. Common is a transformation into voltage waves,

$$a = v + Ri, \qquad b = v - Ri,$$

where a and b represent the incident and reflected wave respectively; R is called the port resistance. When applying these relations in the above topology reformulation step, the connectors that are used to model series and parallel subgraphs get a special name – they are called series adaptor and parallel adaptor respectively. Adaptors have ports where the a/b-equivalents of electrical components or other adaptors can be connected. An adaptor can be understood as a mathematical building block that introduces constraints on the a/b-waves of its connected elements such that in the original circuit Kirchhoff's voltage and current law are fulfilled. Figure 18 shows an electrical circuit and its related wave digital structure containing one series adaptor.

Note that, aside from performance reasons, the reformulated counterpart of circuit S in the a/b-wave domain comes along with superior numerical properties, which simplify the operationalization of S in the form of an integrated circuit.

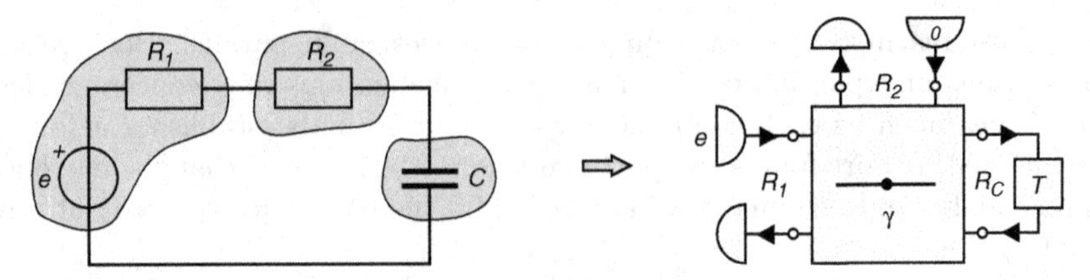

Fig. 18. Electrical circuit with two resistances and a capacity (*left*) and the related wave digital structure (*right*). The *shaded areas* in the circuit indicate the decomposition into three ports

5 Summary

Model construction is an artistic discipline whose overall objective is to find the right model in order to give an answer to an open question. Our research discussed the classical top-down model construction approach and introduced the idea of horizontal model construction. Horizontal model construction starts with an already developed model, which then is improved in different respects, be it efficiency, handling, complexity, or others. Hence, horizontal model construction does not follow a fixed scheme but takes very different forms from which we introduced three important principles: model simplification, model compilation, and model reformulation. For each of these principles we introduced its basic characteristics and exemplified its application within a knowledge-intensive engineering task.

Among experienced engineers and problem solvers horizontal model construction is common practice and may often be applied subconsciously. In this connection our research shall contribute to both a better understanding and a formal theory of model construction within complex and knowledge-intensive tasks.

References

1. M. Abderrahim and A. Whittaker. Mechatronics '98. In J. Adolfsson and J. Karlsen, editors, *Interfacing Autolev to Matlab and Simulink*, pages 897–901. Pergamon, September 1998.
2. Sanjaya Addanki, Roberto Cremonini, and J. Scott Penberthy. Reasoning about assumptions in graphs of models. In *Proceedings of the Eleventh International Joint Conference on Artificial Intelligence (IJCAI 89)*, pages 1324–1330, Detroit, MI, August 1989.
3. Martin Anantharaman, Bodo Fink, Martin Hiller, and Stefan Vogel. Integrated development environment for mechatronic systems. In *Proceedings of the Third Conference on Mechatronics and Robotics*, Paderborn, Germany, 1995.
4. D. Angluin and C. H. Smith. Inductive inference: theory and methods. *Computational Surveys*, 15(3):237–269, 1983.

5. Elisabeth Bradley and Reinhard Stolle. Automatic construction of accurate models of physical systems. *Annals of Mathematics and Artificial Intelligence*, 17(1–2):1–28, 1996.
6. David C. Brown and B. Chandrasekaran. *Design Problem Solving.* Morgan Kaufmann, San Mateo, 1989.
7. François E. Cellier. *Continuous System Simulation.* Springer, Berlin Heidelberg New York, 1991.
8. Daniel J. Clancy and Benjamin Kuipers. Model decomposition and simulation. In Toyoaki Nishida, editor, *Proceedings of the Eighth International Workshop on Qualitative Reasoning about Physical Systems (QR 94)*, pages 45–54, Nara, Japan, June 1994.
9. Rina Dechter and Judea Pearl. The cycle-cutset method for improving search performance in AI applications. In *Proceedings of the Third Conference on Artificial Intelligence Applications*, Orlando, Florida, 1987.
10. Hilding Elmquist. Object-oriented modeling and automated formula manipulation in Dymola. In *SIMS'93*, Kongsberg, Norway, June 1993. Scandinavian Simulation Society.
11. Alfred Fettweis. Wave digital filters: theory and practice. *Proceedings of the IEEE*, 74(2):270–327, February 1986.
12. Paul A. Fishwick. The role of process abstraction in simulation. *IEEE Transactions on Systems, Man, and Cybernetics*, 18:18–39, 1988.
13. Charles L. Forgy. Rete: a fast algorithm for the many pattern/many object pattern match problem. *Artificial Intelligence*, 19:17–37, 1982.
14. Frederick K. Frantz. A taxonomy of model abstraction techniques. In *Proceedings of the 1995 Winter Simulation Conference (WSC 95)*, Proceedings in Artificial Intelligence, pages 1413–1420, Arlington, VA, December 1995.
15. John S. Gero. Design prototypes: a knowledge representation scheme for design. *AI Magazine*, 11:26–36, 1990.
16. Stefan Goldschmidt. Modellbildung am Beispiel von Starrkörpersystemen. Systematische Darstellung und Untersuchung von Software-Unterstützung. Study work, University of Paderborn, November 1996.
17. R. C. Hibbeler. *Engineering Mechanics: Statics and Dynamics*, 6th edition. MacMillan, New York, 1992.
18. Yumi Iwasaki and Alon Y. Levy. Automated model selection for simulation. Knowledge Systems Laboratory (KSL), QS93, 1993.
19. Johan de Kleer. Problem solving with the ATMS. *Artificial Intelligence*, 28:197–224, 1986.
20. G. J. Klir. *Architecture of Systems Complexity.* Sauders, New York, 1985.
21. Granino A. Korn and John V. Wait. *Digital Continuous-System Simulation.* Prentice-Hall, Englewood Cliffs, NJ, 1978.
22. Klaus Ulrich Leweling and Benno Stein. Hybrid constraints in automated model synthesis and model processing. In Susanne Heipcke and Mark Wallace, editors, *5th International Conference on Principles and Practice of Constraint Programming (CP 99), Workshop on Large Scale Combinatorial Optimisation and Constraints*, pages 45–56, Leamington Spa England, October 1999. Dash Associates.
23. R. S. H. Mah. *Chemical Process Structures and Information Flows.* Butterworths, Boston, 1990.
24. Mary Lou Maher and Pearl Pu, editors. *Issues and Applications of Case-Based Reasoning in Design.* Lawrence Erlbaum Associates, Mahwah, NJ, 1997.

25. Sandra Marcus. SALT: a knowledge acquisition language for propose-and-revise systems. In Sandra Marcus, editor, *Automating Knowledge Acquisition for Expert Systems*, pages 81–123. Kluwer Academic, Norwell, MA, 1988.
26. B. T. Messmer and H. Bunke. Subgraph isomorphism in polynomial time. Technical Report AM-95-003, Research Group of Computer Vision and Artificial Intelligence, University of Bern, 1995.
27. Marvin Minsky. Models, minds, machines. In *Proceedings of the IFIP Congress*, pages 45–49, 1965.
28. P. Pandurang Nayak. Causal approximations. *Artificial Intelligence*, 70:277–334, 1994.
29. P. Pandurang Nayak. *Automated Modelling of Physical Systems.* Springer, Berlin Heidelberg New York, 1995.
30. P. C. Piela, T. G. Epperly, K. M. Westerberg, and A. W. Westerberg. ASCEND: an object-oriented computer environment for modeling and analysis: the modeling language. *Computers Chemical Engineering*, 15(1):53–72, 1991.
31. Frank Puppe. *Systematic Introduction to Expert Systems, Knowledge Representations and Problem-Solving Methods.* Springer, Berlin Heidelberg New York, 1993.
32. Lisa Purvis and Pearl Pu. An approach to case combination. In *Proceedings of the Workshop on Adaptation in Case Based Reasoning, European Conference on Artificial Intelligence (ECAI 96)*, Budapest, Hungary, 1996.
33. Olivier Raiman. Order of magnitude reasoning. *Artificial Intelligence*, 51:11–38, 1991.
34. B. Raphael and B. Kumar. Indexing and retrieval of cases in a case-based design system. *Artificial Intelligence for Engineering Design, Analysis, and Manufacturing*, 10:47–63, 1996.
35. Jeff Rickel and Bruce Porter. Automated modeling for answering prediction questions: selecting the time scale and system boundary. In *Proceedings of the Twelfth National Conference on Artificial Intelligence (AAAI 93)*, pages 1191–1198, Cambridge, MA, 1994. AAAI Press.
36. Elisha Sacks. Piecewise linear reasoning. In *Proceedings of the Sixth National Conference on Artificial Intelligence (AAAI 87)*, pages 655–659, Seattle, WA, July 1987. AAAI Press.
37. Munehiko Sasajima, Yoshinobu Kitamura, Mitsuru Ikeda, and Shinji Yoshikawa. An investigation on domain ontology to represent functional models. In Toyoaki Nishida, editor, *Proceedings of Eighth International Workshop on Qualitative Reasoning about Physical Systems (QR 94)*, pages 224–233, Nara, Japan, June 1994.
38. D. B. Schaechter, D. A. Levinson, and T. R. Kane. AUTOLEV *User's Manual*, 1991.
39. L. C. Schmidt and J. Cagan. Configuration design: an integrated approach using grammars. *ASME Journal of Mechanical Design*, 120(1):2–9, 1998.
40. Bart Selman and Henry Kautz. Knowledge compilation using Horn approximations. In *Proceedings of the Ninth National Conference on Artificial Intelligence (AAAI 91)*, pages 904–909, Anaheim, CA, 1991. AAAI Press.
41. Benno Stein. *Functional Models in Configuration Systems.* Dissertation, University of Paderborn, Institute of Computer Science, 1995.
42. Benno Stein. *Model Construction in Analysis and Synthesis Tasks.* Habilitation, Department of Computer Science, University of Paderborn, Germany, June 2001. URL http://ubdata.uni-paderborn.de/ ediss/17/2001/stein/.

43. Benno Stein, Daniel Curatolo, and Hans Kleine Büning. Speeding up the simulation of fluidic systems by a knowledge-based selection of numerical methods. In *Euromech Colloquium 370 Synthesis of Mechatronic Systems*. University of Duisburg, September 1997.
44. Benno Stein, Daniel Curatolo, and Marcus Hoffmann. Simulation in FLUIDSIM. In Helena Szczerbicka, editor, *Workshop on Simulation in Knowledge-Based Systems (SIWIS 98)*, number 61 in ASIM Notes, Bremen, Germany, April 1998. Technical committee 4.5 ASIM of the GI.
45. Peter Struss. Multiple models for diagnosis. SPQR-Workshop on Multiple Models, FRG Karlsruhe, March 1991.
46. Michael Suermann. Wissensbasierte Modellbildung und Simulation von hydraulischen Schaltkreisen. Diploma thesis, University of Paderborn, 1994.
47. J. Wallaschek. Modellierung und Simulation als Beitrag zur Verkürzung der Entwicklungszeiten mechatronischer Produkte. *VDI Berichte, Nr. 1215*, pages 35–50, 1995.
48. Daniel S. Weld. The use of aggregation in causal simulation. *Artificial Intelligence*, 30:1–34, 1986.
49. A. Wayne Wymore. *Systems Engineering for Interdisciplinary Teams*. Wiley, New York, 1976.
50. Kenneth Man-kam Yip. Model simplification by asymptotic order of magnitude reasoning. *Artificial Intelligence*, 80:309–348, 1996.
51. Bernard P. Zeigler, Herbert Praehofer, and Tag Gon Kim. *Theory of Modeling and Simulation*. Academic, New York, 2000.
52. Blaž Zupan. Optimization of rule-based systems using state space graphs. *IEEE Transactions on Knowledge and Data Engineering*, 10(2):238–253, March 1998.

Artificial Intelligence Applied to the Modeling and Implementation of a Virtual Medical Office

Sandro Moretti Correia de Almeida[1], Lourdes Mattos Brasil[1], Edilson Ferneda[1], Hervaldo Sampaio Carvalho[2], and Renata de Paiva Silva[1]

[1] Technology and Science Center, Catholic University of Brasília (UCB), Brasília, Brazil
[2] Medicine Department, University of Brasília (UnB), Brasília, Brazil

This chapter aims at presenting an overview of two Artificial Intelligence (AI) techniques: Case-Based Reasoning (CBR) and Genetic Algorithm (GA). It will present two implementation models, one for each technique, proposed by the Technology and Science Center (CCT) of the Catholic University of Brasília (UCB), in Brazil.

Afterwards, a case study will show the results of the implementation of those models in the context of IACVIRTUAL, which is a project of a Web-based Virtual Medical Office that has been developed by the researchers of CCT.

Therefore, the chapter will provide a practical demonstration of the application of CBR and GA in the field of Health, specifically as to the support to diagnosis.

1 Medical Diagnosis and Knowledge Transfer

The medical profession is very influenced by the scientific advances of the related areas. Like in most of the current professions, medical professionals must be always up-to-date about the latest technologies. Nowadays, knowledge generation and knowledge transfer are key elements for success and a professional challenge as well [1].

According to Davenport and Prusak [2], human beings learn better from histories. The research of Schank [3] and the work of his student Kolodner [4] show that knowledge transfer gets more efficacious when it occurs by means of a convincing, elegant, eloquent narrative. The use of narratives is one of the best ways to learn and to teach complex subjects. Most of the times, it is possible to structure histories to transmit knowledge without substantial loss of the power of communication.

Throughout the diagnosis process, the medical professionals perform several inferences from the body malfunctions. Those inferences derive from

S.M.C. de Almeida et al.: *Artificial Intelligence Applied to the Modeling and Implementation of a Virtual Medical Office*, Studies in Computational Intelligence (SCI) **116**, 169–190 (2008)
www.springerlink.com

observation (the patient's clinical history, signals, symptoms, routine tests, response to manipulation, time elapsed since some events); the clinical, physiological, biochemical, anatomical, pathological knowledge of the doctor; the experience of diagnosing similar cases; and the common sense and intuition [5].

The medical professionals use their past experiences widely in the process of gathering and interpreting information. Those past experiences are essential because they reduce the number of unnecessary questions, avoid superfluous test requisitions, and make the information management more efficient [5].

As can be seen, the medical diagnosis is a very complex process. It demands intuition, reasoning, experience and the analysis of information from distinct sources. Medical experts use several computational techniques to support the decision in the medical diagnosis [6]. Among the computational techniques that aim at retaining, recovering and reusing knowledge, there is case-based reasoning [4].

2 Case-Based Reasoning

Case-Based Reasoning (CBR) is an Artificial Intelligence (AI) technique that solves current problems by making use of past known cases.

Davenport and Prusak [2] suggest that CBR is a way to combine, in the computer context, the power of narrative with knowledge codification. The technique involves knowledge extraction from a series of narratives (or cases) in a knowledge domain. Unlike other expert systems, which demand very structured and non-duplicate rules, CBR allows the case structure to reflect the human way of thinking.

CBR is an approach to learning and to problem solving, and it has numerous applications. In the medical field, there are several reports on the use of CBR [4], such as the use of past medical cases to solve current ones. Like human beings, CBR solves a new problem with the adaptation of a known similar solution. For example, when a doctor is listening to a new patient, he usually remembers the medical history of another one, who presented a similar set of characteristics and symptoms. Therefore he prescribes, to the new patient, a treatment that is analogous to that one successfully applied to a similar case [7].

2.1 The History of CBR

The origin of CBR is in Schank's essays on language understanding [4]. According to his theory, the language is a process based on memory. His theory presupposition is that the human memory adjusts itself in response to every single experience that happens to the person. It implies that learning depends on those memory modifications. The conclusion of such studies is that the language understanding process depends on information previously retained in memory. It means that people can not understand events if they can not associate them with known facts [8].

Another study in the field of CBR, developed by Gentner [9], was related to "analogical reasoning". Gentner says that, in analogical reasoning, changes in a knowledge domain occur along with the transfer of a previous solution to a new problem. For example: "if a fish can swim forward by pushing back the water, I can invent an object called oar to propel a boat in movements pushing back the water". The objective of analogical reasoning is the transformation and the extension of knowledge from a known domain to another one that does not have a totally understood structure.

Later, theories of concept formation, problem solving and experimental learning, all of them derived from Psychology and Philosophy, contributed to CBR. Schank [4] continued exploiting the important role that previous situations in memory and that patterns of situations play in problem solving and in learning.

One of the first computational systems to utilize CBR was CYRUS [4], developed by Kolodner based on the dynamic memory model by Schank. In that model, the memory of cases is a hierarchical structure called episodic memory organization package. The basic idea is to arrange specific cases that have more generic similar properties [10].

Another example is PROTOS, an application to support ear related diseases [7]. It has implemented classification and knowledge acquisition based on cases. PROTOS learned to classify ear related diseases based on symptom descriptions, medical records and test results. The justification for that project was the existence of real restrictions in knowledge representation and learning representation. The base used to train PROTOS was composed of 200 cases of a specialized clinic, placed in 24 categories. After the training, it was reported that the system reached 100% of accuracy.

One of the most famous CBR systems in the health field is CASEY [4]. CASEY is a system to diagnose cardiac dysfunctions. It uses the symptoms of the patient as input to produce an explanation of the patient's cardiopathy. CASEY diagnoses patients by applying a match based on models and heuristic adaptations to the 25 cases in its base. This means that when a new case/problem arrives, CASEY tries to find cases of patients with similar – but not necessarily identical – symptoms. If the new case matches a known one, CASEY adapts the recovered diagnosis according to the differences between the new case and the recovered one.

The commercial applications as well as the development tools available on the market show the current technological maturity of CBR. The first CBR development tools dated the 1990s and nowadays they are even more enhanced.

Among some works using CBR in recent years, we can list the project of Balaa [11] and the study of Perner [12]. Specifically in Cardiology, we must highlight the Heart Disease Program developed under the coordination of Long [13], which aims at supporting the medical diagnosis of patients with heart disease symptoms.

2.2 The CBR Cycle

The main part of knowledge in CBR systems is represented by its cases [7]. Typically, a case represents the description of a situation (problem) along with the acquired experience (solution); and it can be expressed in several ways [14]. Figure 1 shows a sample of case representation.

A case is the formal representation of knowledge, based on a previous practical experience associated to a specific context. The case representation is a complex task and it is very important for the success of the CBR system [15].

Figure 2 is an overview of the CBR architecture. It shows the four phases of the cycle of the CBR process [10], also known as 4R:

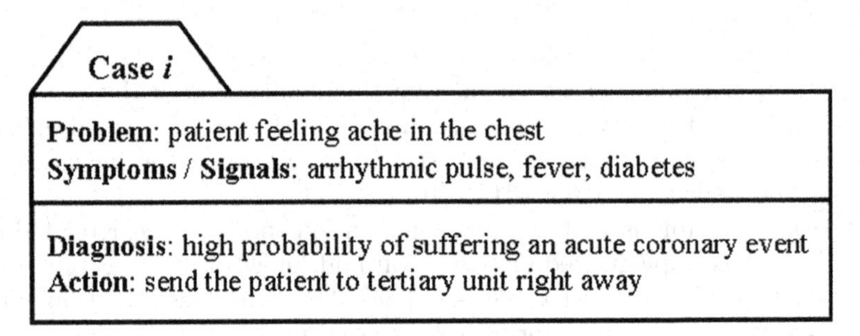

Fig. 1. Case representation

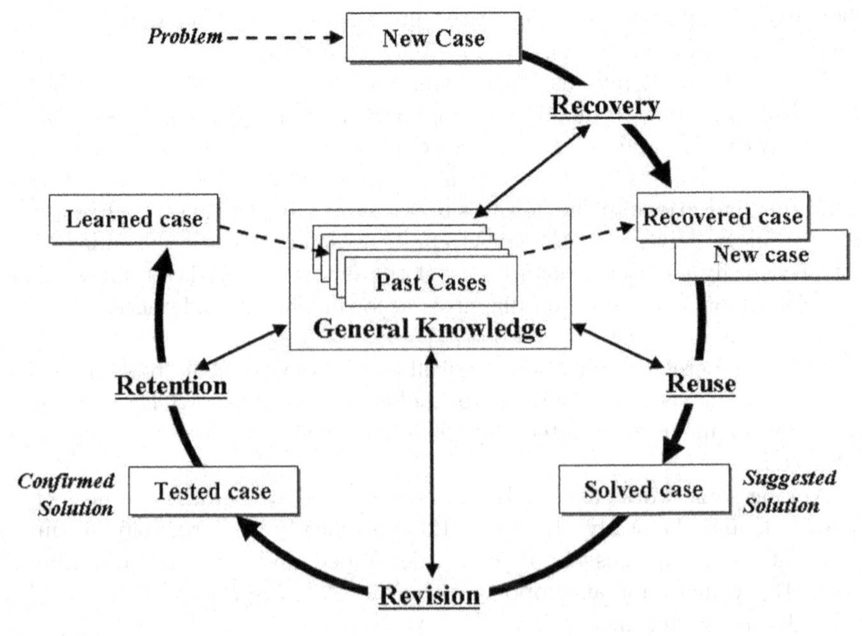

Fig. 2. The CBR cycle [10]

- *Recovery*: search and recovery of the most similar case – or a set of the most similar cases – from the base of cases.
- *Reuse*: reuse of the information that solved the problem in the recovered case by making an association with the context of the current problem.
- *Revision*: if necessary, the solution to the recovered case can be adapted to fit in the new situation (current problem).
- *Retention*: in this phase, all the useful parts of the new problem are recorded in the base of cases. Another decision to be taken is about the correct way to record and how to index the new case to be used in a future recovery.

According to Kolodner [4], it is appropriate to use CBR when it is hard to evaluate objective solutions and the solutions can be justified by cases, or when it is necessary to anticipate potential problems.

The use of CBR has the following advantages: (a) the reduction of time to diagnose; (b) the capture and reuse of the experiences of the medical expert; and (c) continuous learning, because it can feed the system with new experiences [16].

CBR has a lot in common with learning. The human being learns from facing problems that are later solved [17]. As to the system, it is able to automatically generate solutions based on its knowledge domain in a flexible way. Therefore, it can analyze complex, uncommon solutions to problems and recognize and explain errors as well [2].

3 Genetic Algorithm

3.1 Overview

Evolutionary computation is the generic name given to computational methods based on Charles Darwin's Theory of Evolution by Natural Selection. The algorithms used in evolutionary computation are called evolutionary algorithms [18].

The most popular evolutionary algorithms are: genetic algorithms, genetic programming and evolutionary strategies.

According to Goldberg [18], Genetic Algorithm (GA) constitutes an adaptive search method that is based on the analogy between optimization and the natural evolution of species, combining the concepts of selective adaptation with the survival of the most capable individuals.

GA searches for the best hypothesis out of a set of candidate ones. In GA, the best hypothesis is defined as the one that optimizes a predefined numerical measure – called fitness – for a given problem.

3.2 History

The history of GA begins in the 1940s, when computer scientists started to try to get inspiration from nature to create the field of AI. In the 1950s and

1960s several scientists came up with the idea that evolution could be used as an optimization tool for engineering problems [19].

One of the first attempts to associate natural evolution with the optimization problem was made in 1957, when Box presented his evolutionary operation scheme. In the early 1960s Bledsoe and Bremmerman started to work with genes, developing the precursors of the recombination operators [20].

Rechenberg introduced "evolution strategies" in 1965, which is a method he used to optimize real-valued parameters for devices such as airfoils. In 1966, Fogel, Owens and Walsh developed a technique called "evolutionary programming", representing candidate solutions to given tasks like finite-state machines, which evolved by randomly mutating their state-transition diagrams and then selecting the fittest [21].

The theory of genetic-based machine learning was created by Holland in 1962. Finally, GA was developed by John Holland, his colleagues, and his students at the University of Michigan in the 1960s and 1970s [18], and presented in the work entitled "Adaptation in Natural and Artificial Systems" in 1975 [22].

3.3 Biological Terminology in a Simple GA

In this section it will be presented a Simple GA [18] to introduce the terminology used in GA and its respective analogy with biological terms.

Step 1: Generate the Initial Population (Concepts of Population, Chromosome, Gene, Allele and Locus)

In nature: all living organisms consist of cells, and each cell contains the same set of one or more chromosomes – strings of DNA – that serve as a "blueprint" of the organism [21]. *In GA*: both the terms *chromosome* and *individual* are used to represent each member of the population [23] and are often represented by a *bit string.*

In nature: a chromosome can be conceptually divided into genes: functional blocks of DNA, each and every one of them encoding a particular protein. Basically, one may think of a gene as the structure that encodes a characteristic, such as the color of the eye. The different possibilities for each characteristic (e.g., blue, brown, hazel) are called alleles. Each gene is located at a particular locus (position) on the chromosome. *In GA*: each string (or chromosome) is a candidate solution to the problem. In a bit string, the *genes* are either single bits or short blocks of adjacent bits that encode a particular element of the candidate solution (e.g., in the context of multi-parameter function optimization, the bits that encode a particular parameter can be considered as a gene). An *allele* in a bit string is either *0* or *1*; for larger alphabets more alleles (different values) are possible at each *locus* [21].

The simple GA starts with a randomly generated population of n l-*bit* chromosomes (which means that the size of the population and the length of the bit string are determined according to the problem). Now it must be defined a set of operations that take this initial population and generate successive populations which hopefully will improve as time passes by.

Step 2: Calculate the Fitness of Each Member of the Population (Concept of Fitness)

In nature: the fitness of an organism is typically defined either as the probability that the organism will live enough to reproduce (viability) or as a function of the number of offsprings the organism has (fertility). *In GA*: it is used a *fitness* function $f(x)$ to evaluate each chromosome x in the population. Intuitively, it is possible to think of function f as a measure of profit, utility or quality that is intended to be maximized [18].

Step 3: Generate the Next Population (Concepts of Selection, Crossover and Mutation)

The members of the next population can be identical to the members of the current one, or an offspring with the genetic recombination of two parents.

In nature: there are diploid organisms, whose chromosomes are arrayed in pairs (e.g., human beings), and haploid organisms, with unpaired chromosomes. In the haploid sexual reproduction, genes are exchanged between the two parents' single-stranded chromosomes. The offspring are subject to mutation, in which single nucleotides (elementary bits of DNA) are changed, often due to copying errors, while they are transmitted from the father to such offspring [21]. *In GA*: an artificial version of the natural selection is:

Step 3a. Select the parent chromosomes from the current population. Strings with higher fitness value $f(x)$ have a higher probability to be selected [18]. This can be implemented in several ways. Perhaps the easiest one is creating a biased roulette wheel, in which the bracket (probability of selection) is proportional to the fitness of the chromosome (Fig. 3 and Table 1).

To implement such selection, a randomly chosen number, between 1 and 1,200, will indicate the selected parent, which can be chosen more than once in each generation.

Step 3b. Evaluate the probability of crossover. Another parameter of GA is the probability of crossover, which, according to some authors, varies from 60 to 90% [24, 25]. If no crossover takes place, two offsprings are formed as *exact copies* of their parents (both parents are copied for the new population). But if a crossover happens, the two *offsprings* are generated this way: a crossover point is randomly chosen between 1 and the number that corresponds to the total length of the chromosome; the first offspring is formed by the first part of the first parent and the second part of the second parent; and the other offspring is formed by the remaining parts [21].

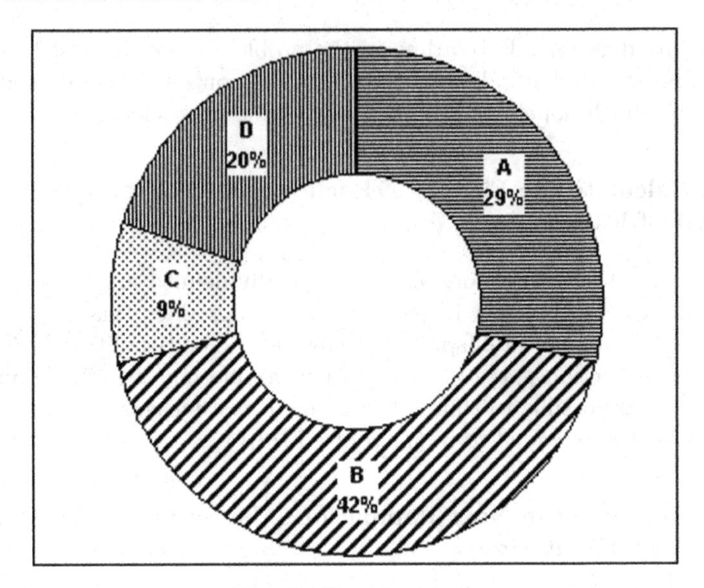

Fig. 3. Biased roulette

Table 1. Sample problem strings and fitness values

No.	String (chromosome)	Fitness	% of the total	Bracket of roulette
A	011011001	348	29	0001–0348
B	110000011	504	42	0349–0852
C	010001010	108	9	0853–0960
D	100110101	240	20	0961–1200
	Total	1,200	100	

That is the single-point crossover (Fig. 4), but there are also multi-point crossovers.

Step 3c. Evaluate the probability of mutation. Usually there is a low mutation rate or mutation probability [24]. Some authors [20,25] suggest values from 0 to 3%. If it takes place, a randomly selected point indicates the bit value (in the example of a bit string) that will be changed (Fig. 5).

Step 3d. Repeat steps to complete the population. Repeat steps 3a to 3c until the new population is complete.

Step 4: Replace the Current Population with the New Population (Concept of Generation)

Each iteration of this process is called *generation*.

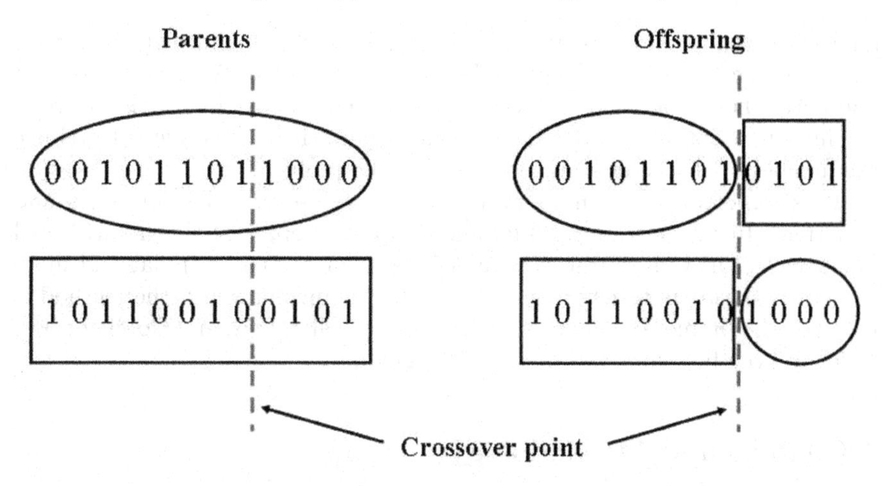

Fig. 4. One-point crossover

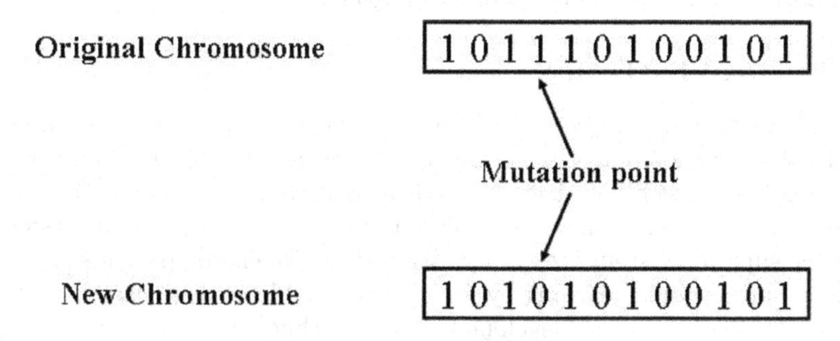

Fig. 5. Mutation in binary strings

Step 5: Go to Step 2 (Concept of Run)

A GA is typically iterated from 50 to 500 times (generations). The entire set of generations is called a *run*. At the end of a run there are often one or more highly fit chromosomes in the population. Since randomness plays an important role in each run, two runs with different random-number seeds will generally produce different detailed behaviors. Researches on GA often report statistics based on the average of many different runs of GA as to the same problem [21].

3.4 The Latest Developments

There are some papers that associate CBR with GA. Wettschereck and Aha [26] studied the nearest-neighbor function, which they say that is the most used to calculate similarity in CBR, but they consider it very sensitive to irrelevant characteristics that may cause distortions in the final result. However, while the present work intends to use GA to find the best weights for each characteristic, their proposal is to detect and eliminate irrelevant characteristics. The recent paper of Beddoe and Petrovic [27] has got a similar approach: they propose to use GA to determine the weights of the nearest-neighbor function used in CBR, but, once again, their objective is to reduce the number of characteristics considered in the problem.

It is important to remark that GA has had several developments and variations. Just for instance, Haupt and Haupt [28] represented genes with real numbers and nowadays tournament selection [25] is more popular and more used than the roulette-wheel selection. However, in this work, the method is based on the simple GA and basic principles by Holland [22] and Goldberg [18], which are the fundamentals of such IA technique.

4 Context and Methodology

This section describes the IACVIRTUAL Project as well as presents a model to implement CBR and another one to implement GA.

4.1 The IACVIRTUAL Project

The IACVIRTUAL Project (Artificial Intelligence Applied to the Modeling and Implementation of a Virtual Medical Office) is a knowledge-based system that simulates a web-based medical office conceived to support (a) patients interested in following their medical reports; (b) professionals interested in decision support systems for diagnosis and treatment of patients; and (c) students interested in learning by following available medical cases.

The project has been developed at the Technology and Science Center (CCT) of the Catholic University of Brasília (UCB), in Brazil, by a group of researchers of the Master's Degree Program in Knowledge and Information Technology Management. IACVIRTUAL aims at: (a) incorporating the latest technologies from Biomedical Engineering; (b) applying Information Technology to the field of Health; (c) contributing to the research of graduation and post-graduation students. The project is supported by several Brazilian and international institutions, medical equipment factories, medical offices and professionals.

The main modules of IACVIRTUAL are: Data Module, User Module, Interface Module, Educational Module and Decision Support Module (Fig. 6).

In the IACVIRTUAL Project, CBR aims at dynamically mapping clinical cases from an electronic medical register to a base of cases. Such past experiences will be available to the professionals in the medical area through the processes of recuperation, reuse and retention of new cases.

4.2 The CBR Model

The knowledge based on experiences can be expressed through a representation and through a reasoning method. Knowledge representation defines the

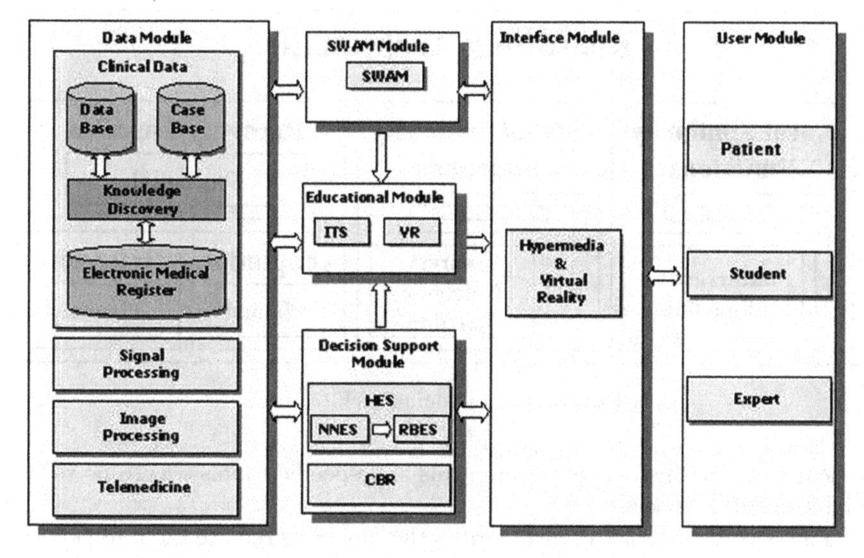

Fig. 6. The architecture of the IACVIRTUAL project

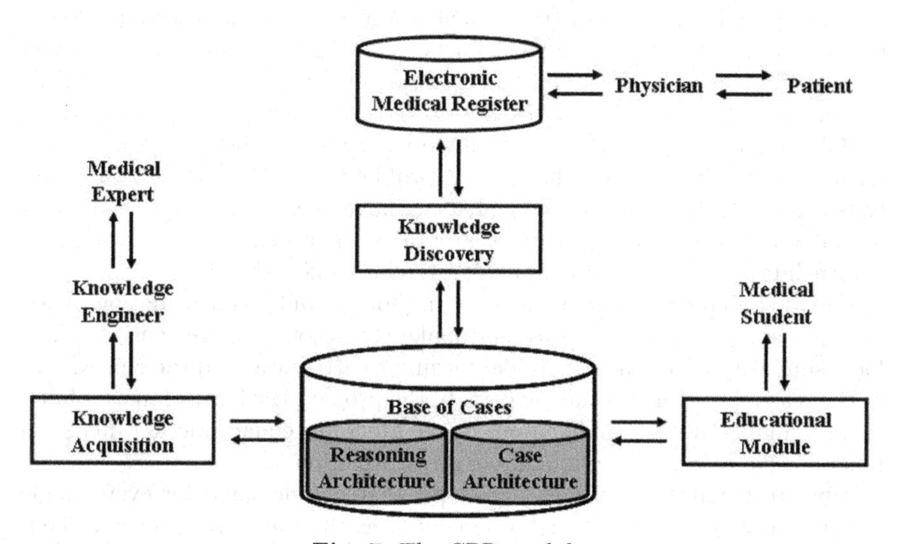

Fig. 7. The CBR module

approach and the organization used to retain and to recover knowledge. The IACVIRTUAL Project aims at supporting the medical diagnosis. Figure 7 shows an overview of the IACVIRTUAL CBR Module [29].

The proposed CBR model describes a global architecture that represents the interaction among the elements of the CBR module.

The *case structure* describes the essential elements to represent cases in a specific domain. In the medical domain, the case structure represents the

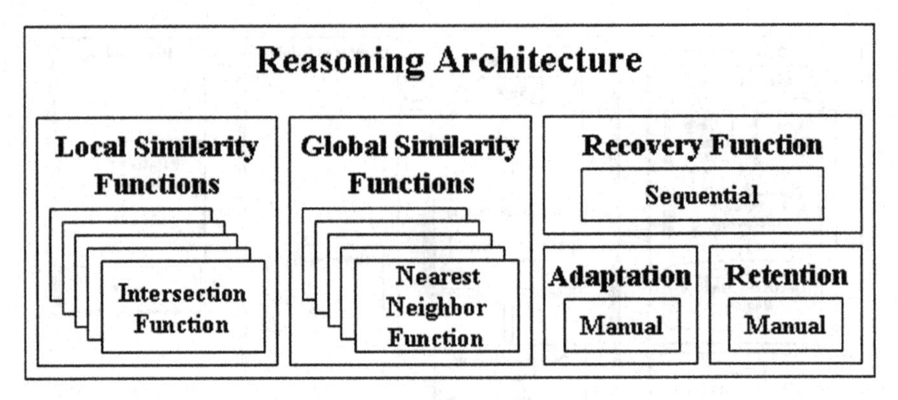

Fig. 8. The reasoning architecture

patient's clinical history, physical exams and specific exams according to the related medical specialty.

The *reasoning architecture* describes the functions required to reuse knowledge based on experiences. It is represented by a functional block that provides the specific functions related to the reasoning methods. In the CBR context, it is composed of functions to implement the indexation, recording, recovery, adaptation and retention mechanisms [30]. Figure 8 shows the reasoning architecture, with the CBR functional components used in the IACVIRTUAL Project.

Global similarity measures determine the similarity between a case with an unknown solution (i.e., the current problem) and the cases of the base, considering all the indexes. The global similarity functions implemented and available to the knowledge engineer are: (a) nearest-neighbor; (b) block-city; (c) Euclidean; (d) Minkowsky; and (e) weight Minkowsky [4].

Local similarity measures are used in global similarity calculation. They compare specific characteristics, and make the recovery more sensitive. The local similarity function is chosen according to attributes and their proximity to the reasoning of the domain expert. In the project, the local similarity functions available are: scalar function, linear function, equal function, maximal function, intersection function and contrast function [7].

Since the similarity of the current problem is calculated for every single case in the base, it is suggested a sequential method for case recovery. Then, the cases shall be sorted by the result of the similarity function [4]. In such process, the adaptation (reuse) and the recording (retention) of cases are manually performed.

The knowledge of the expert is used in the definition of the reasoning architecture and case structure. The expert collaborates with the definition of the most important characteristics and indexes, as well as with the evaluation of local and global similarities. After those definitions, it is made use of the knowledge of the expert once more, to define the weights for global similarity calculation [31].

The knowledge engineer, who prepares knowledge to be inserted in the base of cases, is responsible for developing tools and methods in order to help the expert express his knowledge and also elaborate and get domain models.

In the IACVIRTUAL Project, the knowledge discovery module will dynamically map clinical data from an electronic medical register to a base of cases, as shown in Fig. 7.

The base of cases must have only the relevant information to support the diagnosis decision taking.

In the field of medical education, the evolution of AI techniques and cognitive science researches has brought significant elevation to the intelligence of educational systems. CBR has been applied to those systems satisfactorily, because it presents a rich representation of their domains, allowing the use of knowledge in a manner not explicitly designed by the engineer.

4.3 The GA Model

In a previous section, it has been shown that the CBR module of the IACVIRTUAL Project calculates similarity between a current problem and cases of the base by using a global similarity function. This is a weighted function that uses one weight for each result of the local similarity function. These weights represent the relevance of each particular characteristic or symptom to diagnose the problem. In the CBR model the weights have been defined by a medical expert based only on his experience.

Since the success of the CBR diagnosis primarily depends on the correct selection of the most similar cases [7], it is important to look for a way to optimize the weight definition.

Genetic Algorithm applied to Case-Based ReAsoning SImiLarity

This section will present the GAC-BRASIL model – ***G***enetic ***A***lgorithm applied to ***C***ase-***B***ased ***R***e***A***soning ***SI***mi***L***arity [32]. The GAC-BRASIL model defines a GA whose best solution is a set of weights (one for each local similarity) to support the domain expert in that definition task.

Parameters and Definitions

Each individual (chromosome) is a candidate solution to the weight definition problem. Hence, each one must represent the set of weights to be used in the global similarity function of the CBR module. This way, the number of genes is the same as the number of local similarity functions (one for each symptom or characteristic) of the given CBR system.

The chromosome will be represented by a bit string and each gene will consist of four bits. Therefore, the length of the bit string will be the number of weights (genes) times four.

The initial population must be randomly generated and the number of individuals in each generation must be fixed and between 150 and 250 [32,33].

The criteria to stop the evolution can be the comparison among the genotypes of individuals in the population or among their phenotypes, by using the population fitness average. Each one of those methods can be used in GAC-BRASIL, but the evolution can not stop before 100 generations [33].

The crossover probability is between 60 and 90% and the mutation probability is between 0 and 3%. The selection method will be the already explained biased roulette wheel [18].

These are the steps to generate each next population:

- Two parents must be selected.
- The probability of crossover is tested. If no crossover takes place, both parents are copied for the next generation.
- On the other hand, if a single point crossover happens, the probability of mutation is tested for each offspring. If there is mutation, a randomly selected bit of the offspring is changed.
- These steps are repeated until the expected number of members in the population is complete.

The Fitness Function

Evaluating the fitness of each candidate solution to the problem is a key factor for GA. Since the objective of the GAC-BRASIL model is to define weights for a CBR global similarity function, the ideal fitness function must be an adaptation of it.

The approach suggested by the GAC-BRASIL model is to select two groups of cases from the base of cases. The cases must have a previously known real diagnosis. The first group will represent the "current problems" and the second one will represent the "base of cases". Then, the local similarity between each pair of cases will be calculated. In other words, for each case of the first group it will be calculated the similarity for each characteristic or symptom in each case of the second group. Now, to calculate the global similarity between each one of those pair of cases, it is necessary to use a weight to represent the relevance of each characteristic to the diagnosis.

The fitness function will receive as input parameter the chromosome, in which each gene represents the weight for a determined characteristic.

Then, the fitness function calculates the global similarity between each pair of chromosomes by multiplying the weight times its respective local similarity result and dividing the sum of all the results of the multiplications by the sum of the weights. Since the local similarity results ranges from 0 to 1, the global similarity result is a value ranging from 0 to 100% for each pair of chromosomes.

Each pair of chromosomes is considered similar if that result is above a configurable threshold. The GAC-BRASIL model suggests for this configurable parameter a value from 50 to 70%. Such value must be decided after the tests of the knowledge engineer in accordance with the CBR domain expert.

The function must count the pairs of cases considered similar (*Tsimilar*) and among them, it will count the pairs formed from cases with the same diagnosis (*Tequal*).

Table 2. The fitness function steps

Fitness (chromosome) – the function gets a chromosome as input parameter
Convert each gene of the chromosome from binary to decimal (weights for each characteristic of the case)
Initialize variables (Tsimilar = 0; Tequal = 0)
For each case (x) from group 1 (current problems)
For each case (y) from group 2 (base of cases)
Calculate global similarity (case x, case y) by using received weights
If similarity >= 70% then
Tsimilar = Tsimilar + 1
If the diagnosis of case x = the diagnosis of case y
Tequal = Tequal + 1
End If
End If
Next case y
Next case x
Fitness of the chromosome = Tequal/Tsimilar

The fitness of the set of weights – received as the input parameter – is the percentage of cases with the same diagnosis out of those cases presumed to be similar (Table 2).

It is expected that, with the GAC-BRASIL model approach, the more accuracy an individual (set of weights) brings to the global similarity function, the fitter it will be.

5 Case Study

This section presents a case study in which the CBR module was implemented according to the CBR model proposed. Afterwards, GAC-BRASIL was used as a model to implement a module – which was then incorporated to the CBR module – to help the medical expert define the relevance of the characteristics of the case to the diagnosis.

5.1 Database Preparation

To evaluate the CBR model and the GAC-BRASIL model, it was necessary to use real cases with already known diagnoses.

"Fundação Baiana de Cardiologia" (the Cardiology Foundation of the State of Bahia, in Brazil) [34] has provided a base of cases with 1,052 clinical records collected throughout 2 years, which was the base of cases of the cardiology expert system (SEC) developed by that foundation in collaboration with the Federal University of Rio de Janeiro (COPPE/UFRJ), also in Brazil. As to the probability of suffering a severe coronary event, the base had: 491

cases with low probability; 224 with intermediate probability and 337 with high probability.

In the original base of cases, some cases did not have all the information. With the support of a medical expert, the cases were analyzed and those ones with no essential information have been discarded. Another important decision taken by the medical expert was to discard the cases with intermediate probability, in order to make it possible the calculation of the specificity and sensibility variables.

Therefore, after this phase, 529 cases remained in the CBR base of cases.

5.2 The Implementation of CBR Recovery

The CBR recovery phase was developed according to the CBR model proposed. To determine local similarity for each characteristic or symptom, three different functions were used: (a) function 0 or 1; (b) linear function; and (c) weighted intersection function. Here are some examples:

- For the characteristic "gender" it was used the function "0 or 1": if there was the same gender in both cases, this local similarity was 1, otherwise it was 0.
- For the characteristic "age of the patient" it was used the linear function: there were 10 ranges of ages. If patients were in the same range, the value was 1. If they were in different ranges, the value decreased as the ranges became far from each other. This means that the value of the local similarity for this characteristic varied from 0 to 1, depending on the proximity of the ages.
- For the characteristic "ache location" it was used the intersection function: if two patients were feeling pain exactly at the same points, the value was 1. If they did not have any location in common, the value was 0. The value of the local similarity varied from 0 to 1, depending on the intersection (ache locations in common in the cases).

After calculating the 17 local similarities, it was necessary to calculate the global similarity among the cases. One of the suggested functions of the CBR model was the weight nearest-neighbor function. Such function applied a weight to each partial value (local similarities) and the sum of those multiplications divided by the sum of weights was the global similarity.

Therefore, to determine the global similarity, it was necessary to have a set of 17 weights, one for each characteristic or symptom. In the first version of the CBR module, the weights were defined by a medical expert of the University Hospital of Brasília (HUB). Table 3 of the next section shows those weights, while Fig. 9 shows the form in which the user inserted the characteristics.

5.3 The Implementation of the GA Module

As presented previously, the success of CBR depends mainly on the case recovery, and the case recovery relies on the similarity measure. Furthermore,

Table 3. Comparison as to the relevance of characteristics to the diagnosis (weights suggested by GA × weights suggested by the expert in cardiology)

#	Characteristic or symptom	Expert	GA
1	Gender	0,60	0,27
2	Ache intensity	0,30	0,59
3	Elapsed time since the last ache occurred	0,40	0,27
4	Elapsed time since the beginning of the symptoms	0,60	0,36
5	Heart rate	0,70	0,45
6	Diastolic pressure	0,70	0,39
7	Systolic pressure	0,60	0,51
8	Age of the patient	0,50	0,20
9	Electrocardiogram results	1,00	0,88
10	Antecedents	0,60	0,81
11	Ache characteristic	0,80	0,43
12	Risk factors	0,70	0,75
13	Ache location	1,00	0,85
14	Other related symptoms	0,50	0,75
15	Stethoscope signal	0,70	0,35
16	Other signals observed	0,60	0,76
17	Pulse signal	0,40	0,11

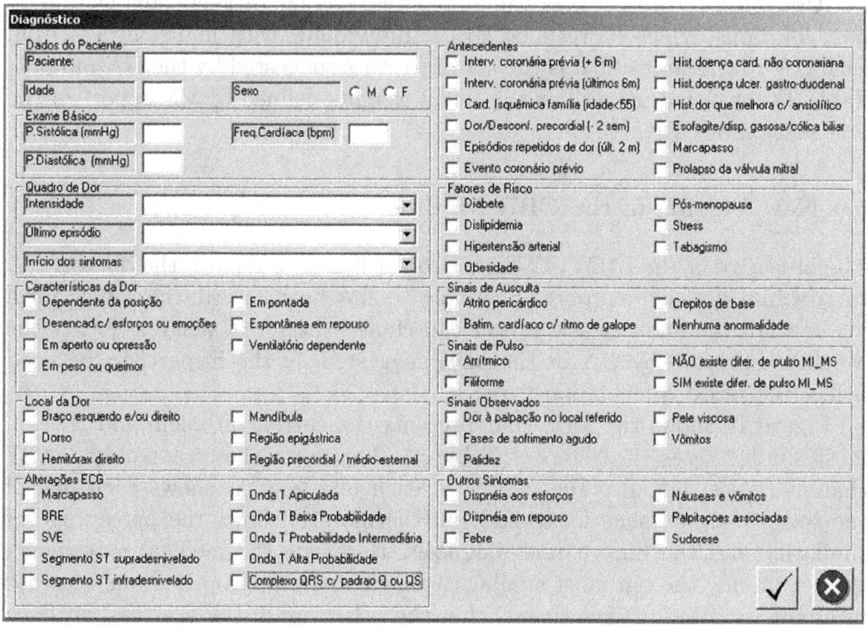

Fig. 9. Form for the characteristics of the case

the results of the global similarity calculation are directly influenced by the weights (relevance) defined for each characteristic. Although those weights can be defined by the professionals themselves, based on their clinical experience, their medical knowledge and the specialized literature, such task can be supported by an automated method based on the recorded clinical cases.

The GAC-BRASIL model aims at helping set the weights for global similarity functions in any CBR system. Therefore, it has been developed a GA module for the IACVIRTUAL Project based on the GAC-BRASIL model to help the medical expert define the relevance of the characteristics of the case.

As shown in the GAC-BRASIL model, the fitness function must be an adaptation of the real global similarity function, by using two groups of cases: one group represents the base of cases and the other one, with fewer cases, represents the current problems. For the first group 64 cases were selected (32 with high and 32 with low probability of suffering a severe coronary event). The second one was composed of 16 cases (8 with high and 8 with low probability of suffering a severe coronary event).

The GA parameters were set according to the ranges suggested by GAC-BRASIL. Such parameters were: binary representation of the chromosome; 17 genes (one for each weight); 4 bits per gene (resulting in a 68 bit length chromosome); 200 individuals in each population; randomized generation of the first population; 100 generations for each end GA run; 75% of probability of crossover; 1% of probability of mutation; biased roulette wheel to select survivors.

After 50 GA runs, it was calculated the average of genes among the 50 best solutions, which correspond to the individuals with higher fitness from each GA run. The result was the set of weights suggested by the GA module. Table 3 shows that, as well as the set of weights defined by the expert in the first version of CBR.

5.4 New Version of the CBR Module

The objective of the IACVIRTUAL Project is not to replace the doctor, but to give him decision support. Therefore, it has been introduced a new form in the CBR module where the user can choose the set of weights (either the weights suggested by GA or the ones suggested by the Expert) to be used before the recovery of similar cases.

Figure 10 shows the form that presents the current problem and the ten most similar cases (last step of Table 4). This form shows not only the 17 characteristics, but also the details of each one (e.g., it shows each one of the seven mapped ache locations). Furthermore it shows the percentage of similarity and the known real diagnosis; this way, the medical expert can choose, among the ten most similar cases, the most appropriate one to reuse or adapt its solution (treatment). On the other hand, the user can go back and run the recovery again, using another set of weights.

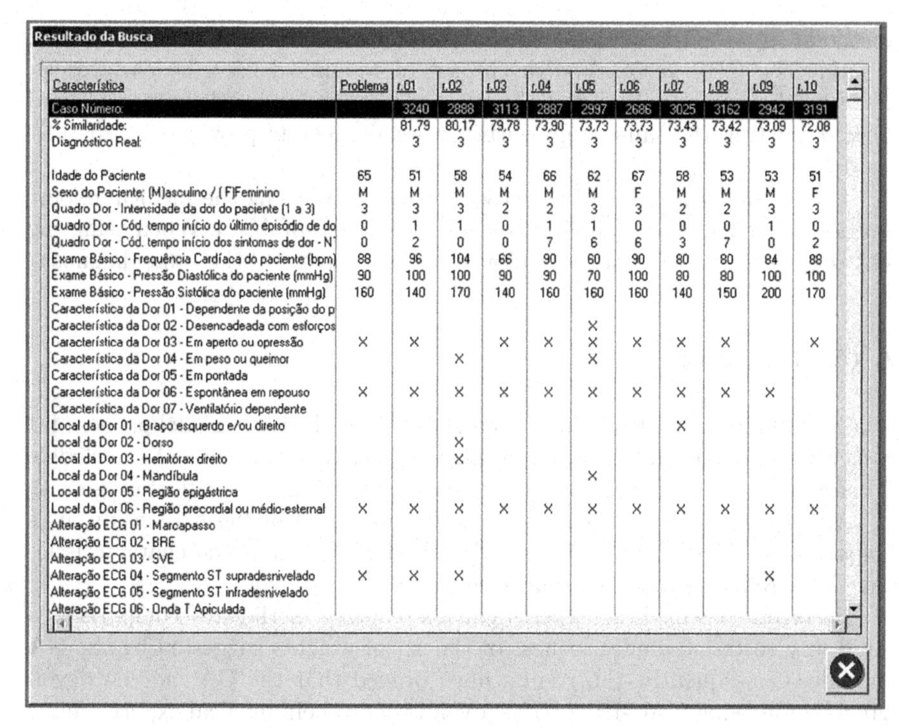

Resultado da Busca

Característica	Problema	r.01	r.02	r.03	r.04	r.05	r.06	r.07	r.08	r.09	r.10
Caso Número:		3240	2888	3113	2887	2997	2686	3025	3162	2942	3191
% Similaridade:		81,79	80,17	79,78	73,90	73,73	73,73	73,43	73,42	73,09	72,08
Diagnóstico Real:		3	3	3	3	3	3	3	3	3	3
Idade do Paciente	65	51	58	54	66	62	67	58	53	53	51
Sexo do Paciente: (M)asculino / (F)Feminino	M	M	M	M	M	M	F	M	M	M	F
Quadro Dor - Intensidade da dor do paciente (1 a 3)	3	3	3	2	2	3	3	2	2	3	3
Quadro Dor - Cód. tempo início do último episódio de do	0	1	1	0	1	1	0	0	0	1	0
Quadro Dor - Cód. tempo início dos sintomas de dor - N	0	2	0	0	7	6	6	3	7	0	2
Exame Básico - Frequência Cardíaca do paciente (bpm)	88	96	104	66	90	60	90	80	80	84	88
Exame Básico - Pressão Diastólica do paciente (mmHg)	90	100	100	90	90	70	100	80	80	100	100
Exame Básico - Pressão Sistólica do paciente (mmHg)	160	140	170	140	160	160	160	140	150	200	170
Característica da Dor 01 - Dependente da posição do p											
Característica da Dor 02 - Desencadeada com esforços						X					
Característica da Dor 03 - Em aperto ou opressão	X	X		X	X	X	X	X	X		X
Característica da Dor 04 - Em peso ou queimor			X			X					
Característica da Dor 05 - Em pontada											
Característica da Dor 06 - Espontânea em repouso	X	X	X	X	X	X	X	X	X	X	
Característica da Dor 07 - Ventilatório dependente											
Local da Dor 01 - Braço esquerdo e/ou direito								X			
Local da Dor 02 - Dorso			X								
Local da Dor 03 - Hemitórax direito			X								
Local da Dor 04 - Mandíbula						X					
Local da Dor 05 - Região epigástrica											
Local da Dor 06 - Região precordial ou médio-esternal	X	X	X	X	X	X	X	X	X	X	X
Alteração ECG 01 - Marcapasso											
Alteração ECG 02 - BRE											
Alteração ECG 03 - SVE											
Alteração ECG 04 - Segmento ST supradesnivelado	X	X	X							X	
Alteração ECG 05 - Segmento ST infradesnivelado											
Alteração ECG 06 - Onda T Apiculada											

Fig. 10. Form for the recovery results

Table 4. New algorithm of the CBR module

Diagnosis support (user = medical expert)

The user informs 17 characteristics of the current problem (case)
The user selects the set of weights to be used for the global similarity calculation
For each case (x) of the base of cases:
  Calculate the local similarity of the 17 characteristics of case "x" and the current case
  Calculate global similarity by applying the selected set of weights
  Save the global similarity
Next case "x"
Select and present the first 10 cases in decreasing order according to the global similarity

5.5 Results

As shown in Table 3, the weights suggested by GA were different from those ones defined by the medical expert. Hence, it was necessary to verify the effectiveness of the weights suggested by that IA technique.

With the help of experts in Cardiology of the University Hospital of Brasília (HUB), current cases were selected to test the system. The methodology was simple: informing the characteristics of the new cases; running the CBR recovery with the set of weights defined by the expert; counting how many cases had the same diagnosis out of the ten recovered ones; repeating the same steps with the set of weights provided by the GA module.

The results were: 85% of the cases recovered by using the medical expert set of weights had the same diagnosis for the current problem. When using the GA set of weights, 90% of the similar cases recovered had the same diagnosis.

6 Conclusions

The CBR module implementation has shown that the CBR model proposed is a methodology that helps the knowledge engineer extract medical knowledge. Therefore, it has achieved the objective of supporting the medical diagnosis.

About the GA implementation results presented in the case study, it is important to notice that real cases with similar symptoms and characteristics may have different diagnoses and vice versa.

However, the medical experts considered this method of comparison a valid way to test the effectiveness of the set of weights suggested by the GA module. Consequently, the results have proved that the GA module developed by following GAC-BRASIL is a valid way to help medical experts define the relevance of the characteristics of cases to the diagnosis. GAC-BRASIL has reached its goal: to be a model to implement weight calibration in CBR systems.

References

1. Portal Médico (2005) Available on: http://www.cremeb.cfm.org.br – accessed on April/18/2005
2. Davenport TH, Prusak L (1998) Conhecimento Empresarial: Como as organizações gerenciam o seu capital intelectual. Ed Campus, Rio de Janeiro
3. Schank RC (1982) Dynamic Memory: A Theory of Learning in Computers e People. Cambridge University Press, New York
4. Kolodner J (1993) Case-Based Reasoning. Morgan Kaufmann, San Francisco
5. Kassirer JP, Gorry GA (1978) Clinical problem solving: a behavioral analysis. Annals of Internal Medicine, 89(1), pp. 245–255
6. Sabbatini RM (2004) Uso do Computador no Apoio ao Diagnóstico Médico. Available on http://www.epub.org.br/informed/decisao.htm – accessed on May/10/2004
7. Wangenheim CH, Wangenheim AV (2003) Raciocínio Baseado em Casos. Manole, São Paulo
8. Silva DJA (2002) Representação e Indexação de Casos em Sistemas de Raciocínio Baseado em Casos para o Domínio da Neurologia. M.Sc. Dissertation. Universidade Federal do Paraíba

9. Gentner K (1983) Structure mapping: a theoretical framework for analogy. Cognitive Science, 7, pp. 155–170
10. Aamodt A, Plaza E (1994) Case based reasoning: foundational issues, methodological variations and system approaches. In: Proceedings of the AI Communications, 7(1), pp. 39–59
11. Balaa ZE, Strauss A, Uziel KM, Traphoner R (2003) Fm-Ultranet: a decision support system using case-based reasoning, Applied to Ultrasonography. In: Workshop on CBR in the Health Sciences
12. Perner P, Günther T, Perner H (2003) Airborne Fungi Identification by Case-Based Reasoning. Workshop on CBR in the Health Sciences
13. Long B (2005) Heart Disease Program. Available on: http://medg.lcs.mit.edu/projects/hdp/hdp-intro-tab.html – accessed on May/18/2005
14. Turban E, Aronson JE (1997) Decision Support Systems and Intelligent Systems. Prentice Hall, Englewood Cliffs, NJ
15. Mattos MM, Wangenheim CGV, Pacheco RC (1999) Aplicação de Raciocínio Baseado em Casos na Fase de Análise de Requisitos para Construção de Abstrações em Lógica de Programação. In: SEMINCO – Seminário de Computação, Blumenau
16. Derere ML (2000) Case-based reasoning: diagnosis of faults in complex systems through reuse of experience. In: Proceedings of the International Test Conference, Atlantic City, NJ, USA, pp. 27–34
17. Almeida SMC, Brasil LM, Carvalho HS, Ferneda E, Silva RP (2006) Proposal of a case-based reasoning model for IACVIRTUAL project. In: Proceedings of the IADIS MCCSIS 2006, pp. 312–314
18. Goldberg DE (1989) Genetic Algorithms in Search, Optimization and Machine Learning. Addison Wesley, CA
19. Mitchell TM (1997) Machine Learning. McGraw-Hill, USA
20. Linden R (2006) Algoritmos Genéticos: uma importante ferramenta da inteligência computacional. Brasport, Rio de Janeiro
21. Mitchell M (1999) An Introducing to Genetic Algorithms. MIT Press, Cambridge
22. Holland J (1975) Adaptation in Natural and Artificial Systems. University of Michigan Press
23. Azevedo FM (1999) Algoritmos Genéticos em Redes Neurais Artificiais. In: Conselho Nacional de Redes Neurais, pp. c091–c121, São José dos Campos - SP
24. Tanomaru J (1995) Motivação, Fundamentos e Aplicações de Algoritmos Genéticos. In: Anais do 2° Congresso Brasileiro de Redes Neurais, Curitiba
25. Lacerda EG, Carvalho ACPLF (1999) Introdução aos Algoritmos Genéticos. In: Anais do XIX Congresso Nacional da Sociedade Brasileira de Computação, vol 2, pp. 51–126
26. Wettschereck D, Aha D (1995) Weighting features. In: Proceeding of the First International Conference on Case-Based Reasoning. Springer, Berlin Heidelberg New York
27. Beddoe GR, Petrovic S (2006) Selecting and weighting features using a genetic algorithm in a case-based reasoning approach to personnel roistering. European Journal of Operational Research, 175(2), pp. 649–671
28. Haupt RL, Haupt SE (1998) Practical Genetic Algorithms. Wiley, New York
29. Silva RP (2005) Modelo de Apoio ao Diagnóstico no Domínio Médico Aplicando Raciocínio Baseado em Casos. M.Sc. Dissertation, Universidade Católica de Brasília

30. Ramos AM (2000) Modelo para Incorporar Conhecimento Baseado em Experiências à Arquitetura TMN. Ph.D. Thesis, Universidade Federal de Santa Catarina
31. Rezende SO (2004) Sistemas Inteligentes: Fundamentos e Aplicações. Manole, Barueri
32. Almeida SMC, Brasil LM, Ferneda E, Carvalho HS, Silva RP (2006) The diagnosis support system for ischemic cardiopathy: a case study in the context of IACVIRTUAL project. In: IFMBE Proceedings, vol 14, World Congress on Medical Physics and Biomedical Engineering, pp. 3541–3544, Seoul
33. Almeida SMC (2006) Modelo GAC-BRASIL – Algoritmo Genético Aplicado à Similaridade para Raciocínio Baseado em Casos. M.Sc. Dissertation, Universidade Católica de Brasília
34. Costa CGA (2001) Desenvolvimento e Avaliação Tecnológica de um Sistema de Prontuário Eletrônico do Paciente. M.Sc. Dissertation, Universidade Estadual de Campinas

DICOM-Based Multidisciplinary Platform for Clinical Decision Support: Needs and Direction

Lawrence Wing-Chi Chan[1], Phoebe Suk-Tak Chan[1], Yongping Zheng[1], Alex Ka-Shing Wong[1], Ying Liu[2], and Iris Frances Forster Benzie[1]

[1] Department of Health Technology and Informatics, Hong Kong Polytechnic University, Hunghom, Hong Kong
[2] Department of Industrial and Systems Engineering, Hong Kong Polytechnic University, Hunghom, Hong Kong

Summary. Multi-disciplinary platform is created to store and integrate DICOM objects from various clinical disciplines. With artificial intelligence, clinical decision support system is built to assess risk of disease complications using the features extracted from the DICOM objects and their interrelationship. Diabetes Mellitus is considered as the disease of interest and the risks of its complications are assessed based on the extracted features. The synergy of the DICOM-based multi-disciplinary platform and the clinical decision support system provides promising functions for extracting and interrelating consistent features of clinical information.

1 Introduction

Clinical decision support (CDS) is of particular importance in providing objective and consistent data analysis to assist the clinicians to make choices in preventive and translational medicine and clinical diagnosis and management through intelligent interactive computer software. The biomedical knowledge base and the intelligent inferencing mechanism are major components of a clinical decision support system (CDSS). A CDSS cannot produce knowledge by itself but have to rely on a biomedical knowledge base incorporating the a priori biomedical knowledge of the experts and the information obtained from historical cases or clinical evidence. To make the CDSS a promising clinical tool, knowledge discovery is an imperative process to discover the useful and novel medical knowledge or the pathogenesis of diseases from the available data.

Nowadays health studies especially of systemic diseases are concerned the synergy of multiple clinical disciplines. More and more biomedical features are considered in these studies to achieve higher significance of the results. In the Dunedin study, the subjects are the 1,037 babies born in Dunedin, New Zealand and they were first followed up at the age from 3 to 32

L.W.-C. Chan et al.: *DICOM-Based Multidisciplinary Platform for Clinical Decision Support: Needs and Direction*, Studies in Computational Intelligence (SCI) **116**, 191–212 (2008)
www.springerlink.com

(http://dunedinstudy.otago.ac.nz/). Examinations on all aspects of the subjects' physical and mental health, including teeth, blood samples, psychological testing, interviews and behaviors were performed in the study. In a study of adolescent idiopathic scoliosis (AIS), magnetic resonance imaging (MRI), somatosensory evoked potentials (SEPs) and radiographs were used to detect the tonsillar ectopia, the disorder in the somatosensory pathway and the curvature in spine respectively and the correlation tests were performed to the measurements [6]. It is easy to observe from these two studies that the data volume of multidisciplinary health studies grows in three dimensions. The first dimension is the number of data types or sources. The second dimension is the number of subjects which is obviously a major consideration in the study. The third dimension is the number of the follow-ups. These three dimensions span orthogonally to form a conceptual block of biomedical data volume. Each point in this data volume represents a biomedical data object obtained from a data source of a subject or patient in one follow-up. The data volume grows with any expansion in these dimensions. The demand for higher computing power and larger system memory is manifest when a large number of biomedical features need to be extracted from a large data volume.

A biomedical feature is in general a numerical value or symbolic label produced by interpretation or computation on values in data objects obtained from more than one data sources of one subject in one follow-up. For example, the anatomical location of a tumor can be determined by the image fusion of computed tomography (CT) and positron emission tomography (PET). Another example is the computation of Reynold number of a blood stream which requires the values of the luminal diameter measured from the ultrasonography, CT Angiography (CTA) or magnetic resonance angiography (MRA) and the blood flow velocity from the Doppler ultrasound waveform. The set of numerical biomedical features transformed from a biomedical data volume can be presented in a multidimensional vector space, referred to as feature space. Of course, the dimensionality of the feature space depends on the complexity of the concerned multidisciplinary health problem. Statistical tests can be used to determine the significant associated features and thus reduce the dimensionality of the feature space.

Of significant importance is the knowledge base produced by the biomedical feature set. The knowledge base could be a rule base describing some relationships between the biomedical features and the development or likelihood of progression of a disease or mathematical analytic model inferencing or predicting the clinical outcomes based on certain conditions. Note that the production of biomedical knowledge base is interactive and the clinicians need to input their a priori knowledge to formulate the framework of the rules or models which will be subjected to the fine adjustment using the available feature set. The CDS can be regarded as the realization of the knowledge base incorporating the integration of the novel knowledge discovered or mined from the data volume and the refined a priori biomedical knowledge.

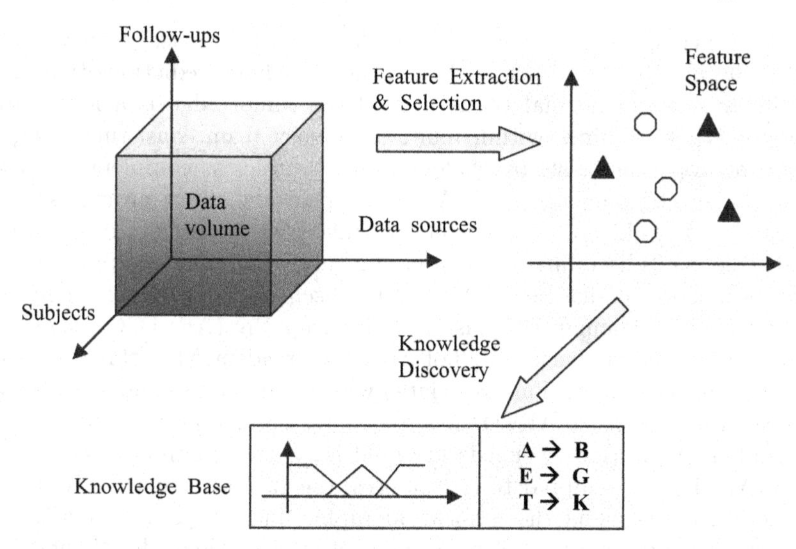

Fig. 1. Concept of information transformations in knowledge discovery

Figure 1 illustrates the conceptual transformations between the data volume, the feature space and the knowledge base.

Exploring the pathogenesis and underlying mechanism of diseases is the major objective of preventive and translational medicine. However, the associations between biomedical features are not sufficient to explain the causes of the diseases. Typical statistical methods could identify some etiological factors of a disease based on a linear regression model but the nonlinear relationship between these factors and the onset, development or progression of the disease cannot be obtained. Artificial intelligent tools can be used to synthesize flexibly adjustable models and meanwhile allows the incorporation of a priori knowledge and the presentation of discovered knowledge in linguistic manner. A number of the existing multidisciplinary health studies will be discussed in Sect. 2 to further explore the solutions in biomedical data management and analysis.

2 Multidisciplinary Health Studies

As many researches are concerned about systemic diseases such as diabetes, vasculatides and hypertension, the investigation and management of these diseases have to be approached in a multidisciplinary fashion. Numerous multidisciplinary health studies have been conducted to investigate the pathogenesis of vascular diseases, dementia and cancer. A study on acid-mediated tumor invasion proposes a hypothetical mathematical model that the increased glycolysis, a phenotypic trait almost consistently observed in human cancers, offers a selective growth advantage on the transformed cells and the in vivo experiments confirm the modeling predictions that the extracellular space of normal tissue around the tumor edge is acidified by the tumors [7]. Disciplines within multidisciplinary team consisting of psychiatrist, neuropsychologist, psychologist and neurologist complement each other in diagnosing dementia [19]. A literature review paper promotes new classification of ocular vasculitis which is linked with systemic diseases to facilitate the multidisciplinary investigation and management approach in a great proportion of its cases [9]. A multidisciplinary study of Arrhythmogenic Right Ventricular Dysplasia/Cardiomyopathy (ARVD/C) suggests determining the genetic background of ARVD/C by identifying chromosomal loci and specific gene mutations associated with the disease in addition to the echocardiography because ARVD/C is frequently familial and the structural abnormalities of the right ventricle are mild [12]. Another multidisciplinary study of ARVD/C attempted to characterize the cardiac structural, clinical and genetic aspects using the echocardiographic findings [20]. A review on the recent discoveries and novel hypotheses of atherosclerosis highlights the essential role of vascular progenitor cells in the homeostasis of the vessel wall and the influence of hemodynamic factors to the distribution of atheroma [8]. Analytical approach was proposed to investigate the atherosclerosis using thermodynamics, hemodynamics and mass transfer physical models [18]. Biochemistry and molecular cell biology were considered together to investigate the mechanism of the vascular diseases in diabetes [3]. Such multidisciplinary studies would face the same issues in the management and analysis of large amounts of highly detailed and heterogeneous biomedical data. An online survey and in-depth interviews were conducted to identify the needs and barriers in management and analysis of dramatically increasing data volume [1]. However, it is rare to find a multidisciplinary research and CDS platform for the management and analysis of biomedical data from different sources and to adopt common data format and communication standard to cater for the needs of data integration in these studies. This chapter proposes a multidisciplinary CDS platform based on the standard of digital imaging and communication in medicine. The feasibility of this proposed platform is validated by a study on the synergistic effect of hematological factors, biomarker profile and hemodynamic factors on the development of atherosclerosis.

3 DICOM Standard

With the introduction of digital medical image sources in the 1970s and the trend of computerized data processing and acquisition, the agreement and promotion of a standard communication method for the biomedical images and their related information was sought to facilitate efficient data exchange. There existed many problems in biomedical research data management and analysis and clinical applications when the biomedical data were stored in the proprietary format and transmitted through proprietary connections or portable media. For the sake of standardization, the American College of Radiology (ACR) and the National Electrical Manufacturers Association (NEMA) formed a joint committee in 1983 and published the ACR-NEMA standard documents. The name of the standard was changed from ACR-NEMA to Digital Imaging and Communications in Medicine (DICOM) during the release of the standard document version 3.0 in 1993. The DICOM committee has long been reviewing and revising the DICOM standard persistently to cater for the needs of continuously evolving clinical and research applications. This section is an informative elaboration of the 2007 DICOM standard (http://medical.nema.org/) for multidisciplinary biomedical researches.

3.1 Initiatives

The DICOM standard is aimed to promote the digital biomedical image communications regardless of the proprietary communication methods developed by the device manufacturers, to facilitate the development and expansion of picture archiving and communication system (PACS) to interface with the other clinical information systems, and to enable the creation of diagnostic information databases for the interrogation of the geographically distributed devices.

Nowadays, the DICOM standard is widely adopted in the manufacturing of medical imaging systems, signaling devices and diagnostic systems, the local and global integration of clinical information systems, medical image data exchange and the biomedical researches. Medical system engineers, software developers, medical informaticians and biomedical researchers tend to use DICOM as their common dictionary or scheme for performing biomedical data analysis and devising the protocols for the biomedical data transfer and exchange between their systems and devices. In interfacing medical imaging or signaling components implemented by different providers, the DICOM conformance statements supplied by the providers constitute very important documentations to ensure promising data compatibility and network connectivity. The Integrating the Healthcare Enterprise (IHE) initiative suggests that the DICOM conformity can streamline the clinical workflow and highly reduce the engineering workload as unnecessary conversions of data format and interfaces of different transmission protocols are no longer required. Due to the unified data format and model, the adoption of DICOM standard avoids the difficulties of reading or decoding the data in proprietary formats, searching for the required files organized in ad hoc indexing methods, and labeling or annotating the data in multiple formats using various software packages.

3.2 DICOM Document

DICOM standard is structured as a multipart document based on the guidelines established in ISO/IEC Directives, 1989 Part 3: Drafting and Presentation of International Standards. As shown in the following list, there are 18

Parts in the document and two of which, Parts 9 and 13, have been retired in the latest version published in 2007.

Part 1: Introduction and Overview
Part 2: Conformance
Part 3: Information Object Definitions
Part 4: Service Class Specifications
Part 5: Data Structures and Encoding
Part 6: Data Dictionary
Part 7: Message Exchange
Part 8: Network Communication Support for Message Exchange
Part 9: Point-to-Point Communication Support for Message Exchange
Part 10: Media Storage and File Format for Media Interchange
Part 11: Media Storage Application Profiles
Part 12: Media Formats and Physical Media for Media Interchange
Part 13: Print Management Point-to-Point Communication Support
Part 14: Grayscale Standard Display Function
Part 15: Security and System Management Profiles
Part 16: Content Mapping Resource
Part 17: Explanatory Information
Part 18: Web Access to DICOM Persistent Objects (WADO)

Generally speaking, the standard facilitates not only the implementations of PACS but also the interoperability of any clinical or research application entities. Figure 2 illustrates the basic building blocks of a general communication model in medical information applications.

As indicated in Fig. 2, the DICOM document Parts specify the conformance scheme for constructing the corresponding building blocks. Two boundaries separate the upper layer service and the basic file service from data processing building blocks of the application entity. The upper layer service represents the online network communication for establishing link between application entities so that agreement is reached on byte order, data types, and operations to be performed. The basic file service represents the offline media storage for secure and reliable data archiving and retrieval. The DICOM document Part 3 "Information Objects Definitions" (IODs) and Part 4 "Service Class Specifications" play very important roles in solving the problems of storing, retrieving, presenting biomedical data which will be further elaborated in Sect. 4.

4 Multidisciplinary DICOM Multimedia Archive

According to the DICOM standard, the IODs specify the common data format not only for medical images but also for graphics, waveforms, reports, audios, videos and printings. Manufacturers attempted to implement DICOM structured reporting for image-associated evidence and findings and also

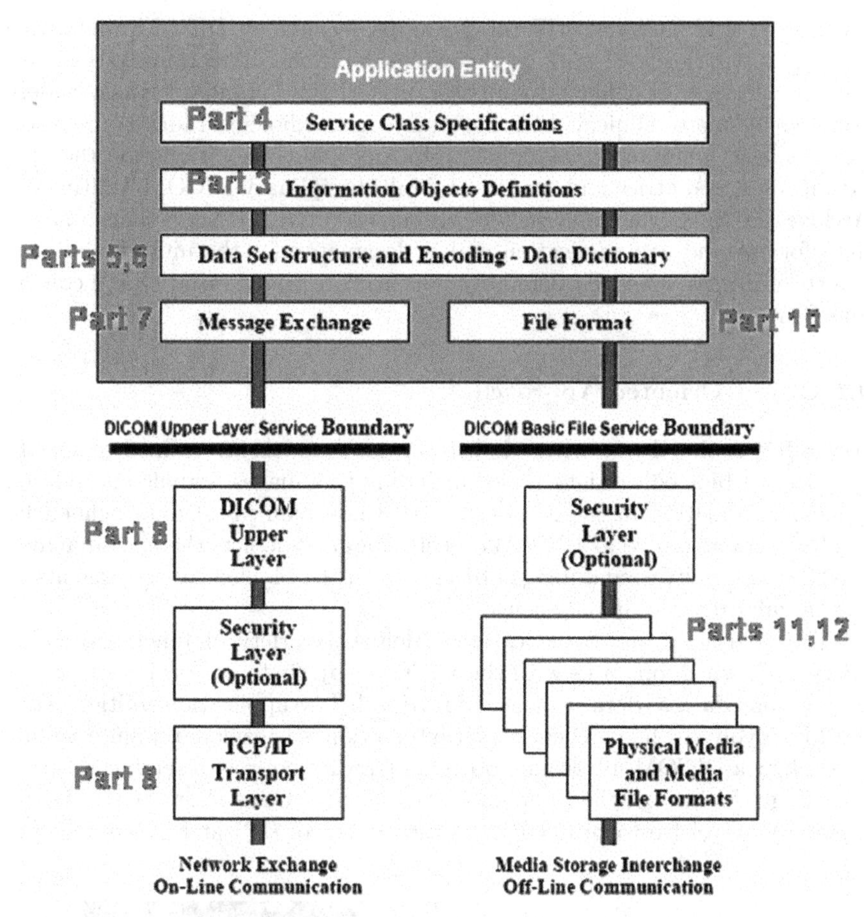

Fig. 2. General communication model for medical information application

the computer-aided diagnosis (CAD) results, so that these structured data can be accessed through the heterogeneous clinical information systems [15]. An object-oriented model was produced to represent the DICOM structured reporting in the web environment using the unified modeling language (UML) [16]. Integration scheme of DICOM objects was proposed to form a consistent framework for information interchange and system interoperability [13]. DICOM protocols have been implemented in a DICOM image storage server and a radiotherapy plan transfer facility to facilitate DICOM messaging and sequencing [10]. Any combination of these multimedia data is possible for the data collection in a multidisciplinary research or clinical application. A biomedical feature may be extracted from the complicated fusion of these data. The data interchange and linkage would be of very high importance in the feature extraction. The DICOM committee oversees continuously the evolution of the standard to fulfill the need for health information integration and exchange. Working groups formed by DICOM committee perform the majority of work on the extension of and corrections to the standard in a variety of clinical disciplines. The DICOM standard is now widely adopted by many clinical disciplines such as cardiology, radiology, pathology, dentistry, endoscopy, mammography, ophthalmology, radiation therapy, orthopedics, pediatrics and surgery. A Multidisciplinary DICOM Multimedia Archive (MDMA) that inherits the properties of DICOM compliance in the data format and communication and is designated for the multidisciplinary health problems would be definitely beneficial to the translational medicine and CDS.

4.1 Object-Oriented Approach

In the IODs, the object-oriented abstract data models are defined to specify a variety of biomedical data types, including the images, signals and related clinical information in the real-world. All these biomedical data defined by the IODs are regarded as DICOM objects. Figure 3 depicts the hemodynamic waveform and ultrasound image objects which the biomedical assessments of the vascular diseases mainly concern.

The "Service Class Specifications" define a variety of functions, called "Services", which operates with the DICOM objects to perform client-server interactions, such as data storage and retrieval. Two application entities (AEs) shall be involved when a client-server interaction is considered. Figure 4 illustrates how a DICOM ultrasound object is transferred from a sending AE to a receiving AE through the layers of storage Service Class User (SCU), Transmission Control Protocol (TCP)/Internet Protocol (IP) at the sending AE,

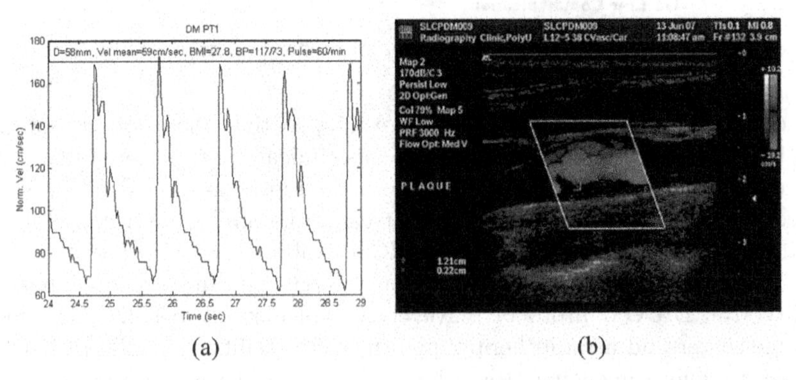

Fig. 3. Two typical DICOM objects for the assessments of vascular diseases: (**a**) Heamodynamic (HD) waveform object stores the cerebral arterial blood flow velocity acquired by Trans-cranial Doppler (TCD) ultrasound; (**b**) Color flow Doppler with B-mode Ultrasound (US) image object stores the blood flow velocity distribution overlay with the B-mode image to highlight the presence of an atheroma

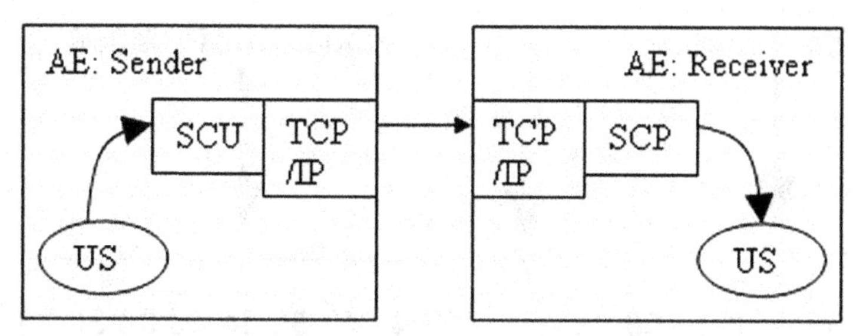

Fig. 4. DICOM storage service class user (SCU) and service class provider (SCP) and their roles in DICOM object transfer

TCP/IP at the receiving AE and then storage Service Class Provider (SCP). The SCU is responsible for getting the DICOM object from locally mounted storage media and then sending the storage request to the SCP of the receiving AE through the TCP/IP layers. The SCP is responsible for accepting association with the authorized sending AE and sending back the acknowledgment message to the SCU to initialize the data transfer. The object is segmented into numerous packets and notifications are required to indicate the completion of transferring one packet and start the transfer of the subsequent packets. The association will continue unless the packets of an object are completely received at the receiving AE. This will ensure the data integrity of the required DICOM object when running a DICOM storage service.

The combination of a service and a particular type of object is referred to as a service-object-pair (SOP). A Service Class Specification defines a group of one or more SOP Classes related to a specific function which is to be accomplished by communicating Application Entities. A specification also defines rules which allow implementations to state some pre-defined level of conformance to one or more SOP Classes. Implementation of SOP exhibits the major features of the object-oriented approach, making the complicated functions, such as image fusion and 3D volume rendering, possible in the clinical applications. Section 4.2 will elaborate more on three important properties of SOP.

4.2 Properties of DICOM Objects and Services

The adoption of DICOM object-oriented approach ensures the data integrity and portability and the quality data management service in biomedical knowledge discovery and CDS. The DICOM objects preserve the properties of closely-packed byte stream, self-contained patient demographic and study information, and cross-referencing between objects. The DICOM services preserve the properties of systematic indexing and seamless data archiving and retrieval.

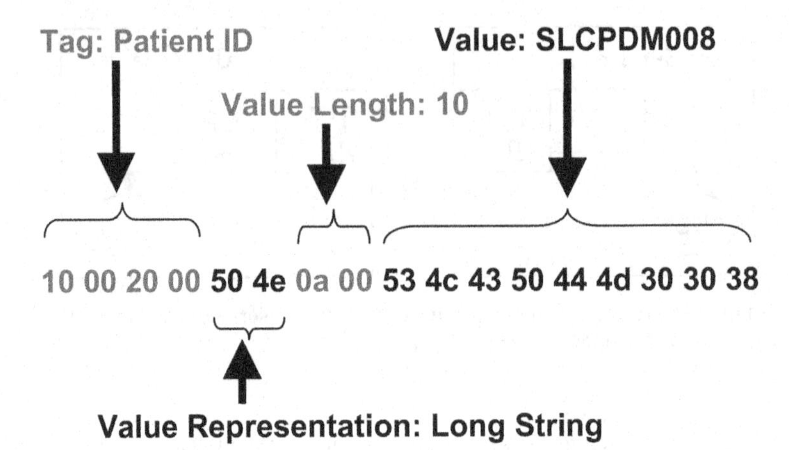

Fig. 5. A sample byte stream in a DICOM objects presenting the Patient ID

Closely-Packed Byte Stream

A DICOM object is concatenation of byte streams presenting data items one by one. Figure 5 shows a sample byte stream presenting the Patient ID in hexadecimal digits. The byte streams are usually displayed in hexadecimal notation where each byte is represented by two hexadecimal digits. To present a data item, a typical byte stream is comprised of DICOM tag, value representation, value length and item value.

Any meaningless redundant bytes don't exist in the byte streams. Moreover, the number of bytes for DICOM tag, value representation and value length is always fixed to 4, 2 and 2, respectively and the length of item value is defined by value length which equals to 10 in this sample data stream. Unlike Extensible Markup Language (XML), spreadsheet and plain text data format, any addition or removal of data in a DICOM object during the transmission can be easily detected. This property guarantees the data integrity within the DICOM objects.

Self-Contained Patient Demographic and Study Information

The data items within a DICOM object are categorized by groups. File Meta Elements (0002), Identifying (0008), Patient Information (0010), Acquisition (0018) and Image (0020) groups are typically incorporated in DICOM objects. The patient demographic profile including patient's name, sex, birth date and time, age, size, weight, address and patient ID is stored in the Patient Information (0010) group. General study information including study ID, description, date and time, and accession number is stored in the Identifying (0008) and Image (0020) groups. Note that nonimage objects, such as DICOM waveforms

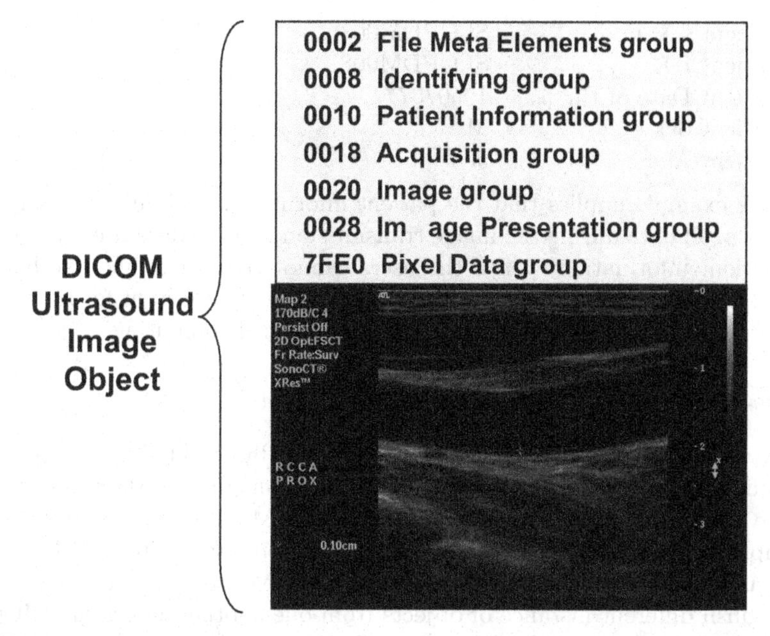

Fig. 6. Basic groups and pixel data in a DICOM ultrasound image object

```
30 21 00 00 00 00 10 00 10 00 0e 00 00 00 53 4c    0!............SL
43 50 44 4d 30 30 38 5e 5e 5e 5e 20 10 00 20 00    CPDM008^^^^ .. .
0a 00 00 00 53 4c 43 50 44 4d 30 30 38 20 10 00    ....SLCPDM008 ..
30 00 08 00 00 00 31 39 35 30 30 37 32 37 10 00    0.....19500727..
40 00 02 00 00 00 4d 20 10 00 00 10 00 00 00 00    @.....M ........
10 00 10 10 04 00 00 00 30 35 36 59 10 00 20 10    ........056Y.. .
```

Fig. 7. Byte streams of the patient information group in a DICOM object

and structured reports, also make use of some elements of the Image (0020) group which specify the signaling study or biomedical analysis result. Figure 6 illustrates a typical format of DICOM ultrasound image object where basic groups are presented in ascending order followed by the pixel data which are the value of the Pixel Data (7FE0) group.

The groups define detailed clinical information about the patient, study, acquisition method concerning the pixel data. This clinical information is self-contained in the DICOM object and independent of any database. The byte streams of the Patient Information group are illustrated in Fig. 7. The hexadecimal digits are shown on the left, with the corresponding ASCII characters on the right.

Note that the dots shown on the right are nonprinting codes of ASCII. These byte streams represent the following patient information.

Patient's Name	SLCPDM008....
Patient ID	SLCPDM008
Patient Date of Birth	1950/07/27
Patient Sex	M
Patient Age	56Y

This example implies that the patient information will automatically follow the pixel data during the image transmission and the data users will know the patient information even if the users cannot connect to the database of the originating information system. This property is important for processing data from different sources using distributed or grid computing.

Cross-Referencing Between DICOM Objects

Every DICOM object is uniquely identified by a SOP Instance UID (Unique Identifier) which is assigned by the imaging modality, signaling device or biomedical system when the DICOM object is produced, for example 1.2.840.113663.1100.156557158.3.1.120070608.1105632. UIDs guarantee uniqueness across multiple countries, sites, vendors and equipment and distinguish different instance of objects from one another across the DICOM universe. To ensure global uniqueness, all UIDs which are being used within the context of the DICOM Standard are registered values according to the definition of ISO 9834-3. The uses of UIDs are defined in the various Parts of the DICOM document.

Each UID consists of two parts, organization and suffix, represented by UID = <org root> . <suffix> where the org root portion of the UID uniquely identifies an organization, such as university, manufacturer, research institute, etc. and is composed of numeric components according to the definition of ISO 8824. The suffix portion of the UID also consists of numeric components, and has to be unique within the scope of the corresponding organization. This implies that the corresponding organization is responsible for guaranteeing the suffix uniqueness by providing registration policies which have to guarantee suffix uniqueness for all UIDs created by that organization.

DICOM structured report also has a special function to itemize the DICOM objects which will undergo observation, analysis or reporting in the same clinical case. Figure 8 illustrates the cross-referencing between the DICOM objects. Note that private UIDs are used in this example. The structured report with UID 1.2.3.00001.0 is regarded as an index of the DICOM objects in the same clinical case. The UIDs in item list are written into the structured report when the respective DICOM objects are created by the equipment and stored into the data archive. From the item list of that structured report, the data integrity of the case can be guaranteed during archiving and retrieval of the DICOM objects. When the data are being processed and analysis using grid or distributed computing, this item list can also act as a portable index to integrate the data irrespective of the database of the originating information system.

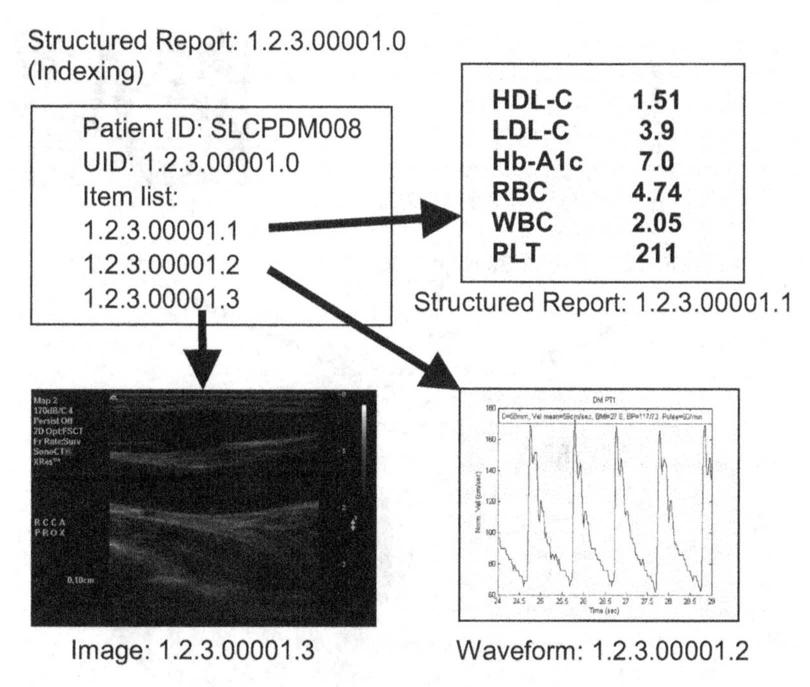

Fig. 8. Cross-referencing between DICOM image, waveform and SR

4.3 Design of MDMA

Through the above mentioned properties of SOP, biomedical data users of different clinical disciplines can obtain, seamlessly and completely integrate and then analyze the biomedical data generated from various data sources anytime and anywhere irrespective of the clinical information systems' databases. A conceptual data archive, referred to as Multidisciplinary DICOM Multimedia Archive (MDMA) is proposed in this chapter to realize the properties of SOP. The schematic diagram of MDMA is illustrated in Fig. 9.

The MDMA adopts the DICOM storage Service Class and query/retrieve Service Class as the elementary blocking blocks. The Storage SCU of the MDMA simply acts as the sender of the data source. Storage SCU is a common software module which has been already incorporated into the imaging modality, such as ultrasound machine. For the other data sources, it is easy to build a storage SCU module by making use of the open source libraries in the web, such as DICOM toolkit. The storage SCP is the receiver of the DICOM objects and triggers the creation of the indexing structured report which list out the UIDs of the archived DICOM objects in the study.

When the data users need to get a desired case from the MDMA, the query/retrieve Service Class will look for the corresponding indexing structured report, find out the UIDs of archived items and then send a complete case to the data users seamlessly. Note that the MDMA can operate without

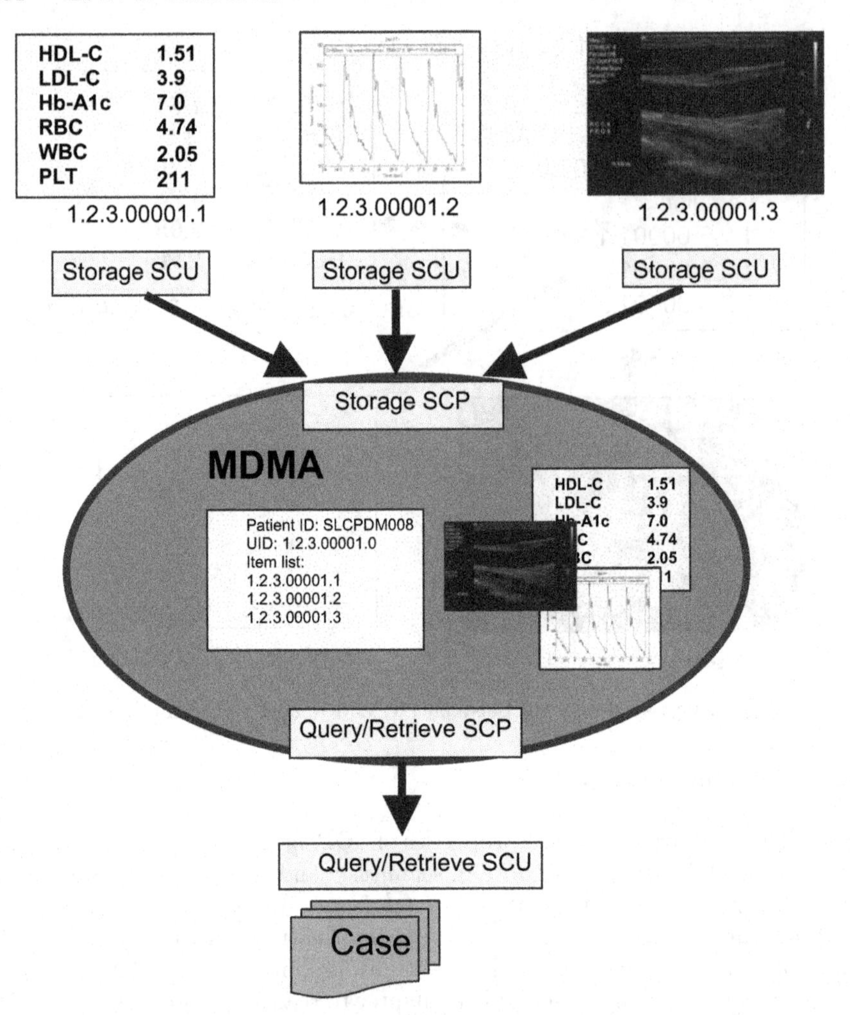

Fig. 9. Schematic diagram of MDMA

the aid of a database and the process is completely automatic without any human intervention.

5 Biomedical Data Processing

Data processing is an essential process in biomedical researches, including noise filtering, missing data handling, feature extraction and feature selection. The quality of data processing greatly affects the results of knowledge discovery and the accuracy of decision support. The extraction and selection of hemodynamic features are considered as an example of biomedical data processing in this section.

5.1 Biomedical Feature Extraction

Two hemodynamic features, Reynold number and percentage of fluid energy loss of blood flow and ratio of end-diastolic velocities, and one morphological feature, percentage of luminal diameter reduction, are derived here from the measurements using B-mode ultrasound, Trans-cranial Doppler ultrasound and manometer.

Reynold Number

Reynold number (Re) is dimensionless feature that indicates the condition of blood flow in arteries given by

$$Re = \frac{\rho d u}{\eta} \tag{1}$$

where ρ is the density of blood ($1\,\mathrm{g\ cm^{-3}}$), η is the viscosity of blood ($3.5 \times 10^{-3}\,\mathrm{Pa\ s}$), d is the diameter of lumen of artery and u is the mean of peak systolic velocity (PSV) across the lumen. Turbulence is most likely established when the Reynold number is greater than 2,000. The occurrences of turbulence can usually be found at the atheroma-prone sites, such as carotid bifurcation, vessel curvature and arterial branching. So, the Reynold number is considered as a hemodynamic feature of atherosclerosis.

Percentage of Energy Loss

According to the Bernoulli's principle, the total fluid energy along the blood vessel is conserved all the time. During systole, the Bernoulli equation is given by

$$P_0 = P_1 + \rho v^2/2 + L, \tag{2}$$

where L accounts for the sum of the inertial, viscous and turbulent losses along the blood vessels from the heart to the middle cerebral artery, P_0 is the systolic pressure, ρ is the density of blood, P_1 is the blood pressure at middle cerebral artery and v is the systolic blood velocity at middle cerebral artery. The value of P_1 is negligible when comparing with P_0. Thus, the percentage of energy loss ($\%EL$) along the blood vessels from the heart to a middle cerebral artery is given by

$$\%EL = \frac{L}{P_0} = 1 - \frac{\rho v^2}{2P_0}. \tag{3}$$

The stenosis caused by atheromas induces substantial energy loss at the artery. The value of $\%EL$ is potentially suitable for indicating the presence of atheromas indirectly and thus also a hemodynamic feature of atherosclerosis.

End Diastolic Velocity Ratio

Symmetry of hemodynamics could be diagnostic criteria for vascular diseases. Asymmetric blood flow implies high possibility of unilateral stenosis. End diastolic velocity (EDV) ratio in CCA was proved to be an additional indicator for predicting unilateral hemodynamically significant stenosis in ICA [11]. EDV ratio is defined as the side-by-side ratio of the end diastolic blood flow velocity at CCA given by

$$EDV\,ratio = \frac{\max\left(EDV_{LCCA}, EDV_{RCCA}\right)}{\min\left(EDV_{LCCA}, EDV_{RCCA}\right)}, \tag{4}$$

where EDV_{LCCA} and EDV_{RCCA} are the EDVs in left and right CCA, respectively.

Percentage of Luminal Diameter Reduction

Vascular laboratories make use of various criteria for the diagnosis of arterial disease, in which luminal diameter reduction is one of their standards [14]. The percentage of luminal diameter reduction ($\%LDR$) is defined by the following equation

$$\%LDR = \frac{h_a}{d}, \tag{5}$$

where h_a and d are the thickness of atheroma causing the reduction of luminal diameter and the radiologist's estimate of the original luminal diameter at the site of atheroma, respectively. If atheroma is not found, $\%LDR$ will be assigned a zero value.

5.2 Biomedical Feature Selection

Multidisciplinary health studies normally consider a large number of biomedical features which may potentially relates to the degree or likelihood of the disease development. In an ongoing study of carotid and cerebral atherosclerosis, over 50 different measured values and derived features are considered and forming a high dimensional feature space. Statistical test among these candidate features can be performed to identify the key biomedical features and reduce the dimensionality of feature space for further analysis.

In the study, two-tailed Pearson's test was applied to find out the correlation (r) and its significance (p) between the extracted hemodynamic, hematological and biochemical features and the percentage of luminal diameter reduction ($\%LDR$). The standard Pearson Product–Moment correlation r [5] is defined by the following formula.

$$r = \sum \frac{\left(X_i - \bar{X}\right)\left(Y_i - \bar{Y}\right)}{nS_XS_Y}, \tag{6}$$

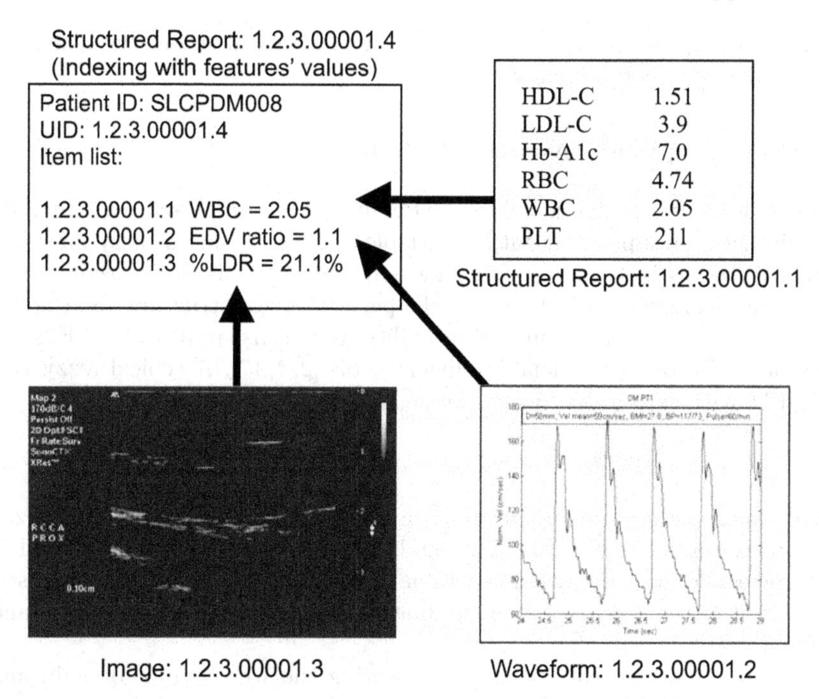

Fig. 10. Indexing DICOM SR for storing the values of the selected features

where $\bar{X}$ and $\bar{Y}$ are the sample mean of X and Y, the two variables to be correlated; S_X and S_Y, the sample standard deviation of X and Y; n is the number of (X, Y) pairs. The Bivariate Correlations procedure was used to computes the coefficient. Biomedical features with high correlation with $\%LDR$ yet reaching high significant levels are considered to be key features. The White Blood Cell (WBC) count and the EDV ratio are selected through the statistical test. The values of the selected features can be further written into the indexing structured report (SR) and then stored into the MDMA as a new DICOM SR object, as shown in Fig. 10.

The mapping between the values of the selected features and the UID of the corresponding DICOM object makes the selected features self-contained in the multidisciplinary case without any explicit linkage with database.

6 Biomedical Knowledge Discovery

The indexing DICOM SR is retrieved from the MDMA to supply the values of the selected features for the process of knowledge discovery. Analytical model can be employed to study the etiology and pathogenesis of diseases. In the study of carotid and cerebral atherosclerosis, Fuzzy Inference System is considered as a tool to establish the analytical model for the formulation of the relationship between the selected biomedical features and the degree of stenosis due to atherosclerosis.

6.1 Multidisciplinary Analytical Model

Fuzzy system is comprised of linguistic fuzzy rules representing the relationship between input and output variables. Adaptive neuro-fuzzy inference system (ANFIS) is a kind of fuzzy system where the input is fuzzified but the output is crisply defined. Due to the parameterized structure and the linguistic representation, the modeling ability and transparency of ANFIS are superior to the other artificial intelligent tools [2, 4, 17]. A typical fuzzy rule in ANFIS with zero-order Sugeno mode has the following form.

$$\text{Rule } i: \text{ If } x \text{ is } A_i \text{ and } y \text{ is } B_i \text{ then } z = k_i$$

where x and y are input variables, A_i and B_i are their corresponding fuzzy sets, for example "Low", "Medium" and "High", z is output variable, and k_i is its corresponding crisply defined value of the ith fuzzy rule. Such linguistic representation can express any nonlinear relationship between input and output variables easily.

By integrating all the fuzzy rules covering the whole input domain and output range, ANFIS can be formulated to a network structure comprised of input and output membership functions and the logical operators between them. The network structure of ANFIS, shown in Fig. 11, inherits the modeling

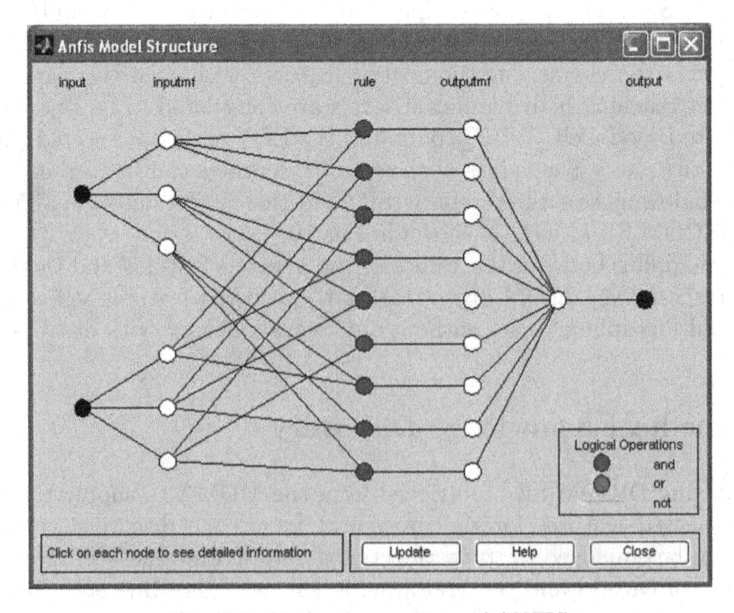

Fig. 11. Network structure of ANFIS

ability of neural networks which are parameterized for further adjustment by training data.

Fuzzy logic toolbox of Matlab® is used to build and train the ANFIS in the study. The key features selected from the biochemical, hematological and hemodynamic features in the statistical test were the input of ANFIS. The output was the percentage of luminal diameter reduction. The ANFIS represents an analytic model formulating the relationship between the selected key features and the degree of atherosclerosis. The following fuzzy rules are obtained by adjusting the parameters of the ANFIS using the available feature values.

1. If WBC is ***Low*** and EDV ratio is ***Low*** Then Stenosis is ***N.A***.
2. If WBC is ***Low*** and EDV ratio is ***Med*** Then Stenosis is ***N.A***.
3. If WBC is ***Low*** and EDV ratio is ***High*** Then Stenosis is ***Mild***
4. If WBC is ***Med*** and EDV ratio is ***Low*** Then Stenosis is ***N.A***.
5. If WBC is ***Med*** and EDV ratio is ***Med*** Then Stenosis is ***N.A***.
6. If WBC is ***Med*** and EDV ratio is ***High*** Then Stenosis is ***N.A***.
8. If WBC is ***High*** and EDV ratio is ***Med*** Then Stenosis is ***N.A***.
9. If WBC is ***High*** and EDV ratio is ***High*** Then Stenosis is ***Mild***

This rule base can be incorporated into the CDS software to assess the risk of atherosclerotic stenosis based on the values of WBC count and EDV ratio. Note that the study is still ongoing and the collected data may not be sufficient to identify the significant associated features for atherosclerosis. However, this would be a good example to demonstrate the use of artificial intelligence to discover knowledge based on the available biomedical data.

7 Synergistic Clinical Decision Support Platform

With the advance of information technology, it is feasible to plan and construct a synergistic platform, referred to as DICOM-based multidisciplinary platform (DMP), to facilitate the archiving and communication of DICOM objects generated from various clinical disciplines. Figure 12 illustrates a conceptual overview and the clinical operations of the platform.

Most of the biomedical imaging modalities are intrinsically able to produce DICOM images. According to the IOD of DICOM standard, proprietary-to-DICOM converters can be built to convert biomedical videos, signals and text documents into DICOM objects. These DICOM objects can be sent to the MDMA through the DICOM storage service class and meanwhile an indexing SR is created by the MDMA. Through the DICOM query/retrieve service class, a complete multidisciplinary case can be retrieved from the MDMA with the aid of the indexing SR.

The biomedical features to be extracted are determined by the a priori knowledge of the clinicians or researchers. The candidate features are computed using the data stored in different DICOM objects and the key features

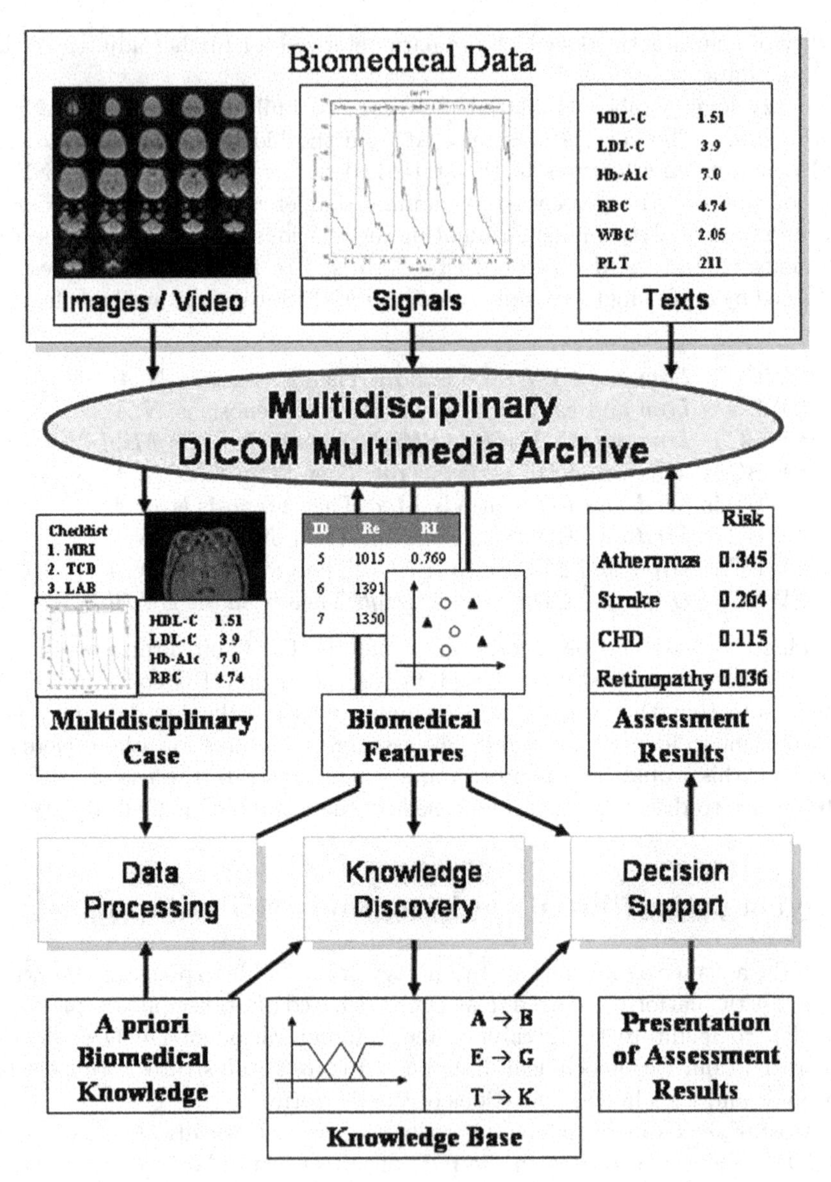

Fig. 12. An overview of the clinical decision support platform

can be selected using the statistical tests. The values of the selected features are then written into the indexing SR and a new version of indexing SR is produced and sent back to the MDMA for record. Analytical model supported by the artificial intelligent tool can be used to formulate the relationship between the biomedical features and the available training dataset can be used to adjust the parameters of the analytical model. The resulting rule base or knowledge base can be employed in the CDS to assess the risk or the likelihood of the development of the disease.

8 Conclusion and New Direction

A DMP for CDS is proposed in this chapter to solve the common problems in conducting multidisciplinary health studies. The unification of data standard allows remarkable improvements in the management of tremendously large data volume and the analysis of multidisciplinary factors for a disease. The proposed platform provides a synergistic framework for knowledge discovery in etiology and pathogenesis. As the explosion of biomedical data volume and feature space becomes very common in the studies of systemic diseases, the grid computing would be a possible solution to strengthen the computing power for the data analysis. Moreover, the data sources could be in different geographic locations which are far apart from each other. The ubiquitous computing technology would be employed for the remote data processing of the portable multidisciplinary cases.

References

1. Anderson NR, Lee ES, Brockenbrough JS, Minie ME, Fuller S, Brinkley J, Tarczy-Hornoch P (2007) Issues in biomedical research data management and analysis: needs and barriers. J. Am. Med. Inform. Assoc. 14:478–488
2. Brown M (1994) Neurofuzzy adaptive modelling and control. New York: Prentice Hall
3. Brownlee M (2001) Biochemistry and molecular cell biology of diabetic complications. Nature 414:813–820
4. Chan CW, Chan WC, Jayawardena AW, Harris CJ (2002) Structure selection of neurofuzzy networks based on support vector regression. International Journal of Systems Science 33(9):715–722
5. Chen PY, Popovich PM (2002) Correlation – parametric and nonparametric measures. Thousand Oaks, CA, Sage
6. Cheng JCY, Guo X, Sher AHL, Chan YL, Metreweli C (1999) Correlation between curve severity, somatosensory evoked potentials, and magnetic resonance Imaging in adolescent idiopathic scoliosis. Spine 24(16):1679–1684
7. Gatenby RA, Gawlinski ET, Gmitro AF, Kaylor B, Gillies RJ (2006) Acid-mediated tumor invasion: a multidisciplinary study. Cancer Res. 66(10):5216–5223
8. Goldschmidt-Clermont PJ, Creager MA, Losordo DW, Lam GKW, Wassef M, Dzau VJ (2005) Atherosclerosis 2005: recent discoveries and novel hypotheses. Circulation 112:3348–3353
9. Herbort CP, Cimino L, El Asrar AMA (2005) Ocular vasculitis: a multidisciplinary approach. Curr. Opin. Rheumatol. 17(1):25–33
10. Kalet IJ, Giansiracusa RS, Jacky J, Avitan D (2003) A declarative implementation of the DICOM-3 network protocol. J. Biomed. Informat. 36:159–176

11. Kamouchi M, Kishikawa K, Okada Y, Inoue T, Ibayashi S, Iida M (2005) Reappraisal of flow velocity ratio in common carotid artery to predict hemodynamic change in carotid stenosis. Am. J. Neuroradiol. 26(4):957–962
12. Marcus F, Towbin JA, Zareba W, Moss A, Calkins H, Brown M, Gear K (2003) Arrhythmogenic right ventricular dysplasia/cardiomyopathy (ARVD/C): a multidisciplinary study: design and protocol. Circulation 107(23):2975–2978
13. Moreno RA, Furuie SS (2002) Integration of heterogeneous systems via DICOM and the clinical image access service. Computers Cardiol. 29:385–388
14. Neumyer MM (2004) Vascular technology. In: Curry RA, Tempkin BB (eds) Sonography: Introduction to normal structure and function. St. Louis, MO, Saunders, 480–496
15. Sluis D, Lee KP, Mankovich N (2002) DICOM SR – integrating structured data into clinical information systems. Medicamundi 46(2):31–36
16. Tirado-Ramos A, Hu J, Lee KP (2002) Information object definition-based Unified Modeling Language representation of DICOM structured reporting: a case study of transcoding DICOM to XML. J. Am. Med. Inform. Assoc. 9(1):63–71
17. Ubeyli ED, Guler I (2005) Adaptive neuro-fuzzy inference systems for analysis of internal carotid arterial Doppler signals. Comput. Biol. Med. 35(8):687–702
18. Wang HH, Wang XF (2006) Analytical methods of atherosclerosis research. In: Schoenhagen M (ed.) Current developments in atherosclerosis research. Hauppauge, NY, Nova Science, 33–66
19. Wolfs CAG, Dirksen CD, Severens JL, Verhey FRJ (2006) The added value of a multidisciplinary approach in diagnosing dementia: a review. Int. J. Geriatr. Psychiatry 21:223–232
20. Yoerger DM, Marcus F, Sherrill D, Calkins H, Towbin JA, Zareba W, Picard MH (2005) Echocardiographic findings in patients meeting task force criteria for arrhythmogenic right ventricular dysplasia: new insights from the multidisciplinary study of right ventricular dysplasia. J. Am. Coll. Cardiol. 45(6):860–865

Improving Neural Network Promoter Prediction by Exploiting the Lengths of Coding and Non-Coding Sequences

Rachel Caldwell[1], Yun Dai[2], Sheenal Srivastava[2], Yan-Xia Lin[2], and Ren Zhang[1]

[1] School of Biological Science, University of Wollongong, Australia
[2] School of Mathematics and Applied Statistics, University of Wollongong, Australia

1 Introduction

Since the release of the first draft of the entire human DNA sequence in 2001, researchers have been inspired to continue with sequencing many other organisms. This has lead to the creation of comprehensive sequencing libraries which are available for intensive study. The most recent genome to be sequenced has been the gray, short-tailed opossum (*Monodelphis domestica*). This metatherian[1] ("marsupial") species, which is the first of its type to be sequenced, may offer researchers an insight not only into the evolution of mammalian genomes in respect to the architecture and functional organization, but may also tender an understanding in the human genome [12].

Much attention within computational biology research has focused on identifying gene products and locations from experimentally obtained DNA sequences. The use of promoter sequence prediction and positions of the transcription start sites can inevitably facilitate the process of gene finding in DNA sequences. This can be more beneficial if the organisms of interest are higher eukaryotes, where the coding regions of the genes are situated in an expanse of non-coding DNA.

With the genomes of numerous organisms now completely sequenced, there is a potential to gain invaluable biological information from these sequences. Computational prediction of promoters from the nucleotide sequences is one of the most attractive topics in sequence analysis today. Current promoter prediction algorithms employ several gene features for prediction. These attributes include homology with known promoters, the presence of particular motifs within the sequence, DNA structural characteristics and the relative signatures of different regions in the sequence.

[1] An infraclass of mammals containing the marsupials.

R. Caldwell et al.: *Improving Neural Network Promoter Prediction by Exploiting the Lengths of Coding and Non-Coding Sequences*, Studies in Computational Intelligence (SCI) **116**, 213–230 (2008)
www.springerlink.com

1.1 Currently Used Algorithms

Different algorithms have been developed which vary in performance and can be categorized into two main groups. The first depends upon recognition of conserved signals such as the TATA box and the CCAAT box as well as the spacing between patterns. This approach uses either the neural network genetic algorithm or the weight matrix methodology. The second relies on identification of promoters within a sequence that may contain the elements. This approach is content-based and distinguishes differences such as triplet base-pair preferences around the transcription start site (TSS), and hexamer frequencies in consecutive 100-bp upstream regions [15]. There are also techniques that combine both these methods, which look for signals and for regions of specific compositions [13].

Many promoter prediction programs are readily available to the scientific community to utilize and explore. The programs that presented relatively high accuracy in their results include the GeneID/Promoter 2.0, TSSW, PromoterInspector and the Neural Network for Promoter Prediction (NNPP) [2,5,9].

Currently, the neural network algorithm is probably the most widely used program in promoter prediction (http://www.fruitfly.org/seq_tools/promoter.html). It is based on a time-delay neural network (TDNN) architecture that originated from speech recognition sequence patterns in time series. This method corresponds to how the brain's learning process operates. What makes this system unique is that it has the advantage of learning to recognize the degenerate patterns that characterize promoter motifs. The algorithm was initially designed for predicting promoters in the *Drosophila* genome and it has been developed to be a common method used to find both eukaryotic and prokaryotic promoters. The NNPP 2.2 algorithm recognizes only the presence and relative location of patterns and motifs within a promoter. It predicts the probability that a tested sequence position $s \pm 3$ base pairs (bp) contains a true TSS denoted by $P(s \in S)$, where S is the class of the true TSS positions [2].

The popularity of NNPP has also been supported by comparative studies. An investigation by Fickett and Hatzigeorgiou (1997) [5] recognized 13 of the 24 promoters (54%) in the test data set by NNPP and 31 false positives (1/1068 bp) were reported. These were significantly better than the outcomes of GeneID/Promoter 1.0 which identified 42% of the promoters and 51 false positives (1/649 bp) and the TSSW program (42% of true promoters and 42 false positives (1/789 bp)). Reese [17] found similar results on the *Drosophila* genome, with a rate of 75% (69/92) of recognition and a rate of 1/547 bases of false positives.

1.2 Further Improvements in Promoter Prediction

Current algorithms to predict promoters are still far from satisfactory. The challenge that occurs in proposing a high level of prediction of promoters, with a reasonable percentage predicted, is that the level of falsely predicted promoters, known as false positives (FPs), is also high when a large percentage of predictions are met.

Another challenge faced which makes prediction difficult is that promoters are very diverse, and even some well-known signals such as TATA box and CCAAT box are not always conserved in all promoters. The TATA box can only be found in ~75% of vertebrates RNA PolII promoters and the CCAAT box is only found in half of vertebrate promoters [16]. Detectable motifs that exist within promoters can also occur randomly throughout the genome creating additional complications [2]. Promoters are defined based on functionality rather than structure, causing major impediments in creating near perfect predictions [14]. The promoter recognition systems for large-scale screening require acceptable ratios of true positives (TPs) and false positive predictions (i.e., those that maximize the TP recognition while minimize the FP recognition).

What currently is required out of these algorithms is the reduction in false positives in respect to promoter prediction. To achieve this it is possible to develop powerful computational methods and to replace current computational promoter prediction procedures. These approaches can be beneficial in increasing the accuracy of promoter prediction, and these changes are not restricted to just computational modifications. One approach in addressing these limitations is to investigate if the outcome of promoter prediction based on current techniques can be improved by incorporating additional information from the underlying DNA sequence.

The influx of DNA sequences, now publicly available, has allowed more and more information to be extracted. This has given computer and mathematical scientists the opportunity to run statistical analysis on this added information. The information gained will increase the understanding of the statistical behavior of promoter positions for different genes across species. While much information can be integrated into any computational promoter prediction algorithms, our approach has been to exploit the distance information between gene elements. Our study on *E. coli* [2] was the first to investigate the use of the distance between TSS and TLS to improve the NNPP2.2 promoter prediction accuracy rate. Analysis and information retrieval performed by computers, particularly when dealing with large data sets has been an important tool for biologists. The information gained by these computations can guide biologists more efficiently in identifying areas of the DNA sequence experimentally infeasible without this data [1].

This paper will summarize the TLS-NNPP approach and further extend the basic idea of the TLS-NNPP to more general circumstances with our more recent research results. The aim of this paper is to firstly demonstrate why and how some measurements in DNA sequences can be used to significantly improve computational promoter prediction. And secondly it is intended to bring researcher's attention to the DNA sequence information which is released through DNA sequence quantitative measurements instead of DNA sequence pattern information. For simplicity reasons, the research will only focus on the NNPP computational method as a reference method and demonstrate how DNA sequence quantitative measurements can be used to improve the promoter prediction of NNPP2.2. The technique discussed in this paper can be easily integrated with other computational promoter prediction algorithms by some minor modifications.

2 Gene Expression

A gene, as illustrated in Fig. 1, is the basic functional unit of genetic material in all living organisms. Any fragment of DNA which can be transcribed in a cell into RNA is called a gene. Those genes coding for proteins are called structural genes and consist of a protein coding region, also known as the coding sequence (CDS) between the translation start site (TLS) and translation stop (TSC), and its 5′ upstream and 3′ downstream non-coding regions. For RNA synthesis, at least one transcription start site (TSS) and terminator site (TTS) are located in the upstream and downstream regions, respectively. The whole region between the TSS and the TTS is transcribed to RNA when a gene is expressed but there may be certain parts (introns[2]) which will be deleted (spliced) during RNA processing in eukaryotes.

All organisms are reliant on the regulation of gene transcription for their function and development. Gene expression is regulated firstly by the initiation of transcription, which generally signifies the most important point of control. In the process of transcription initiation, sets of genes can be turned on or off, determining each cell type, in response to different internal and external cues. The importance of transcriptional control is also associated with all forms of diseases, including cancer which is the improper regulation of the transcription of genes involved in cell growth [7, 8, 15]. Therefore, accurate prediction methods and understanding of these regions can be beneficial in human health in addition to computational biology.

The regulation of gene expression involves a complex molecular network with DNA-binding transcription factors (TFs) being an important element in this network. The level of complexity varies among prokaryotic[3] and eukaryotic[4] organisms. Most prokaryotes are unicellular organisms and promoters are recognized directly by RNA Polymerases, however eukaryotic organisms are more complex with the recognition of promoters consisting of large numbers of transcription control elements. One of the most complex processes found in molecular biology is the function of the promoter in transcription initiation. Promoters contain the nucleotide sequences which indicate the

[2] A nucleotide sequence in a gene that does not code for amino acids.

[3] Any organism belonging to the kingdom Prokaryotae in which genetic material is not enclosed in a cell nucleus.

[4] An organism consisting of cells in which the genetic material is contained within a distinct nucleus.

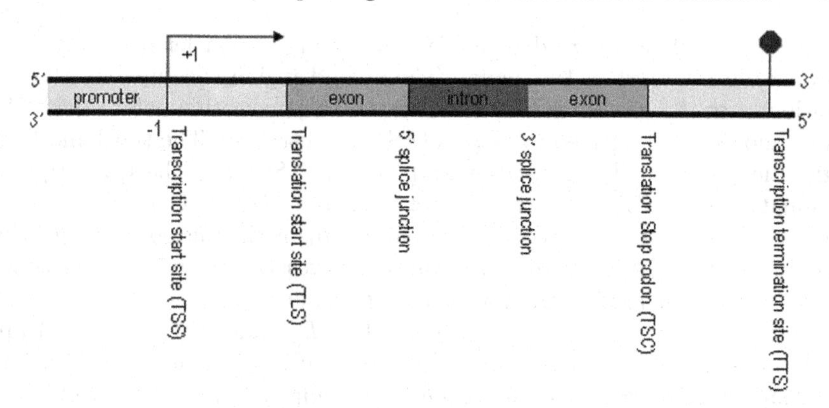

Fig. 1. Diagrammatic illustration of a structural gene

starting point for RNA synthesis. The promoter is positioned within the non-coding region upstream from the transcription start site which is referred to as the +1 position.

Apart from regulatory elements, other attributes of a gene such as its nucleotide composition, length, location (proximity to neighbours) and orientation may also play vital roles in gene expression. Genome size contrasts from organism to organism, and it appears that this divergence correlates with gene length variation. Recent studies examining protein length distributions in several species of eukaryotes and prokaryotes found that there was a positive correlation between average protein lengths and genome complexity [20, 22]. Chiaromonte et al. [3] have found that using distance between neighbouring genes and the lengths of the 3′ untranslated regions, highly expressed genes had greater distances. This research infers that length and distance between genes and their corresponding untranslated regions has important implications in gene expression.

Most studies have just focused on protein length, or the length of its gene coding region (see [10, 20, 21], and reference therein). Little is known about the length distributions of other gene regions and nothing is known of the correlation between gene components among different organisms. Our previous studies have showed that the TSS-TLS distance distribution varies among different organisms and each organism has its own specific character. It seems that, in general, the TSS-TLS distance increases gradually from simple prokaryotic to more complicated eukaryotic organisms [4].

3 Statistical Characteristics on Quantitative Measurements

The gene length can be divided into three sections, and for the purpose of this research the introns were included for each section. As shown in Fig. 1, the first region is situated between the TLS and the translation stop codon (TSC).

This region will be referred to as D_1, or coding region length (TLS-TSC distance) for the rest of this paper. The second region encompasses the +1 position after the promoter (the TSS) to the last nucleotide before the TLS. This region will be referred to as D_2 (TSS-TLS distance). The third and final region lies between the translation stop codon (TSC) and the transcription termination site (TTS), and will be referred to as D_3 (TSC-TTS distance). The distances were measured in base pairs (bp) of the nucleotide sequence. This information is just one of several attributes that could be utilized to improve promoter prediction in a variety of organisms.

The distances, TLS-TSC (D_1), TSS-TLS (D_2) and TSC-TTS (D_3) are varied, and can be considered as random components in gene sequences. The intention of this research is to contend that empirical information of these random components can benefit promoter prediction. Therefore, the aim of this research is to integrate this information with existing computational promoter prediction algorithms, and show that it will provide power to improve the prediction results. To understand why this information might help to improve computational promoter prediction, it is necessary to know the probability structure of these random components and see how much information is involved. Several model species ranging from bacteria to mammals will be used in this research to exploit the statistical information involved in the data. The species involved include *Escherichia coli* and *Bacillus subtilis* (bacterium), *Saccharomyces cerevisiae* (yeast), *Arobidopsis thaliana* (plant), *Mus musculus* and *Homo sapiens* (mammals).

To obtain the TLS-TSC (D_1) and TSS-TLS (D_2) distances, numerous databases were explored to determine absolute TSS, TLS and TSC positions on the various species genomes. The species chosen represent several model organisms that have been studied extensively, and possess a large amount of experimental data available to the public. The species were also chosen as they characterize a range of different classes, ranging from very simple organisms such as bacteria and yeast to the higher organisms such as the mammals.

TSS information was obtained from various databases, depending on the experimental research that had been conducted for each organism. The TSS information for *E. coli* was obtained from RegulonDB [18], *B. subtilis* data were obtained from the DBTBS database [11], SCPD for *S. cerevisiae* [23], TAIR for *A. thaliana* [6] and DBTSS version 5.1.0 for both *M. musculus* and *H. sapiens* [19]. Each of the TSS positions was considered to be positioned at multiple locations in a gene, thereby allowing multiple TSS-TLS distances to be generated. The D_1 data was extracted from protein table files from the NCBI database.

In prokaryotes, the existence of operons is highly common. Therefore, in cases such as this, we regard the genes that are organized in one operon and controlled by the same promoter as separate gene units. Thus a single TSS-TLS distance may correspond to more than one coding region.

The statistical summary on the distances given by the six species was produced by the statistical package SPSS 12.0/15.0 and are presented in Table 1.

Table 1. Statistics of the distances (bp) of D_1 and D_2

Species	TLS-TSC distance (D_1)			TSS-TLS distance (D_2)		
	Sample size	Median	Mean	Sample size	Median	Mean
E. coli	4,237	846	954	1,017	66	164
B. subtilis	4,015	771	896	483	67	93
S. cerevisiae	5,850	1,233	1,503	202	68	110
A. thaliana	30,480	1,623	1,939	20,560	112	213
M. musculus	27,132	10,054	34,552	14,520	378	10,913
H. sapiens	14,796	16,339	45,445	14,588	809	15,291

The mean and median of each species was calculated for the distances between the TSS-TLS and TLS-TSC. The median was used for its simplicity and is not severely affected by extreme values (outliers) as is the mean value. Since the TSS positions have not been experimentally verified for all genes in an organism's genome, the sample size of D_2 is relatively smaller as compared to D_1.

The summary shows that the means of D_1 and D_2 are increasing as the species moves from a relatively simple organism to a more complex organism. The distance between mean and median is also increasing as the species becomes more complex. This denotes that the distribution of both D_1 and D_2 are skewed to the left and exhibits a very long right tail. Accordingly, the data indicates that in the simple organisms such as bacteria, there is a higher likelihood that they have short D_1 and D_2 distances than in the more complex species. It is important to note that even in different species within the eukaryotic and prokaryotic kingdoms there could be differences in the probability distributions for the distance components.

Considering the joint relationship of D_1 and D_2 for the six species, there is more information to be gained. The following two-dimensional scatter plots (Fig. 2) of D_1 vs. D_2 for *E. coli* and *H. sapiens* shows that the correlation between these distances is varied from species to species. The bacterium species D_1 value tended to be smaller and appears that compared with the D_2 value would not change to a great extent. However the plot for the *H. sapiens* illustrates different trends. The D_1 value declined in a different region on the plot and therefore made the distribution of D_2 look different. According to the data of the six species, the research found the more complex a species, the stronger the correlation between D_1 and D_2.

To explorer the relationship between the TSS-TLS and TLS-TSC distances, and to ascertain whether there is a certain level of impact from the D_1 value on the probability distribution of D_2, the complete *H. sapiens* and *M musculus* data sets were used. The higher organisms were chosen due to the higher correlation between these components found in the comparison above. The data was divided into four groups based on the quartiles of the D_1 values. The first group consisted of all D_1 values to the first quartile, the second of

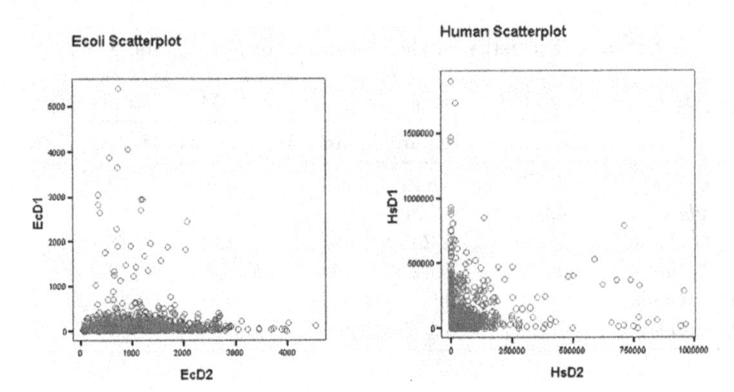

Fig. 2. Scatter plots of D_1 vs. D_2 of *E. coli* and *H. sapiens*

all D_1 values from the first quartile to the median, the third was made up of all D_1 values from the median to the third quartile, and finally the last group consisted of all D_1 values from the third quartile to the maximum D_1 value.

To characterize the location and variability of a data set, the skewness and kurtosis can be used for statistical analysis purposes. Skewness measures the lack of symmetry in a distribution, where the kurtosis describes the data as either peaked or flat relative to a normal distribution.

Table 2 clearly shows that, given D_1 declining into a different region, the associated random component D_2 had significantly different probability distribution. Therefore, the larger the value of D_1 the higher the correlation between D_1 given D_2. Since it is relatively easy to identify D_1 distances from the DNA sequence, with this information, the random component D_2 might show a different portion of information about DNA sequences.

Statistically, there is a great deal of potential to extract information from the D_1 and D_2 data however this research will not delve into every aspect. The purpose of the research is to highlight that different species might have different probability structure on their random components. Therefore, the information of D_1 and D_2 which is related to the distance of the TSS-TLS and could be referenced to the promoter position. Currently many computational promoter predictor algorithms do not take into account the information of D_1 and D_2. This information can be utilized to improve computational promoter prediction results and is discussed below.

4 The Algorithms for TLS-NNPP and TSC-TSS-NNPP

In this section, two algorithms using the information of D_1 and D_2 will be used to demonstrate that these random components can improve the NNPP performance. The first modification will incorporate the D_2 distance values and is called the TLS-NNPP algorithm [2, 4]. The other algorithm is known as the TSC-TSS-NNPP algorithm and uses both D_1 and D_2 values.

Table 2. Statistics of TSS-TLS distances (D_2) given D_1 in different ranges

	Group 1	Group 2	Group 3	Group 4
H. sapiens				
Sample size	2,789	2,789	2,789	2,788
Mean	8,424.28	9,984.15	13,497.72	28,400.62
Median	758	600	816	1,194.5
Std. deviation	40,953.08	48,848.37	45,191.05	73,803.25
Skewness	18.955	12.743	10.812	6.277
Std. error of skewness	0.0464	0.0464	0.0464	0.0464
Kurtosis	470.920	191.664	162.138	54.413
Std. error of kurtosis	0.093	0.093	0.0923	0.093
Minimum	1	1	1	1
Maximum	1,261,540	945,008	967,810	963,680
Pearson correlation	−0.040594	0.0231855	0.01539775	0.15475841
M. musculus				
Sample size	4,717	2,579	1,652	5,640
Mean	6,238.84	6,250.67	8,012.67	13,302.45
Median	422	267	385.5	399.5
Std. deviation	36,156.00	31,896.66	40,082.79	41,833.31
Skewness	18.029	15.958	17.154	9.962
Std. error of skewness	0.036	0.048	0.060	0.036
Kurtosis	396.617	333.241	346.126	150.971
Std. error of kurtosis	0.071	0.096	0.120	0.071
Minimum	4	1	1	1
Maximum	973,006	845,821	906,370	973,292
Pearson correlation	−0.03395	0.007623	−0.03082	0.239054

Reviews conducted on the NNPP algorithm illustrates that it is a competitive tool against several of the other programs available for promoter prediction. However, as with the other programs, this algorithm also suffers from a high instance of false positives. Currently used algorithms are not able to provide highly accurate predictions and the correct prediction promoter rate is only between 13 and 54%. It has been a research challenge to reduce the level of false positive recognition through modifying mathematical modeling and algorithms. Transcription is a complicated process which involves the interactions of promoter *cis*-elements with multiple *trans*-protein factors. The specific interactions rely not only on the specific sequence recognition between the *cis*- and *trans*- factors, but also on some spatial arrangement of them in a complex. Hence, the distance between the TSS and TLS has and can be utilized in promoter prediction.

There are several reasons why the distances between the TSS and TLS (D_2) can be used to improve promoter prediction. For one, the promoter regions are closely associated to the location which in turn will assist in correctly predicting the position of the TSS and will lead to precisely estimating associated promoter regions. Secondly, numerous TSS and TLS experimental data is now accessible by researchers for different species, therefore the empirical probability distribution of TLS-TSS can be obtained. The information of the TLS position can be easily extrapolated from the gene coding region sequence, as it corresponds to the first nucleotide of the coding region. As a result, given a TLS position, and knowing the empirical distribution of the TLS-TSS, the distribution of the TSS can be determined from this distribution. Consequently, improving promoter prediction can be achieved by incorporating this information in the standard NNPP algorithm.

Given a whole DNA sequence of a species, S denotes the set of TSS positions in gene sequences. If position s is a true TSS position in a gene, it will be denoted by $s \in S$; if a range $[s - a_1,\ s + a_2] \subset S$ is used, it means the range $[s - a_1,\ s + a_2]$ covers at least one position which is a real TSS position of the gene. NNPP2.2 will give the probability $P([s - 3,\ s + 3] \subset S)$, sometimes, simply denoted by $P(s \in S)$. The NNPP algorithm has a high instance of false positives, which is due to the estimation of $P(s \in S)$. This probability is not accurate and sometimes overestimates the probability. Therefore, in this paper, we will discuss how to use the information of D_1 and D_2 to adjust the probability given by NNPP.

Two scenarios will be discussed, in the first scenario, only the information of D_2 is considered. But in the second scenario, both information of D_1 and D_2 are take into account for promoter prediction. In the following examples, it always assumes that the position of the TLS and TSC can be easily identified from any given tested gene sequence.

4.1 Scenario 1 – TLS-NNPP Algorithm

In this scenario, it supposes that the NNPP2.2 software has been applied to a tested gene sequence and identified a position s in the sequence with probability of $P(s \in S)$. Given the position s and the tested gene sequence, the distance between s and TLS can be accurately identified. In this circumstance, the probability $P(s \in S, D_2(s) \in [d - a, d + a])$, that is, the probability that s is a TSS position and the distance s and its TLS is between $d - a$ and $d + a$, is used to measure the likelihood of the s being a true TSS position. The higher the probability is the more likely s is a TSS position. In this paper we chose $a = 3$ which is the same value as NNPP employs.

The probability $P(s \in S, D_2(s) \in [d - a, d + a])$ can be evaluated by using the following formula:

$$P(s \in S, D_2(s) \in [d - a, d + a]) = P(s \in S) P(D_2(s) \in [d - a, d + a] | s \in S) \quad (1)$$

Formula (1) is used to adjust the value $P(s \in S)$ given by NNPP2.2. In the formula, $P(D_2(s) \in [d - a, d + a] | s \in S)$ the information is ignored by NNPP2.2. To evaluate the probability the following steps are required:

(1) Collect the position information of the true TSS and its associated TLS for tested species. The larger the sample sizes the superior the output.

(2) Use statistical software to produce the empirical cumulative distribution function $F_{d_2}(d^*)$ for D_2y, $0 \leq d^* < \infty$. Then use a non-parametric method to smooth the empirical cumulative distribution of D_2. Both the above functions can be found from all common statistical software. In this paper, we use SPSS. The empirical cumulative distribution will give the estimation of $P(D_2(s) \leq d^*|s \in S)$ for all $0 \leq d^* < \infty$.
(3) Estimate $P(D_2(s) \in [d-a, d+a]|s \in S)$ by $F_{d_2}(d+a) - F_{d_2}(d-a)$ and substitute it to (1) to evaluate the probability $P(D_2(s) \in [d-a, d+a]| s \in S)$.

The above formula is based on the sample information of D_2 to adjust the probability of s given by NNPP2.2. Sometimes we might consider an alternative way to adjust $P(s \in S)$. From research conducted by Dai et al. [4] it was found that all the density functions of D_2 are positively skewed. For example, considering the histogram plots (Fig. 3), of the *A. thaliana* and *H. sapiens*, the study found when the distance TSS-TLS is large beyond a certain point, say M, the value of the probability density function drops sharply to a very small value.

This offers very little information for the position of the TSS when the distance is beyond that point. Therefore, in such situations, it might be worth considering the probability:

$$\begin{aligned} &P(s \in S, D_2(s) \in [d-a, d+a], D_2(s) \leq M) = \\ &P(s \in S)P(D_2(s) \in [d-a, d+a], D_2(s) \leq M)|s \in S) \end{aligned} \quad (2)$$

instead of (1), while $P(D_2(s) \in [d-a, d+a], D_2(s) \leq M)|s \in S)$ will be evaluated by the empirical probability distribution determined by the entire sample with $D_2 \leq M$ [4].

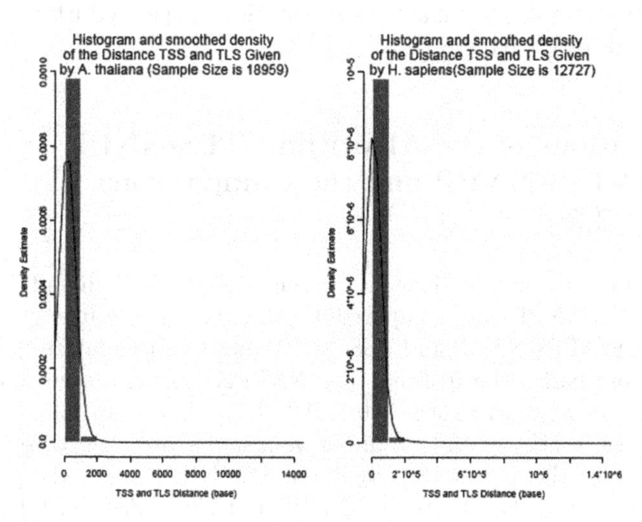

Fig. 3. The histogram and smoothed density of distance TSS-TLS for *A. thaliana* and *H. sapiens*

4.2 Scenario 2 – TSC-TSS-NNPP Algorithm

In this scenario, it is assumed that, the sample information on D_1 and D_2 for tested species is accessible. Under this assumption, given a gene sequence, if the true TSS position is at s; the distance between s and its TLS is D_2(s) and the distance between its TLS and TSC is D_1(s), the following probability will be worth evaluating:

$$P(s \in S, D_2(s) \in [d-a, d+a], D_1(s) \in [b_1, b_2]),$$

where a, b_1, b_2 and d are positive integers, and a is equal to 3 showing that a tested position can differ by plus or minus 3 bp. The probability can be calculated in the following way:

$$\begin{aligned} &P(s \in S, D_2(s) \in [d-a, d+a], D_1(s) \in [b_1, b_2]) = P(s \in S) \\ &\times P(D_1(s) \in [b_1, b_2] | s \in S) P(D_2(s) \in [d-a, d+a] | s \in S, D_1(s) \in [b_1, b_2])^{\cdot} \end{aligned} \tag{3}$$

To evaluate the above probability, the estimation of $P(s \in S)$ is provided by NNPP2.2; following the similar steps listed in Scenario 1, the estimation of $P(D_1(s) \in [b_1, b_2] | s \in S)$ and $P(D_2(s) \in [d-a, d+a] | s \in S, D_1(s) \in [b_1, b_2])$ will be given by the empirical distribution of D_1 and the empirical distribution of D_2 given $D_1 \in [b_1, b_2]$, respectively.

However, if TSS positions are only predicted for gene sequences with $D_1 \in [b_1, b_2]$, the above evaluation can be simplified, and evaluate:

$$P(s \in S) P(D_2(s) \in (d-a, d+a) | s \in S, D_1 \in (b_1, b_2)) \tag{4}$$

instead of (3). In Sect. 5, we only apply (4) to real data.

Figure 4 shows the schematic representation of the algorithms and procedure outlined in this paper.

5 Applications of the Algorithms TLS-NNPP and TSC-TSS-NNPP and the Comparisons to NNPP2.2

In this section, two applications are presented and the results of TLS-NNPP and TSC-TSS-NNPP are compared to the relevant results of NNPP2.2. Using the TSC-TSS-NNPP and TSS-NNPP methods to analyze the data, the adjusted score had to be utilized. The NNPP2.2 algorithm generates scores or cutoff values at tenths such as 0.1, 0.2, 0.3, 0.4, 0.5, up to 0.9. To obtain similar values, tenths of the maximum adjusted score were taken to obtain cutoff values for the TSC-TSS-NNPP and TSS-NNPP methods.

We compare the algorithms TLS-NNPP and TSC-TSS-NNPP to NNPP2.2 in term of the probability of correct prediction. For example, the probability

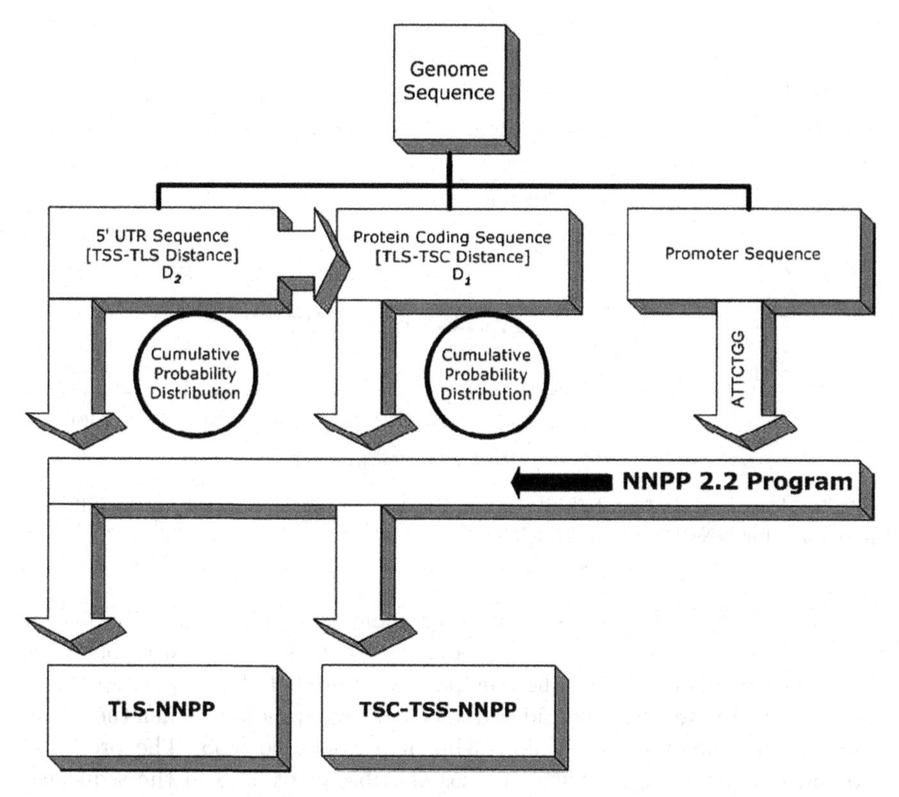

Fig. 4. Schematic representation of promoter prediction using TLS-NNPP and TSC-TSS-NNPP algorithms

of a position which is accepted as TSS position by an algorithm is really a position of TSS.

To save time, the comparison in this paper was done based on a 10% of the gene sample data. This 10% sub-sample is called a testing sample, and is randomly selected from the sample data to reduce the impact of sample error on comparison results. The methods TLS-NNPP, TSC-TSS-NNPP and NNPP2.2 are applied to the sub-sample, respectively. Then, for each cutoff value, the total number of predictions and positive predictions in a range greater than each cutoff value were counted and the probability of correct prediction, denoted by *P(Correct Prediction)* will be evaluated for the TSS-NNPP, TSC-TSS-NNPP and NNPP2.2, respectively. The estimations of *P(Correct Prediction)* are the number of positive predictions divided by the total number of predictions.

5.1 *E. coli* Sequence Study Using the TLS-NNPP Algorithm

We firstly used this technique and modification to the NNPP2.2 algorithm on *Escherichia coli* DNA sequences. The process involved in the implementation

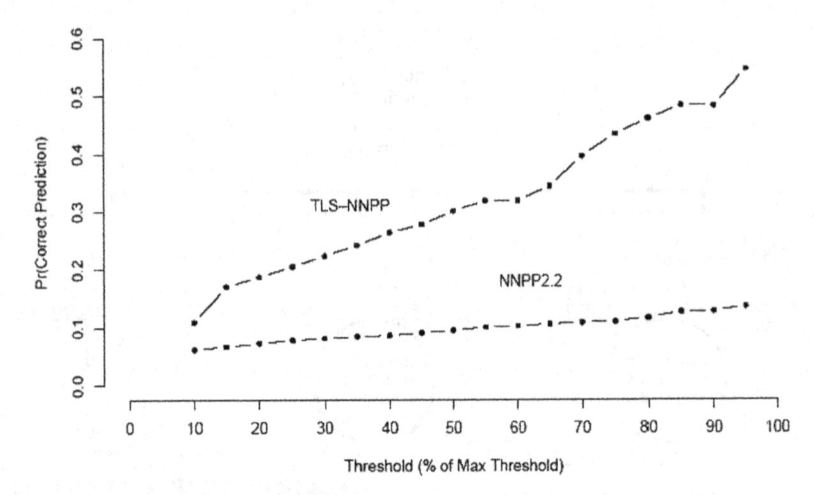

Fig. 5. Comparison of probability of prediction of promoter sequences at different thresholds for NNPP2.2 and TLS-NNPP [2]

took several steps. The steps involved creating an empirical distribution for the TSS-TLS distance, next, DNA sequences (500 bp) were run through the NNPP2.2 program and only the true positively predicted TSS positions were used. The Promoters were considered to be correctly predicted when the actual TSS of the promoter fell within ± 3 bp of a predicted TSS. The predicted promoters must be inline with the closest subsequent TLS in the sequences and the TSS-TLS distance.

The research conducted by Burden et al. [2] showed that by modifying the NNPP2.2 algorithm program by incorporating addition information, such as the TSS-TLS distance, it greatly improved the prediction of promoters and reduced the incidence of false predictions. Figure 5 shows how effective the TLS-NNPP technique was compared to the NNPP2.2 program without the modifications. The number of predictions for this particular species was low due to the training set only containing 293 *E. coli* promoters therefore would not recognize any of the new promoters in the sequences [2].

Further study on a range of species crossing from less complex to more complex organisms also showed that the TLS-NNPP method has power to improve the outcomes of NNPP2.2.

5.2 Human Sequence Study Using the TSS-TSC-NNPP Algorithm

As described in Sect. 4, it is possible to use the TSC-TSS-NNPP approach to improve the performance of NNPP2.2. This is only possible if the data is accessible from databases that could offer large numbers of experimentally defined promoter sequences and start and stop positions for the coding regions and 5′ and 3′ untranslated regions.

In this section the TSC-TSS-NNPP method is applied to human data. Table 2 in Sect. 3 shows that, for human data, D_1 dropped into different regions, and might lead to the variation in the probability distribution of D_2. Since the information of D_2 is related to TSS position, it means that the information of the value of D_1 might have certain level of impact on promoter prediction. In this study, we adopt the four groups, described in Table 2, to group the value of D_1. That is, Group 1 for $D_1 \leq$ 5,583; Group 1 for $5,583 < D_1 \leq$ 17,466; Group 3 for $17,466 < D_1 \leq$ 43,976 and Group 4 for $43,976 < D_1$. The comparisons between the algorithms TSC-TSS-NNPP and NNPP2.2 were done for D_1 in the four groups, respectively.

Our results display that in all four groups of the *H. sapiens* data set, the TSC-TSS-NNPP method achieved better results than both NNPP2.2 and TSS-NNPP, particularly for Group 1. Looking at Fig. 6, 60% seems to be the best cutoff value for the TSC-TSS-NNPP method which has a greater Pr(Correct Prediction) value than the other two methods at this cutoff value. Additionally, within a 10–60% threshold range for Group 1, this showed that the probability of predicting that a sequence is a promoter is highest for TSC-TSS-NNPP.

As shown in Fig. 6, the *P(Correct Prediction)* values for TSC-TSS-NNPP and TSS-NNPP dropped down at large threshold values. This is because time constraints did not allow us to examine a large data set for this research and dividing the data into groups extensively reduced its size so much so that there was no data available and information was exhausted at large threshold levels. Therefore, the data should generally be compared within a range of 0 to around 60%.

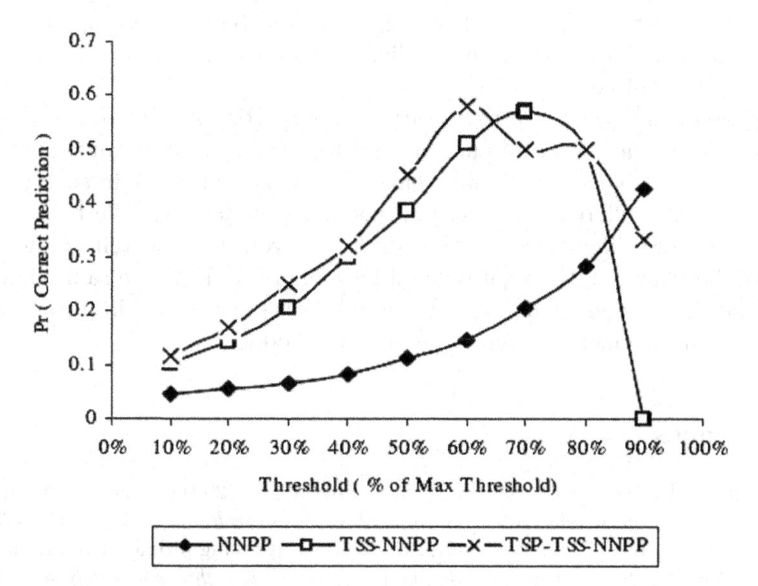

Fig. 6. The comparison of three methods when D_1 in Group 1

The TSC-TSS-NNPP method produced better results compared to the NNPP and TSS-NNPP methods for the *H. sapiens* data set for all four groups. We also apply TSC-TSS-NNPP to *M. musculus* data (The results are omitted from this paper.). For the *M. musculus* data however, our study show that the TSC-TSS-NNPP method is the better choice only for Groups 1 and 2, whereas the NNPP2.2 method is better for Groups 3 and 4. It is interesting to note that the TSC-TSS-NNPP method produced extensively better results than TSS-NNPP and NNPP2.2 for small D_1 values (Group 1) in both species. This is a vital merit for the TSC-TSS-NNPP approach, as generally, shown by 3D histograms of all organisms there is a very high proportion of small D_1 values in the complete data set (3D histograms is omitted from this paper).

Therefore, if the data set consists largely of small D_1 values, this new method will be highly effective in reducing the FP rate for the NNPP2.2 tool, which will then ensure that each promoter that is predicted is associated with a gene coding region.

6 Conclusion

The influx of highly advanced and efficient sequencing technology has made it extremely realistic to complete genomes of a large variety of species. It has become common practice now to sequence new genomes. A big challenge in the post-genome-sequencing era, for deciphering the gene regulation networks, is to improve computational techniques that were lacking in accuracy. It has been shown that using the TSS-TLS and TLS-TSC distances, promoter prediction can be improved with the NNPP2.2 algorithm. However, this new technique does not have to be restricted to this program, but may be applied post-process to many other promoter prediction algorithms that also suffer from a high incidence of false positives.

In delving into the patterns of statistical properties of different gene regions and their correlation we intend to elucidate the spatial organization rules between various gene functional elements and the difference in such organizations among different living organisms and gene families. We believe that these rules and differences are the results of evolution and reflect the complexity differences in the regulation of gene expression. At the same time the outcome of this research has, as presented in this paper, and will provide more leads for improvements in bioinformatics methodology.

References

1. Bajic, V.B., Tan, S.L., Suzuki, Y. and Sugano, S. (2004) Promoter prediction analysis on the whole human genome, *Nature Biotechnology*, 22: 1467–1473.
2. Burden, S., Lin, Y.-X. and Zhang, R. (2005) Improving promoter prediction for the NNPP2.2 algorithm: a case study using *Escherichia coli* DNA sequences, *Bioinformatics*, 21: 601–607.

3. Chiaromonte, F., Miller, W. and Bouhassira, E.E. (2003) Gene length and proximity to neighbors affect genome-wide expression levels, *Genome Research*, 13: 2602–2608.
4. Dai, Y., Zhang, R. and Lin, Y.-X. (2006) The probability distribution of distance TSS-TLS is organism characteristic and can be used for promoter prediction. In: Ali, M. and Daposigny, R. (eds) Advances in Applied Artificial Intelligence – Lecture Notes in Artificial Intelligence (LNAI 4031). Springer, Heidelberg Berlin New York, pp. 927–934.
5. Fickett, J.W. and Hatzigeorgiou, A.G. (1997) Eukaryotic promoter recognition, *Genome Research*, 7: 861–878.
6. Garcia-Hernandez, M., Berardini, T., Chen, G., Crist, D., Doyle, A., Huala, E., Knee, E., Lambrecht, M., Miller, N., Mueller, L.A., Mundodi, S., Reiser, L., Rhee, S.Y., School, R., Tacklind, J., Weems, D.C., Wu, Y., Xu, I., Yoo, D. Yoon J. and Zhang, P. (2002) TAIR: a resource of integrated Arabidopsis data, *Functional & Intergrative Genomics*, 2: 239–253.
7. Gorm Pedersen, A., Baldi, P., Chauvin, Y. and Brunak, S. (1999) The biology of eukaryotic promoter prediction – a review, *Computers & Chemistry*, 23: 191–207.
8. Hughes, T.A. (2006) Regulation of gene expression by alternative untranslated regions, *Trends in Genetics*, 22: 119–122.
9. Knudsen, S. (1999) Promoter2.0: for the recognition of PolII promoter sequences, *Bioinformatics*, 15: 356–361.
10. Lemos, B., Bettencourt, B.R., Meiklejohn, C.D. and Hartl, D.L. (2005) Evolution of proteins and gene expression levels are coupled in *Drosophila* and are independently assocated with mRNA abundance, protein length, and number of protein–protein interactions, *Molecular Biology and Evolution*, 22: 1345–1354.
11. Makita, Y., Nakao, M., Ogasawara, N. and Nakai, K. (2004) DBTBS: database of transcriptional regulation in *Bacillus subtilis* and its contribution to comparative genomics, *Nucleic Acid Research*, 32: D75–D77.
12. Mikkelsen, T.S., Wakefield, M.J., Aken, B., Amemiya, C.T., Chang, J.L., Duke, D., Garber, M., Gentles, A.J., Goodstadt, L., Heger, A., Jurka, J., Kamal, M., Mauceli, E., Searle, S.M.J., Sharpe, T., Baker, M.L., Batzer, M.A., Benos, P.V., Belov, K., Clamp, M., Cook, A., Cuff, J., Das, R., Davidow, L., Deakin, J.E., Fazzari, M.J., Glass, J.L., Grabherr, M., Greally, J.M., Gu, W., Hore, T.A., Huttley, G.A., Kleber, M., Jirtle, R.L., Koina, E., Lee, J.T., Mahony, S., Marra, M.A., Miller, R.D., Nicholls, R.D., Oda, M., Papenfuss, A.T., Parra, Z.E., Pollock, D.D., Ray, D.A., Schein, J.E., Speed, T.P., Thompson, K., VandeBerg, J.L., Wade, C.M., Walker, J.A., Waters, P.D., Webber, C., Weidman, J.R., Xie, X., Zody, M.C., Broad Institute Genome Sequencing Platform, Broad Institute Whole Genome Assembly Team, Marshall Graves, J.A., Ponting, C.P., Breen, M., Samollow, P.B., Lander, E.S. and Lindblad-Toh, K. (2007) Genome of the marsupial *Monodelphis domestica* reveals innovation in non-coding sequences, *Nature*, 447: 167–178.
13. Ohler, U. and Niemann, H. (2001) Identification and analysis of eukaryotic promoters: recent computational approaches, *Trends in Genetics*, 17: 56–60.
14. Pandey, S.P. and Krishnamachari, A. (2006) Computational analysis of plant RNA Pol-II promoters, *BioSystems*, 83: 38–50.
15. Qui, P. (2003a) Recent advances in computational promoter analysis in understanding the transcriptional regulatory network, *Biochemical and Biophysical Research Communications*, 309: 495–501.

16. Qui, P. (2003b) Computational approaches for deciphering the transcriptional regulatory network by promoter analysis, *Biosilico*, 4: 125–133.
17. Reese, M.G. (2001) Application of a time-delay neural network to promoter annotation in the *Drosophila Melanogaster* genome, *Computers and Chemistry*, 26: 51–56.
18. Salgado, H., Cama-Castro, S., Peralta-Gil, M., Daz-Peredo, E., Snchez-Solano, F., Santo-Zavaleta, A., Martnez-Flores, I., Jimnez-Jacinto, V., Bonavides-Martnez, C., Segura-Salazar, J., Martnez-Antonio, A., and Collado-Vides, J. (2006) RegulonDB (version 5.0): *Escherichia coli* K-12 transcriptional regulatory network, operon organization, and growth conditions, *Nucleic Acids Research*, 34: D394–D397.
19. Suzuki, Y., Yamashita, R., Sugano, S. and Nakai, K. (2004) DBTSS, DataBase of transcriptional start sites: progress report 2004, *Nucleic Acids Research*, 32: D78–D81.
20. Tan, T., Frenkel, D., Gupta, V. and Deem, M.W. (2005) Length, protein–protein interactions, and complexity, *Physica A*, 350: 52–62.
21. Wang, D., Hsieh, M. and Li, W. (2005) A general tendency for conservation of protein length across eukaryotic kingdoms, *Molecular Biology and Evolution*, 22: 142–147.
22. Zhang, J. (2000) Protein-length distributions for the three domains of life, *Trends in Genetics*, 16: 107–109.
23. Zhu, J. and Zhang, M.Q. (1998) SCPD: a promoter database of the yeast *Saccharomyces cerevisiae, Bioinformatics*, 15: 607–611.

Appendix

cDNA	Complementary deoxyribonucleic acid
d_1	Coding region length
d_2	Distance between transcription start site and translation start site
d_3	Distance between stop codon and terminator site
D_1	Coding region length with introns
D_2	Distance between transcription start site and translation start site with introns
D_3	Distance between stop codon and terminator site with introns
DNA	Deoxyribonucleic acid
RNA	Ribonucleic acid
TLS	Translation start site
TSS	Transcription start site
TSC	Translation stop codon
TTS	Transcription termination site
TF	Transcription factor
TLS-NNPP	D_2 distance used in conjunction with NNPP algorithm
TSC-TSS-NNPP	D_1 and D_2 distances used in conjunction with NNPP algorithm

Artificial Immune Systems for Self-Nonself Discrimination: Application to Anomaly Detection

Sanjoy Das, Min Gui, and Anil Pahwa

Electrical and Computer Engineering Department, Kansas State University, Manhattan, KS, USA, sdas@ksu.edu, mingui@ksu.edu, pahwa@ksu.edu

Summary. Self-nonself discrimination is the ability of the vertebrate immune systems to distinguish between foreign objects and the body's own self. It provides the basis for several biologically inspired approachs for classification. The negative selection algorithm, which is one way to implement self-nonself discrimination, is becoming increasingly popular for anomaly detection applications. Negative selection makes use of a set of detectors to detect anomalies in input data. This chapter describes two very successful negative selection algorithms, the self-organizing RNS algorithm and the V-detectors algorithm, which are useful with real valued data. It also proposes two new approaches, the single and the multistage proliferating V-detector algorithms to create such detectors. Comparisons with artificial fractal data as well as with real data pertaining to power distribution failure rates, shows that while the RNS and the V-detector algorithms can perform anomaly detection quite well, the proposed mechanism of proliferation entails a significant improvement over them, and can be very useful in anomaly detection tasks.

1 Introduction

One of the fundamental characteristics of the mammalian immune system is the ability to recognize the presence of foreign bodies called *pathogens* [1]. This process is carried out by means of *antibodies*, which are receptor sites attached to lymphocytes and capable of binding to pathogens. There are two kinds of antibody producing cells, the *T-cells* and the *B-cells*. T-cell antibodies possess the ability to distinguish invading antigens from the body's own cells. This phenomenon is termed as *self-nonself discrimination*. Self-nonself discrimination proceeds within a specimen's immune system by generating a large variety of detector cells, systematically culling out those that erroneously categorize native cells (self) as foreign (nonself) while retaining the rest. This principle is called *negative selection* [2].

Self-nonself discrimination is becoming increasingly popular as a paradigm in industrial and other applications. An algorithm based on this approach typically works in the following manner. During the training phase, a large amount

S. Das et al.: *Artificial Immune Systems for Self-Nonself Discrimination: Application to Anomaly Detection*, Studies in Computational Intelligence (SCI) **116**, 231–248 (2008)
www.springerlink.com

of samples of normal data are presented to the algorithm, allowing it to adapt. These samples are referred to as self samples. In the input space, the region containing all possible self samples is called the *self region*, or simply self (**S**). Its complement is the nonself region, or nonself. (In an example that we have considered later in this chapter, the entire self space will be defined a priori, instead of through self samples.) A fully trained algorithm, when presented any sample from the input space should be capable of correctly classifying it as either self or nonself. This feat is not readily accomplished with a conventional machine learning classification scheme as proper training requires the availability of enough samples of both classes, self as well as nonself. Data pertaining to nonself is usually sparse in most industrial applications, where nonself is linked to abnormal conditions.

Negative selection algorithms use a set, **D**, of detectors, which also correspond to points in the input space. These detectors are the biological analogues of the B-cell antibodies. During the training phase, a sufficient number of detectors are generated to fill the detector set. In each iteration, a candidate detector is generated, usually at random. It is then cross-checked with the self sample that are available for training to see if it can detect any such point. If it does, the detector is discarded, otherwise it is inserted into the detector set. Training is terminated only after a large number of detectors are found. During the detection phase, the input is checked to see if it can be detected by any existing detector. When it does, the input is classified as a nonself sample. On the other hand, an input that cannot be detected is labeled as self. Figures 1 and 2 depict the overall scheme used during the training and detection phases of the negative selection algorithm.

Negative selection algorithms have been successfully applied to many anomaly detection tasks in engineering [3–5]. In these applications, the goal is to spot irregular patterns in data, classifying them as anomalies, which is equivalent to nonself. Recently, it was also shown that the performance of negative selection algorithms are at least as good as a one class support vector machine classifier for a variety of tasks [6]. In one application, negative

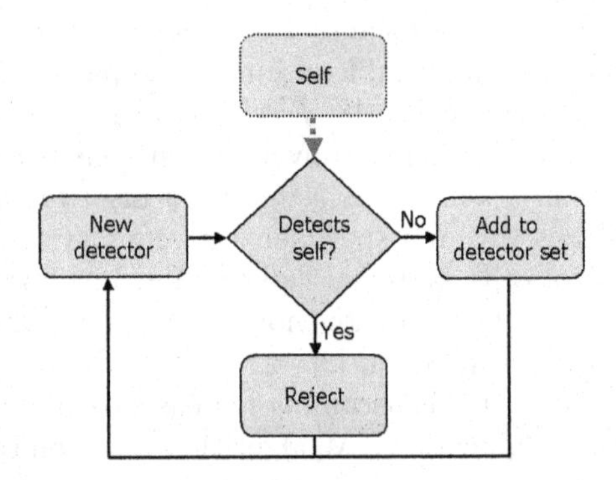

Fig. 1. Creating detectors using negative selection

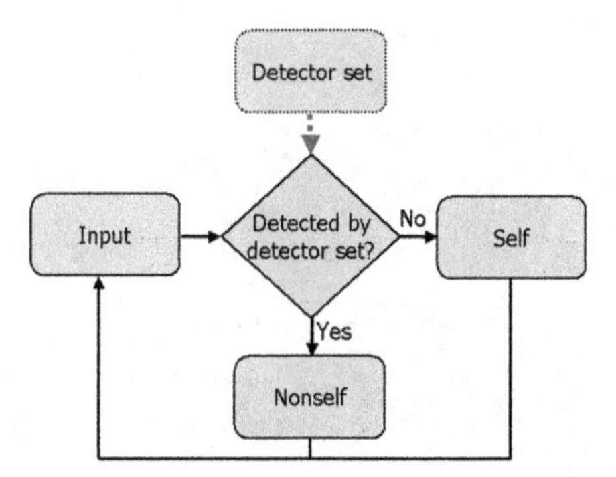

Fig. 2. Input classification using negative selection

selection used to detect faults in squirrel cage induction motors [7]. Applying negative selection to detect anomalies in time series data was proposed in [5,8,9]. Computer security is another important application where negative selection algorithms have met with great success [10–12]. They have been used in a self-nonself discrimination application involving ball-bearing data as well as another one to recognize Indian Telugu characters in [6, 13].

2 Real Valued Negative Selection

Real valued negative selection is a special case of negative selection, where the input samples, and consequently, also the detector set, are represented as real valued, fixed size vectors. Under these circumstances, the detectors' task of detecting an input sample is relatively easy. A distance measure, such as the Euclidean distance, $\text{dist}(\mathbf{x}, \mathbf{z})$, between an input $\mathbf{z}$ and a detector at location $\mathbf{x}$ in the input space, can be used to measure the proximity (or affinity) between an input and a detector. Each detector $\mathbf{x}$, has a radius of coverage, r and is said to *cover* the region defined by a hypersphere with radius r and centered at $\mathbf{x}$. The detector at $\mathbf{x}$ *detects* the input $\mathbf{z}$ if and only if $\mathbf{z}$ lies in the covered region.

During the training phase of the algorithm, only discrete samples of self data are provided, and the algorithm has to construct its own model of the continuous self region $\mathbf{S}$. Here, a simplified model of $\mathbf{S}$ that is commonly used, is the region formed by the union of all hyperspheres centered around the self samples. The hyperspheres have a radius ρ. This model of $\mathbf{S}$ is shown in Fig. 3, where the shaded region is self.

2.1 Recent Approaches

In recent years, several algorithms for real valued negative selection have been proposed. Some of them have been shown to be very effective approaches for

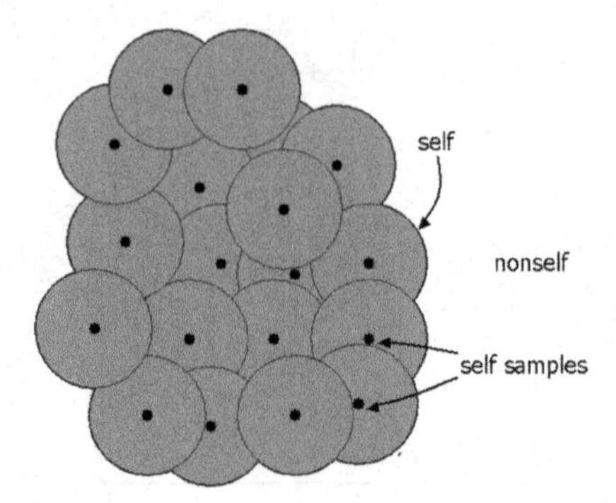

Fig. 3. A model of the self **S** reconstructed from self samples

self-nonself discrimination. Many of them have been applied to engineering anomaly detection tasks, where they have met with a great deal of success. In this chapter, we have selected two successful approaches for real valued negative selection for further investigation, namely the *RNS algorithm*, and the *V-detectors* algorithm. While the RNS algorithm rearranges fixed size detectors within nonself, the latter uses variable sized detectors. A comparison of the two methods would show their relative effectivenesses in covering the nonself region. We also consider two very recent algorithms, the *proliferating V-detectors* and the *multistage proliferating V-detectors* algorithms. Although derived from the idea of V-detectors, they incorporate new features that greatly enhance the original algorithm's performance.

The remainder of this section is used to outline these four aproaches for self-nonself discrimination.

RNS Algorithm

The self-organizing RNS (real-valued negative selection) algorithm places detectors inside the nonself region [14]. The region covered by these detectors are hyperspheres of a fixed radius, r. The algorithms innovation lies in its clever use of self-organization to achieve better placement among the detectors.

The RNS algorithm begins with a set of detectors, $\mathbf{D}$, whose locations $\mathbf{x}$ in the input space are generated randomly. Then the algorithm invokes self-organization to push those detectors that are inside $\mathbf{S}$ to the nonself region, and to space the detectors already outside $\mathbf{S}$ away from each other. This process, which is implemented to maximize nonself coverage, is carried out in an iterative manner. In the ith iteration, the detector whose location will be denoted as $\mathbf{x}_i$ is picked from $\mathbf{D}$. In order to check whether it overlaps $\mathbf{S}$, a sorted list of the k nearest neighbors of $\mathbf{x}_i$, $\mathbf{N} \subseteq \mathbf{S}$ is computed from self samples. If the distance between the median in $\mathbf{N}$ and $\mathbf{x}_i$ is below a certain limiting value, then $\mathbf{x}_i$ is considered to overlap self, and must be moved outside it. Since the median of the nearest neighbors are considered, this method of determining overlap makes the algorithm more robust to noise. It allows some peripherally located self points (presumably noise) to remain within the hypersphere of radius r around the location $\mathbf{x}_i$. Otherwise, only the nearest neighbor of $\mathbf{x}_i$ in $\mathbf{S}$ can be considered. This simplified version has been used for comparisons in the experiments described in this chapter.

Adjusting the location $\mathbf{x}_i$ of each detector that overlaps the self region is done by first computing a direction, $\mathbf{d}$, as follows:

$$\mathbf{d} = \frac{\sum_{\mathbf{n} \in \mathbf{N}} (\mathbf{x}_i - \mathbf{n})}{|\mathbf{N}|}. \tag{1}$$

In this equation, $|\mathbf{N}|$ is the cardinality of $\mathbf{N}$. The detectors location $\mathbf{x}_i$ is nudged along the direction $\mathbf{d}$ in the following manner

$$\mathbf{x}_i = \mathbf{x}_i + \eta \mathbf{d}. \tag{2}$$

The quantity η in the above equation is a parameter of the RNS algorithm, called the adaptation rate.

On the other hand, when the detector $\mathbf{x}_i$ is determined to be outside $\mathbf{S}$ based on the above procedure, then the location $\mathbf{x}_i$ is adjusted as follows:

$$\mathbf{d} = \frac{\sum_{j \neq i} (\mathbf{x}_i - \mathbf{x}_j) w_j}{\sum_{j \neq i} w_j}. \tag{3}$$

where each w_j is a weight assigned to the detectors at location $\mathbf{x}_j$. These weights are assigned to the detectors in such a manner that those already far away from $\mathbf{x}_i$ do not influence its movement. The weight associated with the jth detector in the above equation is determined in the following manner:

$$w_j = \exp(-\alpha \times \mathrm{dist}(\mathbf{x}_i, \mathbf{x}_j)), \tag{4}$$

where α is a constant.

Once the direction $\mathbf{d}$ is determined, the location $\mathbf{x}_i$ is updated as previously in (2) so that the detector at $\mathbf{x}_j$ moves away from the other detectors.

Each detector also has its own age, t_i. This quantity is used to prevent using up wasteful iterations trying to relocate detectors that are placed well inside the self region, and far way from the self-nonself boundary. Initialized to zero, the age is incremented each time the location $\mathbf{x}_i$ of such an overlapping detector is moved. It is reset to zero each time a replacement occurs, or when a detectors location is updated away from other detectors.

The algorithm is run multiple times within an outer loop until the locations of all the detectors in $\mathbf{D}$ approach desired values, also ensuring that none remain within $\mathbf{S}$. Although not shown, the adaptation rate η is allowed to drop progressively with each iteration of the outermost loop. This is done to ensure that the detectors eventually settle down to stable locations towards the end of the algorithm. The pseudocode for the RNS algorithm is provided below:

```
D = rns-generate(S, N)
begin
    D = rand (N)
    while (termination) do
        for i = 1 to N
            if t_i ≥ t_max
                x_i = rand (), t_i = 0
            else
                N = near-nbr(x_i, S)
                if overlap(x_i, S)
                    d = direction(N, x_i)
                else
                    d = weighted-direction(D, x_i), t_i = 0
                end
            endif
            x_i = x_i + ηd, t_i = t_i + 1
        endfor
    endwhile
end
```

V-Detectors Algorithm

V-detectors (variable-sized detectors) are detectors whose radii are not fixed, but are assigned by the algorithm based on their distance from self. Allowing the on-line computation of the radii provides additional flexibility to the algorithm. For each detector in the nonself, the V-detector algorithm adjusts its radius to maximize its coverage, without allowing it to overlap with **S**. This obviates the need to adjust their locations as was the case of the RNS algorithm discussed earlier.

In the V-detectors algorithm, the detector set **D** is initially an empty set. In each iteration a new candidate detector is generated and assigned a location $\mathbf{x}_i$. It is then tested for possible overlap with the self region **S**. The detector is considered to overlap only if its location $\mathbf{x}_i$ lies within a range r of any **S** sample. Any detector that is found to overlap **S** is promptly discarded. No attempt is made to adjust its location. The radius of detection of each detector is set to be equal to the distance from the detectors location to self, dist($\mathbf{x}_i$, **S**), which is the shortest Euclidean distance between the location $\mathbf{x}_i$ and selfs boundary. It is easily computed by measuring the distance from the detector to its closest self sample, and subtracting the latters radius, ρ.

The V-detectors algorithm is repeated until the termination condition is satisfied. In a simple version that we have considered, the algorithm is run for a fixed number of iterations, N. The pseudocode for this algorithm shown below. An additional parameter, θ, has been included as an argument to be passed by the calling procedure. This parameter is used to create a gap between the self and nonself, classifying any detector whose boundary is only within θ of that of self, as an overlap. The reason for doing so is explained subsequently, but to implement the V-detectors algorithm of [13], θ must be set to zero.

$\mathbf{D}$ = generate($\mathbf{S}, N, \theta$)
begin
  $i = 1$
  for $j = 1$ to N
    $\mathbf{x}_i$ = rand ()
    if not overlap($\mathbf{x}_i, \mathbf{S}, \theta$)
      $r_i = \text{dist}(\mathbf{x}_i, \mathbf{S})$, $\mathbf{D} = \mathbf{D} \cup \{(\mathbf{x}_i, r_i)\}$, $i = i + 1$
    endif
  endfor
end

One of the successful improvements of V-detectors is the boundary-aware V-detector algorithm [15]. The boundary-aware algorithm is an improvisation over the basic V-detectors algorithm. Although it places detectors centered at distances at least equal to the radius, ρ, of the self samples, it allows the circumference around each detector to touch the self samples. The boundary between $\mathbf{S}$ and its exterior, the nonself region, in this algorithms reconstruction is therefore unlike Fig. 3, as it does not include an edge of thickness r from the boundary self samples.

Proliferating V-Detectors Algorithm

This method is very recent, having being first suggested in [5]. It consists of a *generation* stage followed by a *proliferation* stage. During the generation stage, the detector set $\mathbf{D}$ is filled with an initial set of detectors in the same manner as that in V-detectors. The novelty of this approach lies in its exploitation of the information gained from previous detectors to rapidly fill interstices in nonself that remain uncovered.

During the proliferation stage the algorithm proliferates the remaining portions of nonself with new detectors which are created from old ones in $\mathbf{D}$. We will refer to these detectors as *offspring*. At the beginning of the proliferation stage, the algorithm already has a set of detectors $\mathbf{D}$ from the previous generation stage. In the ith iteration, it selects one of those detectors whose location and detection radius are $\mathbf{x}_i$ and r_i from the set $\mathbf{D}$, and creates new offspring detectors located at a distance r_i from $\mathbf{x}_i$. If an original detector in $\mathbf{D}$ is regarded as a sphere of radius r_i in the nonself region centered around $\mathbf{x}_i$, the offspring detectors will be located along the sphere's circumference, i.e., at a location $\mathbf{x}_i + \hat{\mathbf{u}}r_i$ where $\hat{\mathbf{u}}$ is some unit vector. Its radius is set to be equal to the minimum of the distance from its center to the self region $\mathbf{S}$. An offspring detector that found to cover a region that is entirely covered by another existing detector in $\mathbf{D}$ is said to be *subsumed* by $\mathbf{D}$. Since a new detector has additional coverage value only when another detector does not subsume it, only those detectors that are not subsumed can enter $\mathbf{D}_o$, the new set of offspring detectors. The algorithm is stopped when all the original detectors in $\mathbf{D}$ have been given the opportunity to proliferate a few times. The detectors in $\mathbf{D}$ are selected for proliferation in a sequential manner. In the version used in this chapter, the unit vectors $\hat{\mathbf{u}}$ are kept to be either parallel or antiparallel to each dimension. Hence, in a two dimensional input space, there will be four possible $\hat{\mathbf{u}}$, whereas if the input space is three dimensional, there are six such vectors.

The procedure for the proliferation stage is provided below. The argument for computing distances includes the threshold θ, which needs to be subtracted from the actual Euclidean distance between $\mathbf{x}_j$ and $\mathbf{S}$.

```
Dₒ = proliferate(D, S, θ)
begin
    j = 1, Dₒ = {}
    for each detector (xᵢ, rᵢ) ∈ D do
        for each unit vector û do
            xⱼ = xⱼ + ûrᵢ, rⱼ = dist(xⱼ, S, θ)
            if not subsume(xⱼ, D)
                Dₒ = Dₒ ∪ {(xⱼ, rⱼ)}, j = j + 1
            endif
        endfor
    endfor
end
```

Maintaining the threshold θ that is initially high during the generation stage, and lowering it towards zero in a stepwise manner during subsequent stages of detector proliferation can result in significantly faster creation of detectors. This is because decrementing the threshold θ at the end of each stage creates a gap between the self-nonself boundary. This gap can then be filled by the proliferating offspring of the next stage. Steadily decreasing it by lowering θ will result in increasingly smaller, but strategically placed offspring to proliferate around the self-nonself boundary region. To ensure full coverage of nonself, the threshold must be zero during the last proliferating stage.

The complete procedure to proliferate in stages, which makes use of decreasing thresholds θ (i.e. $\theta_0 \geq \theta_1 \geq \cdots \theta_{M-1} \geq \theta_M = 0$), is provided below:

$\mathbf{D}$ = multistage-proliferate($\mathbf{D}, \mathbf{S}$)
begin
determine θ_0 thru θ_{M-1}, $\theta = \theta_0$
$\mathbf{D}_0$ = generate($\mathbf{S}, N, \theta$)
$\mathbf{D} = \mathbf{D}_0$
for $s = 1$ to M do
$\theta = \theta_s$
$\mathbf{D}_s$ = proliferate($\mathbf{D}_{s-1}, \mathbf{S}, \theta$)
$\mathbf{D} = \mathbf{D} \cup \mathbf{D}_s$
endfor
end

Once negative selection has been employed to obtain an effective set of detectors, nonself is fully covered. Therefore, any input that falls within the boundaries of any detector must be treated as an anomaly.

3 Results with Koch Curve

The first set of experiments was carried out with a well known fractal, the Koch curve. One side of this curve was (arbitrarily) designated as self, while the other side was considered as nonself. Although the Koch curve is relatively easy to generate, being a fractal it has infinite length, providing arbitrarily small regions to be filled with detectors. The Koch curve is defined recursively, starting from a straight line. At each depth within the recursive procedure, each straight line is replaced with four lines of $\frac{1}{3}$ length. The new lines at the middle are rotated by angles $\pm 60°$ from the others. As the overall length gets increased by a factor of $\frac{4}{3}$ in every step, the total length approaches $(\frac{4}{3})^{\infty} = \infty$. In the simulations described below, the recursion was limited until a depth of six. From visual inspection, it was found that increasing it beyond six produced no difference to the curve, but increased the computation substantially. An earlier study showing how proliferating V-detectors improves the performance of the V-detectors algorithm was reported in [5].

In Fig. 4 is shown the detector placed by applying the RNS algorithm. Note that the bottom right side of the Koch curve is self. The detectors are shown as circles. As the RNS algorithm produces only detectors with fixed radii r, the circles have the same size. The optimal value of this radius was found to be $r = 0.02$. Higher or lower values produced lesser coverage of nonself. The result shown is for N = 5,500. Based on a Monte Carlo simulation the percentage of the nonself covered by this algorithm was only 90.021%.

The next figure, Fig. 5 shows the placement of detectors obtained through the V-detectors algorithm. As before the value of the parameter N was kept at N = 5,500. As this algorithm discards detectors overlap with self instead of reorganizing them, the actual number of detectors in the detector set $\mathbf{D}$ produced by this method is lower than 5,500. In spite of less detectors, this

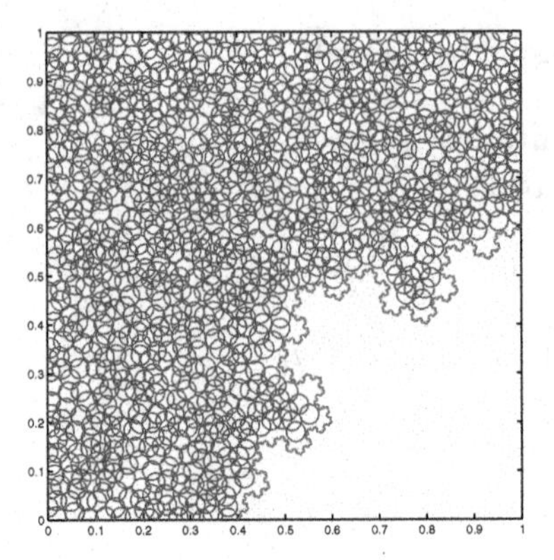

Fig. 4. Application of the RNS algorithm to the Koch curve after 5,500 iterations

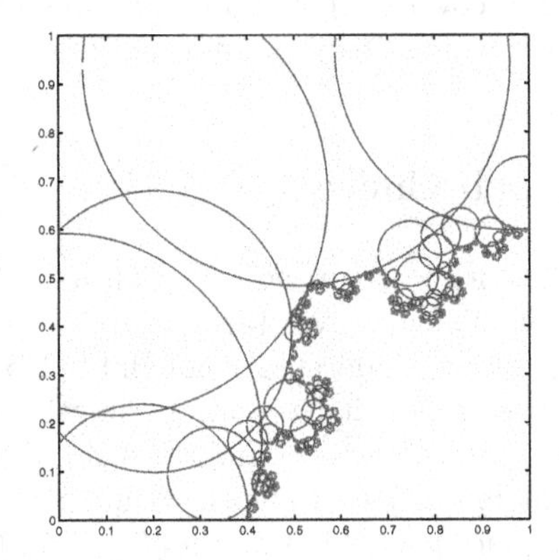

Fig. 5. Application of the V-detectors algorithm to the Koch curve after 5,500 iterations

method produced an estimated coverage of 98.368%, which is much better than the previous method.

Figure 6 shows the performance of the proliferating V-detectors algorithm, where only a single stage was used. Again, a total of $N = 5{,}500$ random detectors were tried, out of which a fraction entered the detector set. This algorithm's coverage of nonself is 99.962%, which is better than the previous method.

Multistage V-detectors was also applied in four stages. The resulting detectors are shown in Figs. 7–10. The threshold, θ was kept at the following values: $\theta = 0.05$ for the first 1,000 iterations when an initial set of detectors were generated using the original V-detectors algorithm, $\theta = 0.025$ for the first stage of proliferation until 2,500 iterations, $\theta = 0.0125$, for the next, which was until 4,000 iterations, and finally, $\theta = 0$ until the algorithm completed the last stage of proliferation at 5,500 iterations. In spite of the already substantial coverage of the single stage proliferation method, using four stages could accomplish

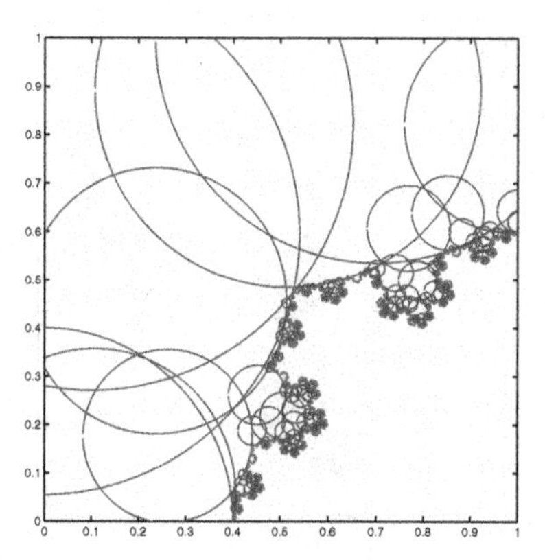

Fig. 6. Application of the proliferating V-detectors algorithm to the Koch curve after 5,500 iterations

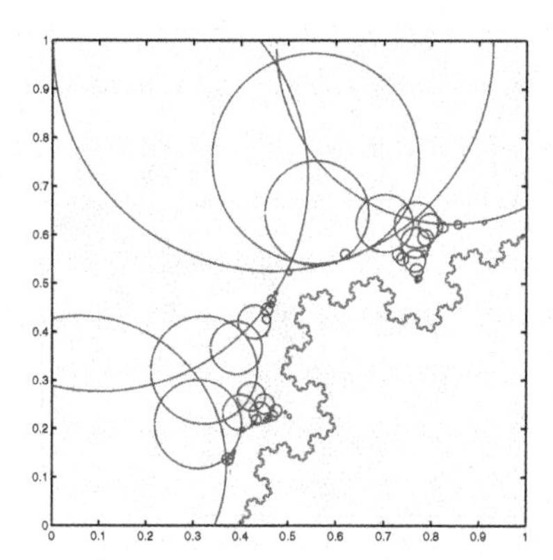

Fig. 7. Application of the multistage proliferating V-detectors algorithm to the Koch curve after 1,000 iterations

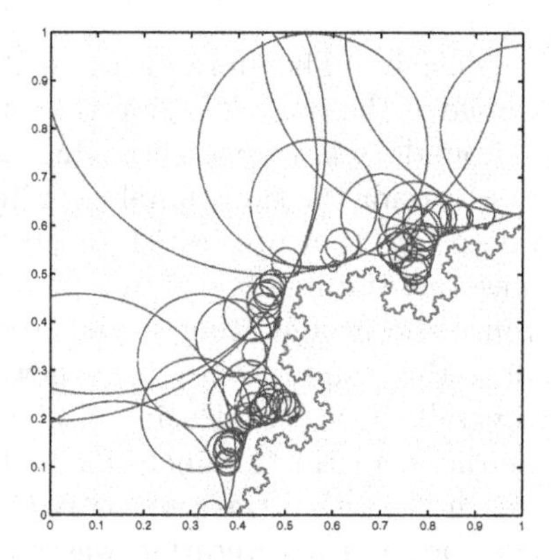

Fig. 8. Application of the multistage proliferating V-detectors algorithm to the Koch curve after 2,500 iterations

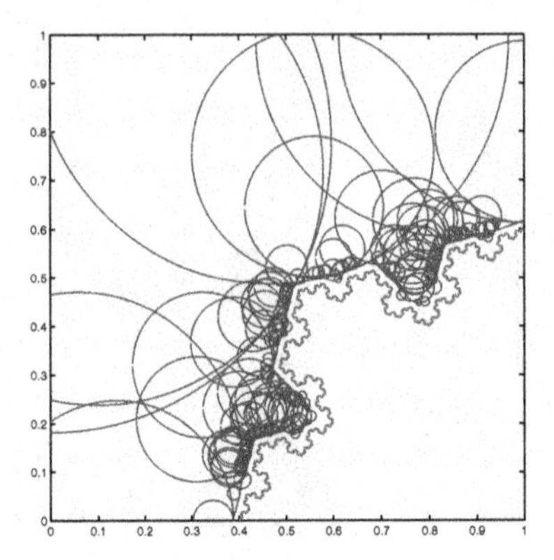

Fig. 9. Application of the multistage proliferating V-detectors algorithm to the Koch curve after 4,000 iterations

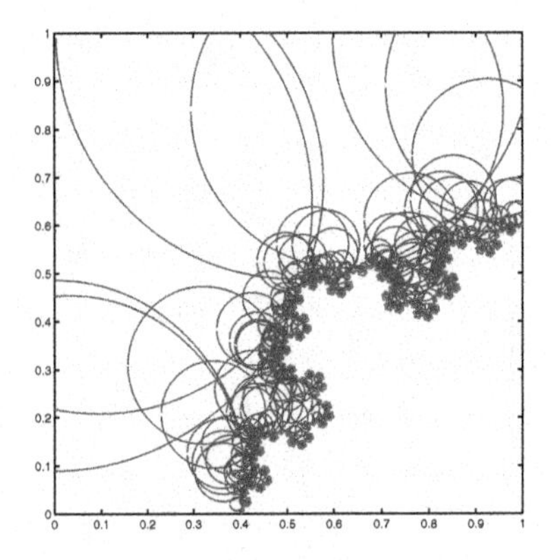

Fig. 10. Application of the multistage proliferating V-detectors algorithm to the Koch curve after 5,500 iterations

even better coverage, at 99.994%. Therefore, using multiple stages of proliferation, almost full coverage of the nonself region is achieved, even when the shape of the self-nonself boundary is very long, having the shape of a fractal.

Since each algorithm is run for the same number of iterations, the number of randomly generated detectors that are tested for possible inclusion in the detector set is fixed. However, the number of accepted detectors would differ from one another. Since the proliferating stages grow offspring detectors from old ones, instead of testing random ones, these new detectors are always created outside **S**. The overall effect is that using proliferation, the frequency of accepted detectors by the algorithm for inclusion in **D** is higher. In order to test this hypothesis, each algorithm was run for 10,000 iterations, and the number of accepted detectors at each iteration was recorded. The result is shown in Fig. 11. Clearly, using the proliferating mechanism, detectors are accepted more frequently than the V-detectors method. It can also be seen

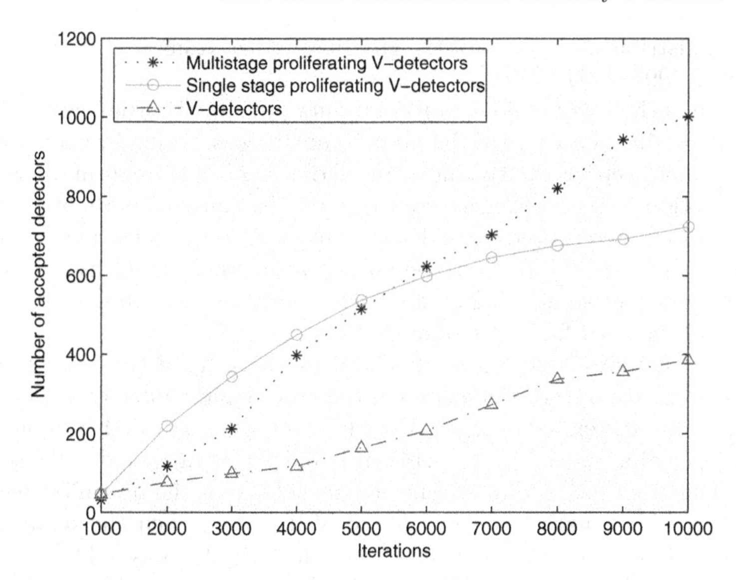

Fig. 11. Detectors accepted using the thee V-detector approaches for real valued negative selection

that initially the single stage of proliferation is better than a multistage one. We theorize that this is because of the presence of a nonzero θ that limits the size of the region within nonself, where detectors can be placed. At around 5,500 iterations, both methods produce an equal number of detectors. However, increasing the number of iterations beyond this value clearly reveals that proliferating in stages produces a higher success rate.

It must be noted that Fig. 11 does not show the performance of the RNS algorithm. This is because that approach does not discard as many detectors as the other ones. Eventually most detectors are accepted after the algorithm reorganizes their locations in the input space, discarding only a small fraction of those, whose ages exceed the limit. This however does not make the RNS algorithm more efficient than the others in terms of computer time, as a substantial amount of computing is used up for self-organization, whose need is obviated in the other methods.

4 An Application to Anomaly Detection in Distribution Systems

This section presents results of simulations where the RNS, V-detectors, and the single and multistage proliferating V-detectors algorithms were applied to real world distribution system data.

The data for these experiments was obtained from various sources, including a local utility company. It pertains to weekly animal related failure rates of electrical distribution system of the city of Wichita, Kansas, collected during the period 1998–2004.

Although there are various causes for outages in distribution systems, animal related failures are one of the most common ones. Animal related failures are those that occur explicitly due to the activity of live animals in the vicinity of distribution lines. In Kansas, squirrels are the cause of most of such failures. An automated procedure to detect unusual patterns in the power outages caused by squirrels will prove to be an important tool for utility companies. For instance, consistently high rates in any location may indicate excessive foliage and the need for tree trimming.

Squirrel activity is strongly influenced by the time of the year, squirrels being most active during summers. On the other hand winter conditions produce little animal related outages. Therefore, our data included an input field, *month type*, which was a value between 1 and 3 (for further details, see [16] or [5]). The other factor that affects squirrel activity is the prevailing weather conditions. It is usually between 40 and 85°F. Days when the temperature range is within these limits are classified as fair weather days. Therefore, the other input field that we have included in this study is the *fair weather days per week*. The *total number of failures* seen during each week was the third and last input. Strong deviations in either direction, from typically observed rates for a given month type and fair weather days, are considered anomalies. Hence a self-nonself detection approach will not only classify excessively high failure rates as anomalies, but unusually low ones too. Since all three input fields were normalized prior to training, the input space for the anomaly detection is the three-dimensional cube, $[0, 1]^3$.

In order to investigate the performance of the four methods, training was carried out removing one year's data from the self samples, enabling the algorithms to place detectors from the remaining data. The algorithm was then tested with the data belonging to the year that was excluded, to see what fraction of it would be classified as anomalous. In Fig. 12 is shown the percentage of inputs that were detected as anomalies by each algorithm when each year was left out. The top curve pertains to multistage proliferation with four stages. Clearly, the multistage proliferation method was able to identify the highest number of anomalies. This is closely followed by an algorithm with a single stage of proliferation. The RNS algorithm (bottom-most curve) could detect the least number of anomalies, and the basic V-detectors algorithm could perform marginally better than it.

In addition to speed, the multistage proliferating V-detectors method can also be used for a graded detection of anomalies, where anomalous inputs closer to self are considered to be lesser anomalies than those further away, and are assigned a higher *level*. This is because the application of multiple stages of proliferation with constantly decreasing θ causes offspring detectors to be placed closer to self than their parents from the previous stage, thus breaking the boundary around self into discrete levels (see Fig. 13). Whenever detectors from the sth and subsequent stages are able to detect an input, which

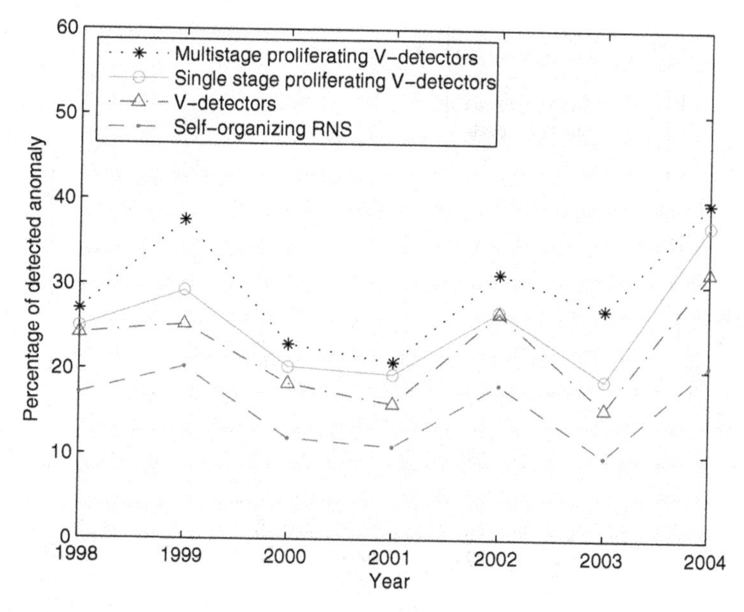

Fig. 12. Performance of the algorithms with the Wichita data

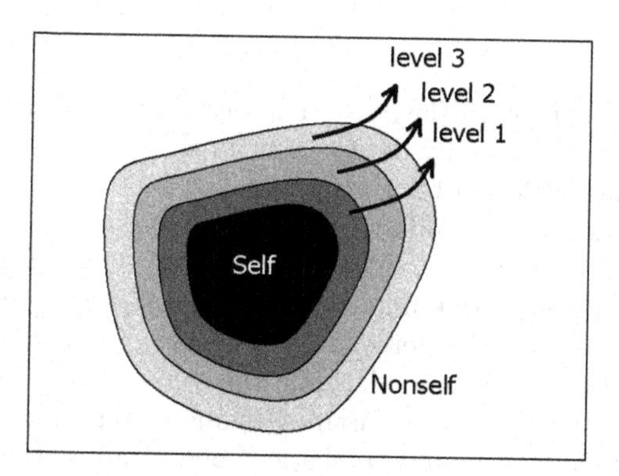

Fig. 13. Graded nonself in multistage detection

none of the previous ones can, the input can be assigned a discrete level of s. Graded detection of anomalies was considered elsewhere before [10], and can be put to good use in industrial applications.

Figure 14 illustrates how graded anomaly detection is done using a generation stage followed by three separate stages of proliferation for each year. The topmost curve corresponds to inputs that were detected by a detector from any stage (i.e. level 1). The curve below it (level 2) pertains to those that could only be detected by detectors drawn from the first three stages alone (generation as well as the first two stages of proliferation). Those that could be identified as anomalies by the detectors from the generation or their immediate offspring

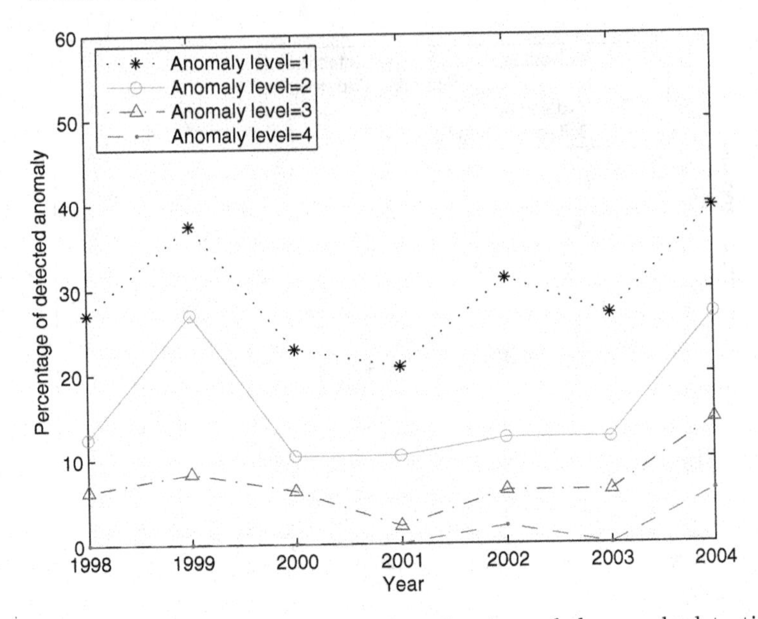

Fig. 14. Performance of the multistage algorithm in graded anomaly detection for the Wichita data

appears beneath that curve (level 3). The last curve is associated with data that could be detected by only detectors from the generation stage, and thus represents the highest level of anomaly (level 4).

In order to examine the influence of the number of fair days during each week, the algorithms were again applied to produce detectors using data pertaining to six years, and excluding that of one year. The outcome is shown in Fig. 15. The x-axis shows the percentage of fair days that were present during the entire year that was excluded, while the y-axis shows the percentage of inputs from that year that were classified as anomalous. The curves produced by all four algorithms are concave shaped, but attained their highest levels when the fair weather days was very high. In other words, anomalous data is more likely to be produced during weeks when the fair weather days were either too low or too high, but highest in case of the latter.

The results that was obtained with the present Wichita data are quite similar to an earlier study, where the V-detectors, and the single and multistage proliferating V-detectors methods were compared for failure data obtained from the city of Manhattan [5]. This earlier study also showed that the proliferation of additional detectors could significantly increase the performance of negative selection in detecting anomalies from distribution system failure patterns. The results reported in this article not only corroborate the findings in [5], but also establish the overall superiority of V-detectors based methods over the self-organizing RNS approach.

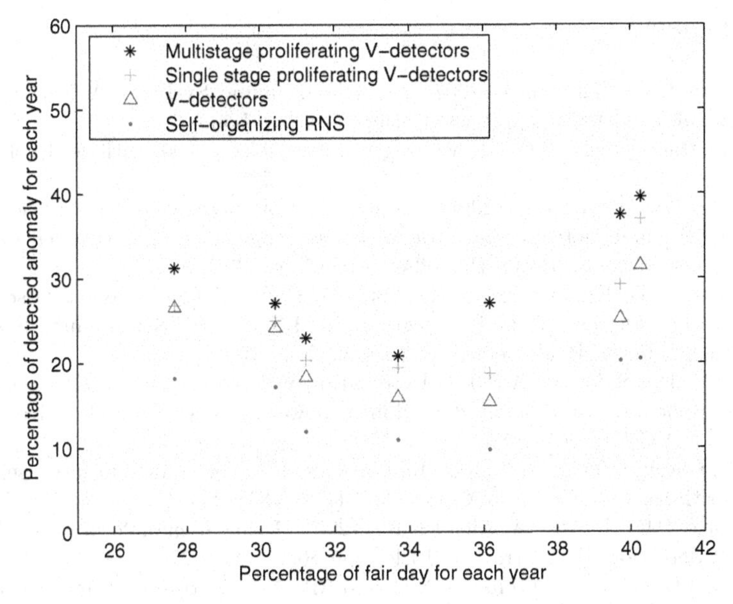

Fig. 15. Fraction of detected anomalies as a function of fair days per year in the Wichita data

5 Conclusion and Further Research

In this article, two earlier algorithms for real valued negative selection, the RNS and the V-detectors approaches, as well as the newer single and multistage proliferation methods are discussed in details. When applied to the Koch curve, proliferating detectors to fill small interstices in the nonself region greatly increases the coverage of nonself. This study also showed that proliferating detectors from existing ones instead of generating new ones also reduces the total number of detectors required to be tested. When applied to failure rate data, the results show that when detectors are produced through proliferation, the anomaly detection capability of the algorithm improves. Using more than a single stage of proliferation further enhances the overall performance.

The method of proliferation shows some promise in addressing scalability issues of negative selection algorithms. More investigation would be needed to establish this claim. One of the possible extensions that the authors are looking at is a subspace proliferation technique where V-detectors are allowed to populate subspaces of the multidimensional input space. They are also considering other implementations of the idea of proliferation when the input space is not real valued. Such a technique would be immensely useful for computer security when the bulk of the input stream is non-numeric. Lastly, the applicability of proliferating detectors to algorithms based on danger theory [17] – a new immune systems paradigm upon which other anomaly detection algorithms can be built upon – may be taken up for further investigation.

References

1. Castro LND, Timmis J (2002) Artificial Immune Systems: A New Computational Intelligence Approach. Springer-Verlag, London
2. Ji Z, Dasgupta D (2007) Revisiting negative selection algorithms. Evol Comp 15(2): 223–251
3. Taylor DW, Corne DW (2003) An investigation of the negative selection algorithm for fault detection in refrigeration systems. In: LNCS Artificial Immune Systems. Springer, Berlin Heidelberg New York 2787: 34–45
4. Dasgupta D, KrishnaKumar K, Wong D, Berry M (2004) Negative selection algorithm for aircraft fault detection. In: LNCS Artificial Immune Systems. Springer, Berlin Heidelberg New York 3239: 1–13
5. Gui M, Das S, Pahwa A (2007) Procreating V-detectors for nonself recognition: An application to anomaly detection in power systems. Proc Gen Evol Comp Conf (GECCO): 261–268
6. Ji Z, Dasgupta D (2006) Applicability issues of the real-valued negative selection algorithms. Proc Gen Evol Comp Conf (GECCO): 111–118
7. Branco PJC, Dente JA, Mendes RV (2003) Using immunology principles for fault detection. IEEE Trans Ind Electron 50(2): 362–373
8. Dasgupta D, Yu S, Majumdar N (2003) MILA – Multi-level Immune Learning Algorithm and its application to anomaly detection. Soft Comp J 9(3): 172–184
9. Nunn I, White T (2005) The application of antigenic search techniques to time series forecasting. Proc Gen Evol Comp Conf (GECCO): 353–360
10. Dasgupta D, Gonzalez F (2002) An immunity-based technique to characterize intrusions in computer networks. IEEE Trans Evol Comp 6(3): 281–291
11. Harmer PK, Williams PD, Gunsch GH, Lamont GB (2002) An artificial immune system architecture for computer security applications. IEEE Trans Evol Comp 6(3): 252–280
12. Nio F, Gmez D, Wong D, Vejar R (2003) A novel immune anomaly detection technique based on negative selection. In: LNCS Artificial Immune Systems. Springer, Berlin Heidelberg New York 2723: 204
13. Ji Z, Dasgupta D (2004) Real valued negative selection algorithm using variable-sized detectors. Proc Gen Evol Comp Conf (GECCO) 3102: 287–298
14. Dasgupta D, Gonzalez F (2003) Anomaly detection using real-valued negative selection. Gen Prog Evol Mach 4(4): 383–403
15. Ji Z (2005) A boundary aware negative selection algorithm. Proc 9th IASTED Int Conf Art Intell Soft Comp
16. Sahai S, Pahwa A (2004) Failures of overhead distribution system lines due to animals. Proc N Am Power Symp Moscow Idaho
17. Aickelin U, Cayzer S (2002) The danger theory and its application to artificial immune systems. In TimmisJ, Bentley PJ (eds), Proc Int Conf Art Immune Syst (ICARIS). University of Kent at Canterbury 141–148

Computational Intelligence Applied to the Automatic Monitoring of Dressing Operations in an Industrial CNC Machine

Arthur Plinio de Souza Braga[1], André Carlos Ponce de Leon Ferreira de Carvalho[2], and João Fernando Gomes de Oliveira[3]

[1] Department of Electrical Engineering, Federal University of Ceará – CT, Caixa Postal 6001 – CEP - 60450-760, Fortaleza, Ceará, Brazil
[2] Department of Computer Science, University of São Paulo, Caixa Postal 668,13560-970, São Carlos, SP, Brazil
[3] Department of Production Engineering, University of São Paulo, Av. Trabalhador Sãocarlense, 400, 13566-590, São Carlos, SP, Brazil

1 Introduction

In manufacturing, *grinding* is the process that shapes very hard work pieces with a high degree of dimensional accuracy and surface finish. The efficiency of the grinding process is regarded as a very important issue in the modern and competitive metal-mechanic industry, since it usually represents the major portion of processing costs [1–8]. The grinding process is strongly dependent on the topography surface of the *grinding wheel*, an expendable wheel that carries an abrasive compound on its periphery, since it is the cutting tool in grinding operations [1–3, 9–11]. The process responsible for preparing the topography surface of the grinding wheel is named *dressing* [2, 9, 12]. This process removes the current layer of abrasive, leading to a fresh and sharp surface. Thus, the dressing process has an important effect on the efficiency of the grinding process, because the quality of the cutting tool directly affects the quality of the final product [2].

This research targets a common problem in industrial grinding operations. Every time a grinding wheel is changed, or after an automatic balancing procedure, it is necessary to perform dressing operations in order to get a concentric working surface. The number of dressing strokes is normally determined by trial and error. If an excessive number of dressing strokes is executed, the grinding wheel's life-cycle is reduced and that implies an economic loss. If an insufficient number of dressing strokes is executed, the quality of the grinding process will be compromised and it also leads to an economic loss caused by work pieces. The solution proposed in this paper is to develop an *automatic monitoring dressing system* that provides support to execute the just-right

A.P. de Souza Braga et al.: *Computational Intelligence Applied to the Automatic Monitoring of Dressing Operations in an Industrial CNC Machine*, Studies in Computational Intelligence (SCI) **116**, 249–268 (2008)
www.springerlink.com

number of dressing strokes. As a result, it guarantees the high quality of final products, its efficient use of grinding wheel's life-cycle and eventually a lower cost [1, 3, 9, 10].

An automatic monitoring dressing system is often regarded as a nontrivial issue [2, 3, 6, 9, 12]. The particular nature of the grinding wheel, which has its abrasive grains randomly spaced on the work surface, makes it more difficult than a pure mathematics problem. A model-based approach for the monitoring problem has to include many functional parameters that could not be easily measured [4]. Thus, we propose to monitor dressing operations through a parameter that measures, indirectly, the grinding wheel's working condition. Computational intelligence (CI) algorithms use these measurements to induce classifiers which are able to decide if the wheel needs another dressing stroke.

The results obtained in the performed simulations are very promising (Sect. 7), with 100% of correct matches with the best classifiers, resulting in a sophisticated monitoring system that is able to either replace or provide support to skilled operators.

This chapter is organized as follows: Section 2 describes the parameter adopted to indirectly monitor the grinding wheel; Sections 3 and 4 discuss how the measurement of the parameter chosen is structured, in this work, to represent the grinding wheel's condition; Section 5 presents the CI algorithms used in this research to decide if a new dressing operation is necessary; Section 6 summarizes the proposed monitoring system followed by its initial results obtained in the simulations described in Sect. 7. A general discussion of the experimental results and final conclusions are presented in Sect. 8.

2 Acoustic Emission in Grinding and Dressing

Several methods for the monitoring of the grinding wheel surface have been developed in the last few years, most of them based on the analysis of measures that indirectly monitor parameters of the wheel [1–3, 9–11]. In recent years, acoustic emission (AE) instruments and systems have been developed for the monitoring, nondestructive testing of the structural integrity, and general quality of a variety of manufacturing processes [9]. Li [9] defines AE as: "*The class of phenomena whereby transient elastic waves are generated by the rapid release of energy from a localized source or sources within a material, or the transient elastic wave(s) so generated*". Cutting process produces such kind of waves that propagate through the machine structure [12], and the AE generated during a grinding process has been proven to contain information strongly related to the condition changes in the grinding zone [4]. The major advantage of using AE to monitor tool condition is that the frequency range of the AE signal is much higher than that of the machine vibrations and environmental noises, and does not interfere with the cutting operation [9].

As occurs in the grinding process during the contact between the wheel and the workpiece [13], in the dressing process every single contact of an abrasive grain from the grinding wheel with the dressing tool is assumed to generate a stress pulse. During operation, the properties of each single grain and their overall distribution on the circumference of the grinding wheel will change. Hence, many different sources of AE have to be considered, since every single effect has to be regarded as the origin of a wave propagating through a solid-state body [8].

The application of AE sensors has become very popular in many kinds of machining process over the last two decades [1–5, 9, 10, 13, 14]. To avoid disturbances like vibrations in the machine [6,13], AE sensors should be placed as close as possible to the tool. These sensors are very advantageous in a monitoring system, since they combine the following requirements [13]: (a) relatively low costs, (b) no negative influence on the stiffness of the machine tool and (c) they are easy to mount.

However, the AE signals will generate between 10^6 and 10^7 data points per second during monitoring if the full dynamic range of the signal is captured [12]. Thus, the "curse of dimensionality", which plagues so many patterns recognition procedures, reaches here an exceptional intensity. A variety of methods for dimensionality reduction have been proposed for AE signal [13]: root-mean-square value, raw AE signals and frequency analysis. In the area of AE in grinding and dressing, the root-mean-square value $U_{AE,\ RMS}$ of the time domain signal recorded from the AE sensor is one of the most important quantities to characterize the process state [6], [12], and [13].

3 Acoustic Maps

As mentioned in Sect. 2, the traditional AE-based monitoring systems generally use the following parameters for acquired AE signals [3, 4, 10]: AE RMS peak value, peak value of a chosen frequency in the FFT of the AE signal, number of threshold detections on AE signal and standard deviation of the AE signal. However, these parameters have limitations imposed by the unpredictable changes in AE signal linearity and amplitude with time [5].

The nonlinearity between the AE and the dressing or grinding quantities can be caused by signal saturation in the first stage of the signal pre-amplification – Oliveira and Dornfeld [5] observe that, in dressing operation, when the sensor output is positioned close to the diamond dressing tool, the sensor output can reach values higher than the power supply voltage. Another source of problem is the RMS level instability. The AE level may fluctuate up to 50% of its value after some hours. One of the main reasons for this signal oscillation is the variable performance of the machine interfaces in transmitting the AE signal along the period of operation [5]. These influences make the conventional AE RMS solutions difficult to apply in production lines, leading to frequent need of human interference for adjustments.

Oliveira and Dornfeld [5] proposed the representation of the AE signal by an image, the *acoustic map* [5], as an alternative to avoid some of the

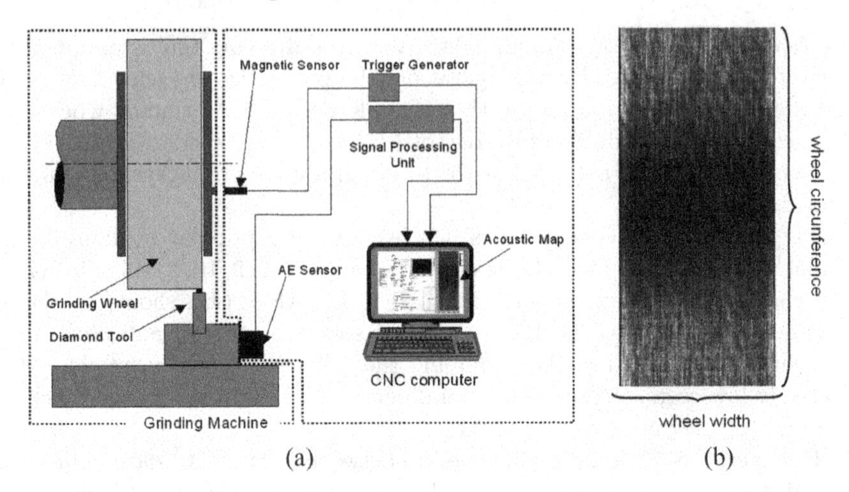

Fig. 1. Acoustic emission (AE) in dressing operation: (**a**) the diagram of the experimental setup for acquiring the AE signals during dressing (based on [11]), and constructing the (**b**) acoustic map, where each line corresponds to a full rotation of the grinding wheel

limitations present in most of the currently AE based monitoring systems in industrial grinding. The acoustic maps are constructed in real time by adding columns in an array as the dressing tool travels along the wheel surface [11]. The vertical and horizontal directions in the acoustic map are the wheel circumference and width, respectively. The gray intensity shows the acoustic emission RMS value measured from the interaction between the dressing tool and the abrasives grains. Figure 1b shows an acoustic map produced during a dressing operation with the experimental setup of Fig. 1a.

The data acquisition is carried out in data arrays corresponding to a full rotation of the grinding wheel. It is triggered by a magnetic sensor and a target positioned on the spindle. In the acoustic maps, the lack of contact between dressing tool and grinding wheel appears as dark areas in the map (Fig. 2). The gray intensity shows the acoustic emission RMS value measured from the interaction between the dressing tool and the abrasives grains. By using acoustic maps, the limitations on AE-based monitoring are reduced, since the patterns present in this graph are not influenced by the RMS fluctuation along the time (normally much longer times).

Figure 3 shows that the fluctuations on AE RMS can lead to changes in the acoustic maps intensity, but the pattern (the relationship between pixels) is preserved. Therefore, the diagnosis based on the acoustic maps should not be influenced by temperature or other long-term AE disturbances [5].

The approach of monitoring grinding processes through acoustic maps allows human operators to perform the monitoring of a dressing operation by simple visual inspection. The use of image patterns allows an automatic monitoring by using CI techniques [5–23] for the classification of the acoustic

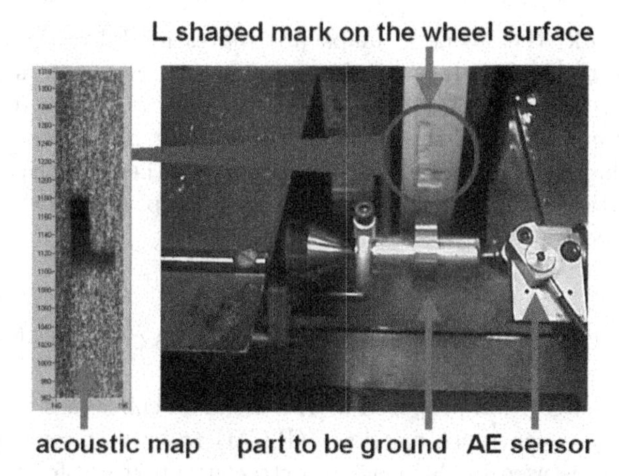

Fig. 2. An real experiment that shows how efficient is an acoustic map to represent the topography surface of an grinding wheel with an L-shaped mark – see the dark region on acoustic map at the left side, and the L-shaped mark in detail on the right side

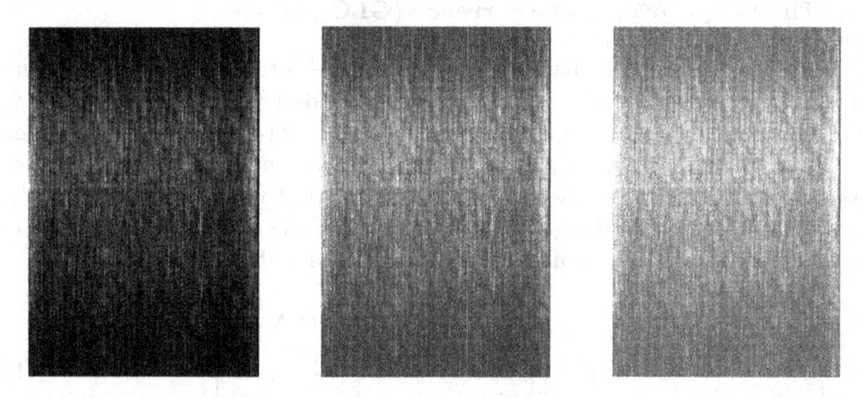

Fig. 3. An example of how acoustic maps deal with fluctuations on AE RMS: the three acoustic maps are associated with the same grinding wheel, but their AE RMS levels are different. Despite the AE RMS fluctuations, the acoustic maps present the same pattern of darker up and down regions and a brighter central region

maps. However, the success of a pattern classifier depends on the selection of a proper representation that should satisfy some desirable characteristics [24]:

1. It should consider a small number of features, which might lead to simpler decision regions, and facilitate a classifier induction.
2. It should have features that are robust, i.e., relatively insensitive to noise or other errors. In practical applications, a classifier may need to either act fast or use less memory or processing steps [17].

In this work, the acoustic maps are represented by a vector of texture descriptors [24–26]. In acoustic maps, texture descriptors can provide measures, like smoothness, coarseness, and regularity, which define mutual relationships among intensity values of neighboring pixels repeated over an area larger than the size of the relationship. Thus, these descriptors can preserve information about the patterns in the map, despite AE RMS oscillations. Section 4 describes the texture descriptors adopted.

4 Extracting Textural Features from Acoustic Maps

The statistical algorithm proposed by Haralick and colleagues in [26] is based on gray level co-occurrence matrices (GLCM) to extract texture measures. This algorithm is used in this work to extract the textural features from the acoustic maps. The following subsections describe the co-occurrence matrix and the Haralick's textural descriptors computed from this matrix.

4.1 The Gray-Level Co-Occurrence (GLC) Matrix

Considering the acoustic map as a matrix $M(x, y)$ with N discrete values, the spatial gray level dependency matrix $\mathbf{P}(d, \phi)$ is described for each d and ϕ. The element $p_{i,j}$ is defined as the relative number of times a gray level pair (i, j) occurs when pixels separated by the distance d along the angle ϕ are compared. Each element is normalized by the total number of occurrences, producing the *gray-level co-occurrence (GLC)* matrix $\mathbf{P}$. A spatial gray level dependency matrix is also named a co-occurrence matrix. It is given by

$$\mathbf{P}(d, \phi) = \begin{bmatrix} p_{0,0} & p_{0,1} & \cdots & p_{0,N-1} \\ p_{1,0} & p_{1,1} & \cdots & p_{1,N-1} \\ \vdots & \vdots & \ddots & \vdots \\ p_{N-1,0} & p_{N-1,1} & \cdots & p_{N-1,N-1} \end{bmatrix}, \quad (1)$$

where $p_{i,j}$ is defined by

$$p_{i,j} = \frac{\text{number of pixel pairs with intensity } (i, j)}{\text{total number of pairs considered}}. \quad (2)$$

The number of operations required to process the matrix is proportional to the number of resolution cells N present in the image. In comparison, the number of operations has complexity of order $N. \log N$ if Fourier or Hadamard transform is used to extract texture information [26]. Besides, to compute the entries in the gray-tone spatial-dependence matrices, it is necessary to keep only two lines of the image data in core at each time instant. Thus, no severe storage constraints are imposed [26].

4.2 Haralick's Textural Descriptors

The frequent assumption in characterizing image texture is that all the texture information is contained in the gray-tone spatial-dependence matrices (1). However, these matrices are typically large and sparse, and metrics of these matrices are taken to get textural descriptors. Haralick [26] proposed a set of textural descriptors obtained from the GLC matrix: angular second moment, contrast, correlation, variance, inverse diference moment, sum average, sum entropy, entropy, difference variance, difference entropy, correlation and maximal correlation coeficient. The following equations describe the Haralick's textural descriptors [26]:

$$f_1 = \sum_i \sum_j P(i,j)^2_{\Delta x, \Delta y}, \tag{3}$$

$$f_2 = \sum_i \sum_j (i-j)^2 P(i,j)_{\Delta x, \Delta y}, \tag{4}$$

$$f_3 = \frac{\sum_i \sum_j i \cdot j \cdot P(i,j)_{\Delta x, \Delta y} - \mu_l \cdot \mu_c}{\sigma_l \cdot \sigma_c}, \tag{5}$$

$$f_4 = \sum_k k P^D(k)_{\Delta x, \Delta y}, \tag{6}$$

$$f_5 = \sum_i \sum_j \frac{P(i,j)_{\Delta x, \Delta y}}{1 + (i-j)^2}, \tag{7}$$

$$f_6 = \sum_k k P^S(k)_{\Delta x, \Delta y}, \tag{8}$$

$$f_7 = \sum_k \left(k - MS_{\Delta x, \Delta y}\right)^2 \cdot P^S(k)_{\Delta x, \Delta y}, \tag{9}$$

$$f_8 = -\sum_k P^S(k)_{\Delta x, \Delta y} \cdot \log P^S(k)_{\Delta x, \Delta y}, \tag{10}$$

$$f_9 = -\sum_i \sum_j P(i,j)_{\Delta x, \Delta y} \log P(i,j)_{\Delta x, \Delta y}, \tag{11}$$

$$f_{10} = \sum_k (k - MD_{\Delta x \Delta y})^2 \cdot P^D(k)_{\Delta x \Delta y}, \tag{12}$$

$$f_{11} = -\sum_k P^D(k)_{\Delta x, \Delta y} \cdot \log P^D(k)_{\Delta x, \Delta y}, \tag{13}$$

$$f_{12} = \sum_k P^S(k)^2_{\Delta x, \Delta y}, \tag{14}$$

$$f_{13} = \sum_k P^D(k)^2_{\Delta x,\Delta y}, \tag{15}$$

$$f_{14} = \sum_i \sum_j \frac{P(i,j)^2_{\Delta x,\Delta y}}{P(i,^*)_{\Delta x,\Delta y} \cdot P(^*,j)_{\Delta x,\Delta y}}, \tag{16}$$

where:

$$\mu_l = \sum_i i \cdot P(i,^*)_{\Delta x,\Delta y}, \qquad \mu_c = \sum_j j \cdot P(^*,j)_{\Delta x,\Delta y},$$

$$\sigma_l = \sqrt{\sum_i (i-\mu_l)^2 \cdot P(i,^*)_{\Delta x,\Delta y}}, \qquad \sigma_c = \sqrt{\sum_j (j-\mu_c)^2 \cdot P(^*,j)_{\Delta x,\Delta y}},$$

$$P(i,^*)_{\Delta x,\Delta y} = \sum_j P(i,j)_{\Delta x,\Delta y}, \qquad P(^*,j)_{\Delta x,\Delta y} = \sum_i P(i,j)_{\Delta x,\Delta y}.$$

Based on these features (3)–(16), a vector representing an acoustic map can be created. Gotlieb and Kreyszig [27] divide the Haralick's features into four groups: (a) features that express visual textural characteristics, (b) features based on statistics, (c) features based on information theory and (d) features based on information measures of correlation. They show that not necessarily all 14 Haralick's descriptors should be used to represent an image – a subset of these features can be enough. In Sects. 6 and 7, this representation is used to perform the monitoring of the dressing operation by classifying the textural descriptors vector of the acoustic maps. Section 5 discusses how the classification is performed.

5 Pattern Classification

The problem now is to obtain a partition of the Haralick textural descriptors feature space into different regions, each one associated to a different decision – dressing or not dressing the grinding wheel. In this work, pattern classification techniques [17] are employed to obtain the decision boundary between these patterns. For such, a classifier associates a feature vector to a label related with the action to be performed.

The central aim of a classifier is to label data according to some grouping even when presented to novel patterns [17] – this is the issue of generalization. In dressing monitoring, this characteristic is particularly important, since it is impossible to collect beforehand all possible acoustic maps to design the classifier. Four classifiers, with different learning[1] capabilities, are investigated in our work: (a) a multi-layer perceptron network, (b) a radial-basis function network, (c) a support vector machine and (d) a decision tree. The

[1] Learning refers to some form of algorithm for reducing the error on a set of training data. In the broadest sense, any method that incorporates information from training samples in the design of a classifier employs learning [15].

following subsections discuss the general principles of these classifiers (for a more detailed discussion, see [15–17, 19, 20, 22]).

5.1 Multi-Layer Perceptron (MLP) Networks

A multi-layer perceptron (MLP) network has at least one hidden layer of neurons[2]. Usually, each neuron in the hidden layer is connected to all the neurons in the previous and in the next layer. The input signal is propagated through the network in a forward direction, on a layer-by-layer basis [16]. MLP networks can be seen as universal approximators, i.e., they can be viewed as a nonlinear input–output mapping of a general nature [16]. These networks have their free parameters adjusted by being trained in a supervised learning process, using input–output examples. The most frequently used training algorithm is known as *error back-propagation* [16]. Its basic idea is to efficiently compute partial derivatives of an approximating function $F(\mathbf{w};\mathbf{x})$ – the relationship between input–output – with respect to all the elements of the adjustable weight matrix $\mathbf{w}$ for a given value of input vector $\mathbf{x}$. The actual response of the network is subtracted from a desired response to produce an error, and this error is propagated backward through the network being used as a target to adjust the network's weights.

5.2 Radial-Basis Function (RBF) Networks

The learning of the relationship between the feature vector and the desired outputs can be carried out in different ways. While this association in MLPs is based on optimization methods known in statistics as *stochastic approximation* [16], the *Radial-Basis Function* networks take the relationship as a *curve-fitting (approximation) problem* in a high-dimensional space. The RBF technique consists of choosing a function F that has the following form [28]:

$$F(x) = \sum_{i=1}^{N} w_i \varphi\left(\|x - x_i\|\right), \tag{17}$$

where: $\{\varphi(\|x - x_i\|),\ i = 1, 2, \ldots, \mathrm{N}\}$ is a set of N arbitrary (generally nonlinear) functions, known as *radial-basis functions*, and $\|.\|$ denotes a *norm* that is usually taken to be Euclidean [16]. The principle for the approximation of the desired association $F(\mathrm{x})$, (17), is to solve, for the unknown weights $\{w_i\}$ – vector $\mathbf{w}$, the following set of simultaneous linear equations [16]:

$$\underbrace{\begin{bmatrix} \varphi_{11} & \varphi_{12} & \cdots & \varphi_{1N} \\ \varphi_{21} & \varphi_{22} & \cdots & \varphi_{2N} \\ \vdots & \vdots & \vdots & \vdots \\ \varphi_{N1} & \varphi_{N2} & \cdots & \varphi_{NN} \end{bmatrix}}_{\Phi} \cdot \underbrace{\begin{bmatrix} w_1 \\ w_2 \\ \vdots \\ w_N \end{bmatrix}}_{\mathbf{w}} = \underbrace{\begin{bmatrix} d_1 \\ d_2 \\ \vdots \\ d_N \end{bmatrix}}_{\mathbf{d}}, \tag{18}$$

[2] Neuron refers to a mathematical model of the biological neuron (see [14]).

where Φ is the *interpolation matrix* and $\mathbf{d}$ is the desired response vector. Haykin [16] argues that Φ is positive definite, and cites [29] to justify that the interpolation matrix is nonsingular - thus, (18) can be solved by:

$$\mathbf{w} = \Phi^{-1} \cdot \mathbf{d} \tag{19}$$

where Φ^{-1} is the inverse of the interpolation matrix Φ.

5.3 Support Vector Machine (SVM)

SVMs represent a classification technique based on statistical learning theory [18]. Given a training set of instance-label pairs $(\mathrm{x}_i,\ \mathrm{y}_i)$, a SVM is trained by solving optimization problem that finds a linear hyperplane $\mathbf{w} \cdot \mathbf{x}_i + b$ able of separating data with a maximal margin. The recent interest in SVMs is motivated by [19]: (a) strong generalization capability; (b) capability to deal with high dimensional patterns; (c) use of a convex objective function; as a SVM solution implies in the optimization of a quadratic function, there is only one global minimum - what represents a clear advantage in comparison with neural networks with multiple global minimums; (iv) a solid theoretical background based on learning statistics theory [20]. This is accomplished by solving the following optimization problem [20] and [21]:

$$\mathbf{Minimize}: \|\mathbf{w}\|^2 + C\sum_{i=1}^{n} \varepsilon_i,\ \mathbf{Restricted\ to}: \begin{cases} \varepsilon_i \geq 0 \\ y_i\left(\mathbf{w} \cdot \mathbf{x}_i + b\right) \geq 1 - \varepsilon_i \end{cases} \tag{20}$$

where $\mathbf{x}_i \in \Re^m$, C is a constant that imposes a tradeoff between training error and generalization and ε_i are the slack variables, which increase its tolerance to noise. The decision frontier obtained is given by [20] and [21]:

$$F(\mathbf{x}) = \sum_{\mathbf{x}_i \in SV} y_i \alpha_i \mathbf{x}_i \cdot \mathbf{x} + b \tag{21}$$

where α_i are Lagrange multipliers determined in the optimization process and SV corresponds to the set of support vectors for which the associated Lagrange multipliers are larger than zero.

5.4 Decision Trees (DT)

In general, a DT is an arrangement of tests that prescribes the most appropriate test at every step in an analysis [22]. Each decision outcome at a node is called a split, since it corresponds to splitting a subset of the training data. Each node is connected to a set of possible answers. Each ***nonleaf node*** is connected to a test that splits its set of possible answers into subsets corresponding to different test results. Each branch carries a particular test result's subset to another node. Usually, DTs represent a disjunction of conjunctions of constraints on the attribute-values of instances. Each path from the tree root to a leaf corresponds to a conjunction of attribute tests, and the tree itself to a disjunction of these conjunctions [15].

More specifically, DTs classify ***instances*** by sorting them down the tree from the ***root node*** to one of the ***leaf nodes***, which provides the classification of the instance. Much of the work in designing trees focuses on deciding which property test or query should be performed at each nonleaf node [17].

The fundamental principle underlying tree creation is that of simplicity: preferring decisions that lead to a simple, compact tree with few nodes. To this end, a property test T is looked for at each node n that makes the data reaching the immediate descendent nodes as pure as possible. In formalizing this notion, it turns out to be more convenient to define the impurity, rather than the purity of a node. Several different mathematical measures of impurity have been proposed, all of which have basically the same behavior [17] and [23]. Let $i(n)$ denote the impurity of a node n. In all cases, $i(n)$ should be closed to 0 if all of the patterns that reach the node bear the same category label, and should be large if the categories are equally represented.

The most popular measure is the entropy impurity (or, occasionally, information impurity) [17]:

$$i(n) = -\sum_{j} P(c_j).\log_2 P(c_j) \tag{22}$$

where $P(c_j)$ is the fraction of patterns at node n that are in the category c_j. By the well-known properties of entropy, if all the patterns are of the same category, the impurity is 0; otherwise it is positive, with the largest value occurring when the different classes are equally likely.

6 Intelligent Monitoring of Dressing Operations

The flowchart of the proposed monitoring strategy of dressing operations is sketched in Fig. 4: the AE signals acquired along a pre-defined period of time are stored in an acoustic map; at the end of this period, the co-occurrence matrix of the map is computed and, based on this matrix, a vector with a set of Haralick's descriptors is created. This vector is the input of the classifiers, whose binary output indicates if the dressing operation must either stop (output = 0) or continue (output = 1). All four classifiers learn a mapping between the vector of Haralick's descriptors, representing the acoustic map, and a label ("**stop dressing operation**" or "**continue dressing operation**").

Based on Gotlieb and Kreyszig (see [27]), a subset of the Haralick's features can be enough to represent the acoustic map, instead of the set of all Haralick's features. Thus, the time spent calculating these features can be reduced by choosing a just a subset of features to compose the elements of

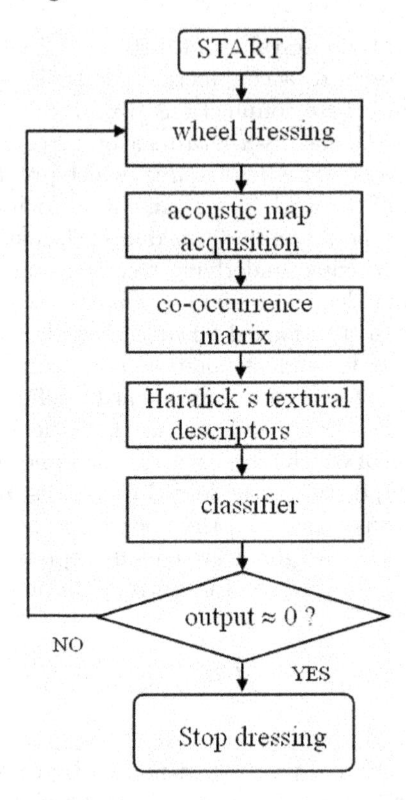

Fig. 4. Flowchart of the proposed automatic monitoring system of dressing operations. The system classifies the acoustic map in sense to determine if the grinding wheel should be dressed or not

the input vector for the classifiers. The empirical approach adopted to select the subset of features to represent the acoustic map had began with the first three Haralick's features listed in Sect. 4.2, and the systematic inclusion of the subsequent features on the representation if necessary. The validation of each representation was obtained through the convergence or not of the classifiers trained with each current subset of Haralick's descriptors. A vector with the first six features (3)–(8) was the shortest representation found that satisfied the chosen validation. Section 7 presents the results obtained with such compact representation of the acoustic maps.

7 Experiments and Results

The four classifiers described in Sect. 5 were employed in the experiments. The main goal of the experiments is the separation, in the Haralick's descriptors space, of the maps associated with concentric and excentric grinding wheels. A grinding wheel is concentric, and the dressing operation is satisfactory, when the dressing tool touches its entire circumference during dressing. Two classes were used: maps indicating that the wheel needs to be re-dressed, and maps indicating good dressing. The acoustic maps were acquired on-line using a CNC open architecture from the Laboratory for Optimization of Manufacturing Processes (OPF) of the University of São Paulo (USP). The CNC platform is based on the GE-FANUC platform 180i dual processor and a Pentium IV based interface. In this system, all the mapping software and analysis run in the CNC interface, allowing its use in an industrial environment.

The computation of (a) the co-occurrence matrix, (b) the textural descriptors and (c) the classifiers output were performed off-line using MATLAB® routines developed by the authors. The following results present an analysis of classifiers performance in two main aspects: (a) comparative number of correct classifications, and (b) comparative behavior in the presence of noise.

7.1 Experimental Setup

The signal processing, the monitoring of the acoustic maps and the execution of controlling routines were performed through a graphical interface developed by OPF in LabView®. The data was collected in real time during dressing operations with different wheel balancing conditions at a sampling rate of 100 K samples/second. The scanning process takes about 10 s and, depending on the desired resolution, can generate an acoustic map with a resolution of 1,000 × 700 pixels. The AE signal intensity is represented by a 12-bit word.

The balancing was changed by using a fluid based balancer installed in the wheel hub. Each time the wheel is balanced or unbalanced in a differrent level, a concentric/excentric dressing condition was achieved. The dataset was visually divided into two classes: (a) concentric (Fig. 5a) and (b) excentric grinding wheels (Fig. 5b).

The experimental conditions used in collecting the acoustic maps are listed in Table 1.

After a large number of experiments using different values for the free parameters, the authors decided to use the following configurations for the techniques investigated in this work: a MLP network with a 6-6-1 one-hidden-layer architecture. This network was trained with the backpropagation algorithm using a learning rate of 0.08; a RBF network with ten Gaussian radial basis functions. Their centers were chosen during the training phase, based on the largest reduction of the network error by a supervised selection [16], and standard deviation of 0.001; a SVM with Gaussian radial basis function kernel, parameter C equal to 10, and the DTs whose impure nodes must have five or more observations to be spitted, generated using the CART algorithm [23].

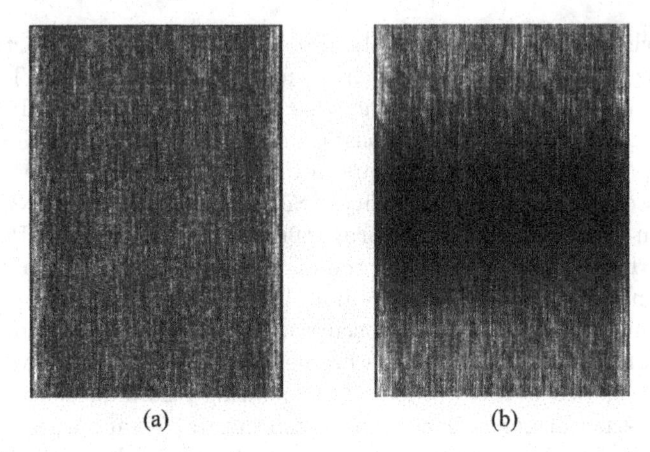

Fig. 5. Examples of acoustic maps collected during dressing operations. The acoustic map (**a**) corresponds to a sharp and concentric grinding wheel. The acoustic map (**b**) corresponds to a nonconcentric grinding wheel that needs to be dressed to correct its surface

Table 1. Experimental specifications and conditions

Grinding wheel	SGB 60 L VHB
Cutting fluid	Integral oil
Grinding wheel speed	1,770 rpm
Dressing depth	0.02 mm
Grinding wheel width	54 mm
Dressing speed	200 mm min^{-1}

7.2 Simulation Results

The training and test sets used in the experiments have 40 patterns for each class. In the training phase, the maps were randomly presented to the classifiers. The images were classified as belonging to one of two classes, related with the balance of the wheel grinding:

1. Class with output 0 indicates a concentric wheel and that no more dressing operations are necessary;
2. Class with output 1 indicates a nonconcentric wheel and further dressing operations are necessary.

In order to obtain more statistically significant estimates for the classifiers generalization performance, the authors employed the ten-fold cross-validation methodology [23]. According to this methodology, each experiment is performed ten times, each time using a different subset as the validation set and combining the other subsets for the training set. Thus, each example is used in the validation set for one of the experiments and in the training set for the other nine experiments. The final results are the average of the results obtained

Table 2. Accuracy using tenfold cross-validation

	Training				Test			
Fold	MLP	RBF	SVM	DT	MLP	RBF	SVM	DT
01	100	100	100	100	100	100	100	100
02	100	100	100	100	100	100	100	100
03	100	100	100	100	100	100	100	100
04	100	100	100	100	100	100	100	87.5
05	100	100	100	100	87.5	100	87.5	87.5
06	100	100	100	100	100	100	100	100
07	100	100	100	100	100	100	100	100
08	100	100	100	87.5	87.5	100	100	100
09	100	100	100	100	100	100	87.5	87.5
10	100	100	100	100	100	100	100	100

Table 3. Average squared error using tenfold cross-validation

Classifier	Training	Test
MLP	0.0001 ± 0.0000	0.0059 ± 0.0122
RBF	0.0042 ± 0.0018	0.0050 ± 0.0042
SVM	0.0000 ± 0.0000	0.0250 ± 0.0527
DT	0.0028 ± 0.0059	0.0250 ± 0.0527

in these ten experiments. Thus, the 80 available examples were portioned into ten disjoint subsets, each of size 8.

Table 2 shows the accuracies[3] observed for each partition by each classifier. It can be seen that the four classifiers presented good generalization: all of them showed at least 87.5% (seven correct classifications and one incorrect classification in the fold) of correct classifications for the test set patterns (Table 2). Table 3 shows the mean squared error rates on the total acoustic map dataset: all classifiers produced output values very close to the desired values, what suggests a good potential for the monitoring of the dressing operation.

Despite these good results, one important question is how meaningful, robust and relevant they are for a real monitoring system. Although the results seem good, a few patterns (Table 2), produced under controlled situation, were misclassified by most of the classifiers. Thus, how the performance would be affected by the presence of noise, commonly found in industrial plants? To answer this question, this dataset was expanded with the addition of artificial noise data.

These additional experiments took into account the uncertainty regarding the estimation of the AE RMS used to compose the acoustic maps. It was assumed an addition of noise where the new images would be in

[3] Here, "accuracy" refers to the percentage of correct classifications.

the format $g(x, y) = f(x, y) + n(x, y)$, where $n(x, y)$ would be the noise. In the experiments, noise was considered independent of spatial coordinates and uncorrelated with respect to the original image $f(x, y)$ (there was no correlation between pixel values and the values of noise components). The noise was modeled as a Gaussian random variable, with probability density function (PDF) given by [30]

$$p(n(x, y)) = \frac{1}{\sqrt{2\pi}\sigma} e^{-(n(x,y)-\mu)^2/2\sigma^2}, \tag{23}$$

where μ is the noise mean value, σ is its standard deviation, and σ^2 the *variance* of $n(x, y)$. When $n(x, y)$ is described by (23), approximately 70% of its values will be in the range $[(\mu - \sigma), (\mu + \sigma)]$, and about 95% will be in the range $[(\mu - 2\sigma), (\mu + 2\sigma)]$ [30].

The noise simulations were performed by adding the noise to the collected images through the MATLAB® function *imnoise*, which was used to generate a Gaussian white noise with zero mean and variable variance. The main goal of using different variances was to see how the performance of each classifier would be affected by the increment of noise intensity in the input data. The variance is a measure of statistical dispersion, so the increment of variance makes the probability density of the noise closer to a uniform distribution – the worst case, when the collected AE signal is totally random and has no correlation with the dressing process (Fig. 6).

The percentage of correct classifications, accuracy, was adopted to measure the classifiers' performance. The curves in Fig. 7 show that MLP, RBF and SVM had a similar behavior in the presence of noise: with 0 variance, all present a very good performance, but as soon as the noise intensity increased,

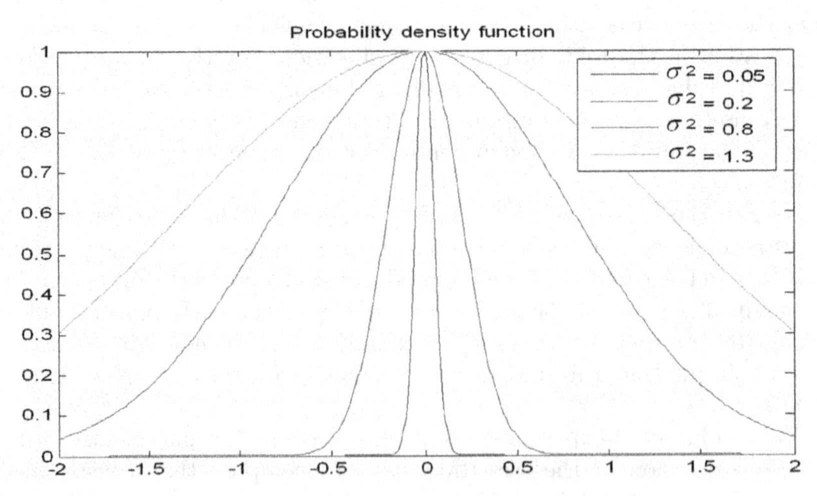

Fig. 6. The variance is a measure of statistical dispersion. The increment of the variance makes the PDF closer to a uniform distribution

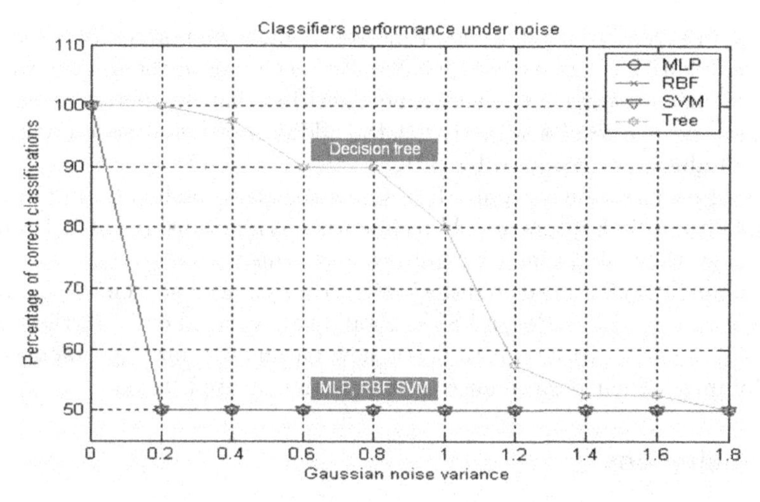

Fig. 7. Performance of the classifiers in presence of a zero mean Gaussian noise with different variances. The MLP, RBF and SVM techniques presented a similar behavior with the increment of the noise variance. The decision tree was less affect by the change on noise variance

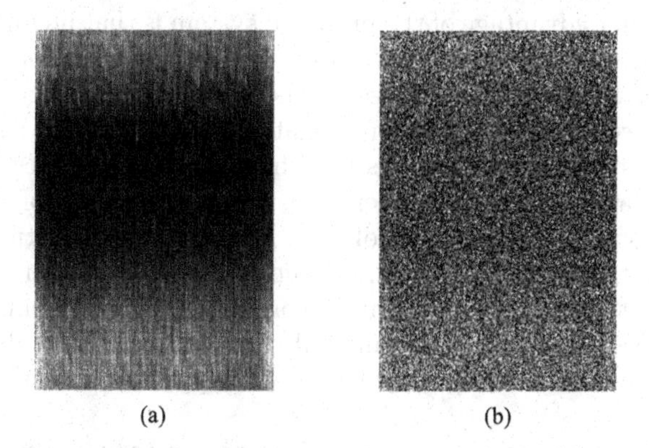

Fig. 8. Examples of the same acoustic maps: (**a**) in the absence of noise, and (**b**) with the inclusion of Gaussian noise with zero mean and different unitary variance

their outputs saturated and they classified all patterns with output 1 – what explain the percentage of 50% in the graph. In defense of MLP, RBF and SVM results, Fig. 8 shows an example of how the same map can have its appearance affected by the presence of noise – it is easy to see that even a human operator could make some of the misclassifications performed by these classifiers. However, different from the other classifiers, the decision tree showed a surprising performance (Fig. 7). Even with a zero mean, variance 1, Gaussian white noise (an example in Fig. 8b), the decision tree correctly classified up to 70% of the new input patterns. The superior performance obtained by the DTs might no be expected, since SVMs have been superior to them in several other datasets. However, it must be pointed out that each learning algorithm has its bias, what makes each one of them more suitable for particular datasets [23]. There is no algorithm that is superior to others in any dataset. The results suggest that the addition of noise made the dataset more suitable to be modeled by DTs.

Based on the results reported, it seems that the combination of acoustic maps, textural descriptors and AI techniques has a promising potential in real implementations of automatic monitoring systems.

Although the DT presented the better performance between the simulated CI algorithms, MLP, RBF and SVM should not be ruled out – further experiments should be carried out to define new parameters for these models that could improve their performance in this particular application.

8 Conclusions

In this paper, an automatic monitoring system for the dressing operations is described. Different from previous work that adopted CI methods to classify AE measurements in grinding operations, the proposed system is less sensitive to the AE RMS fluctuation along the time series (normally much more time needed). Another advantage of the proposed system is that its implementation is less expensive than solutions based on sensor fusion or Fourier spectrum analysis, since no additional hardware is needed.

Four CI-based classifiers were investigated: MLP networks, RBF networks, SVMs and DTs. The experimental study showed that accuracies could achieve over 87% for all classifiers. Furthermore, the performance of the proposed monitoring system in the presence of noise had also been investigated (Fig. 7). The results obtained revealed that, depending on the classifier, the proposed monitoring system could deal with noise conditions where the simulated noises were reproduced to match those commonly existed in industry plants. Overall, in the experiments, DT represented the best performance in the presence of noise.

The performed experiments suggest that this strategy is very promising for a future implementation of an on-line monitoring system. At its current stage, the experiments only considered the wheel concentricity problem. However, they can be applied to many other dressing anomalies that result in pattern changes in the acoustic map, like, for example, profile loss, vibration in dressing, lack of diamond in feed, diamond breakage and so on. It awaits our further exploration.

Acknowledgement

The authors would like to thank the support from FAPESP, CNPq, Zema Industrial Grinders, GE Fanuc, who provided financial support and donation of the equipment, and to the laboratory for Optimization of Manufacturing Processes (OPF) at University of São Paulo. They would also like to thank Dr. Ana Patrocinio for her helpful comments regarding the Haralick's descriptors.

References

1. Tönshoff, H.K., Jung, M., Männel, S., Rietz, W. (2000), Using acoustic signals for monitoring production processes. Ultrasonics 37, 681–686.
2. Maksoud, T.M.A., Atia, M.R. (2004), Review of intelligent grinding and dressing operations. Machining Science and Technology 8(2), 263–276.
3. Kwak, J.-S., Ha, M.-K. (2004b), Intelligent diagnostic technique of machining state for grinding. International Journal of Advacned Manufacturing Technology 23, 436–443.
4. Kwak, J.-S., Song, J.-B. (2001), Trouble diagnosis of the grinding process by using acoustic emission signals. International Journal of Machine Tools & Manufacture 41, 899–913.
5. Oliveira, J.F.G., Dornfeld, D.A. (2001), Application of AE contact sensing in reliable grinding monitoring. Annals of CIRP 2001, Berne, Switzerland 51(1), 217–220.
6. Toenshoff, H.K., Friemuth, T., Becker, J.C. (2002), Process monitoring in grinding. Annals of CIRP 51/2, 551–571.
7. Inasaki, I. (1999), Sensor fusion for monitoring and controlling grinding processes. International Journal of Advanced Manufacturing Technology 15, 730–736.
8. Oliveira, J.F.G., Dornfeld, D.A., Winter, B. (1994), Dimensional characterization of grinding wheel surface through acoustic emission. Annals of CIRP 1994, Berne, Switzerland 44(1), 291–294.
9. Li, X. (2002), A brief review: acoustic emission method for tool wear monitoring during turning. International Journal of Machine Tools & Manufacture 42, 157–165.
10. Kwak, J.-S., Ha, M.-K. (2004a), Neural network approach for diagnosis of grinding operation by acoustic emission and power signals. Journal of Materials Processing Technology 147, 65–71.
11. Oliveira, J.F.G., Silva, Eraldo Jannone Da, Biffi, Marcelo et al. (2002), New architecture control system for an intelligent high speed grinder. Abrasives Magazine, USA, no. 06, pp. 4–11.
12. Byrne, G., Dornfeld, D., Inasaki, I., Ketteler, G., König, W., Teti, R., (1995), Tool condition monitoring (TCM) – the status of research and industrial application. Annals of the CIRP 44/2: 541–567.
13. Tönshoff, H.K., Inasaki, I. (ed.) (2001), Sensors in Manufacturing. Wiley, New York.
14. Iwata, K., Moriwaki, T. (1977), Application of acoustic emission measurement to in-process sensing of tool wear. Ann. CIRP 26 (1–2), 19–23.
15. Winston, P.H. (1993), Artificial Intelligence. Addison-Wesley, Reading, MA.
16. Haykin, S. (1999), Neural Networks – A Comprehensive Foundation. Prentice-Hall, Englewood Cliffs, NJ.
17. Duda, R.O., Hart, P.E., Stork, D.G. (2000), Pattern Classification, 2nd Edn. Wiley, New York.
18. Hsu, C.-W., Chang, C.-C., Lin, C.-J. (2003), A practical guide to support vector classification. (http://www.csie.ntu.edu.tw/~cjlin/papers.html).
19. Smola, A.J., Barlett, P., Schölkopf, B. and Schurmmans, D. (1999), Advances in Large Margin Classifiers. MIT Press, Cambridge, MA.
20. Vapnik, V.N. (1995), The Nature of Statistical Learning Theory. Springer-Verlag, Berlin.

21. Lorena, A.C., Carvalho, A.C.P.L.F. (2003), Introdução às Máquinas de Vetores Suporte. Relatório Técnico do Instituto de Ciências Matemáticas e de Computação (USP/São Carlos), No. 192.
22. Quinlan, J.R. (1986), Induction of decision trees, Machine Learning 1, 81–106.
23. Mitchell, T.M. (1997), Machine Learning. McGraw-Hill, NY.
24. Kulkarni, A.D. (1994), Artificial Neural Networks for Image Understanding. Van Nostrand Reinhold - An International Thomson Publishing Company, New York.
25. Egmont-Petersen, M., Ridder, D., Handels, H. (2002), Image Processing with Neural Networks - A Review. Pattern Recognition 35, 2279–2301.
26. Haralick, R.M., Shanmugam, K., Dinstein, I. (1973), Textural features for image classification. IEEE Transactions on Systems, Man, and Cybernatics, SMC-3, 610–621.
27. Gotlieb, C.C., Kreyszig, H. (1990), Texture descriptors based on co-occurrence matrices. Computer Vision, Graphics, and Image Processing, 51, 70–86.
28. Powell, M.J.D. (1988), Radial basis function approximations to polynomials. Numerical Analysis 1987 Proceedings, 223–241, Dundee, UK.
29. Miccheli, C.A. (1986), Interpolation of scattered data: distances matrices and conditionally positive definite functions. Constructive Approximation 2, 11–22.
30. Gonzalez, R.C., Woods, R.E. (2002), Digital Image Processing. Prentice Hall, Englewood Cliffs, NJ.

Automated Novelty Detection in Industrial Systems

David A. Clifton[1,2], Lei A. Clifton[3], and Peter R. Bannister[1], and Lionel Tarassenko[1]

[1] Department of Engineering Science, University of Oxford, Oxford, UK, davidc@robots.ox.ac.uk, prb@robots.ox.ac.uk, lionel@robots.ox.ac.uk
[2] Oxford BioSignals Ltd., Abingdon, Oxfordshire, UK
[3] Nuffield Department of Anaesthetics, University of Oxford, Oxford, UK, lei.clifton@nda.ox.ac.uk

1 Introduction

1.1 Novelty Detection

Novelty detection is the identification of abnormal system behaviour, in which a model of normality is constructed, with deviations from the model identified as "abnormal". Complex high-integrity systems typically operate normally for the majority of their service lives, and so examples of abnormal data may be rare in comparison to the amount of available normal data. Given the complexity of such systems, the number of possible failure modes is large, many of which may not be characterised sufficiently to construct a traditional multi-class classifier [22]. Thus, novelty detection is particularly suited to such cases, which allows previously-unseen or poorly-understood modes of failure to be correctly identified.

Manufacturers of such high-integrity systems are changing the focus of their business such that they take on responsibility for provision of system maintenance [23]. Intelligent data analysis techniques are required to assess the "health" of system components, aiming to identify potential precursors of component failure in advance of actual system failure. This prognostic approach to condition monitoring is useful for types of fault that can be avoided if identified sufficiently early. In addition to providing early warning of critical failure, such techniques enable a "needs-based" approach to system maintenance, in contrast to the traditional dependence on maintenance scheduled at fixed intervals. Typically such warning systems require highly robust alarming mechanisms, with a minimal number of false positive activations, due to the cost involved of decommissioning and examining equipment following an alarm.

D.A. Clifton et al.: *Automated Novelty Detection in Industrial Systems*, Studies in Computational Intelligence (SCI) **116**, 269–296 (2008)
www.springerlink.com

In order to provide early warning of this large set of potentially ill-defined possible failures, a novelty detection approach may be adopted. Novelty detection is alternatively known as "one-class classification" [24] or "outlier detection" [25].

This chapter describes recent advances in the application of novelty detection techniques to the analysis of data from gas-turbine engines. Whole-engine vibration-based analysis will be illustrated, using data measured from case-mounted sensors, followed by the application of similar techniques to the combustor component. In each case, the investigation described by this chapter shows how advances in prognostic condition monitoring are being made possible in a principled manner using novelty detection techniques.

1.2 Chapter Overview

Novelty detection theory is introduced in Sect. 2, in which data evaluation techniques, modelling methods, and setting of a decision boundary are described. Complementary on-line and off-line analysis of jet engine vibration data are presented in Sect. 3, in which different novelty detection methods are investigated for their benefit in providing advance warning of failure, in comparison to conventional techniques. Section 4 presents an analysis of novelty detection techniques applied to a gas-turbine combustor, showing that single-component analysis can provide indication of failure. Finally, conclusions are drawn in Sect. 5, providing recommendations for future novelty detection implementations.

2 Novelty Detection for Industrial Systems

This section describes a framework for novelty detection divided into the following stages:

- Existing Methods
- Pre-processing
- Visualisation
- Construction of a model of normality
- Setting novelty thresholds to detect abnormality

2.1 Existing Methods

Novelty detection techniques are typically divided into statistical (or probabilistic) methods, and machine learning (or neural network) methods. The former, being based upon probabilistic foundations, often claim to be the more principled approaches, in which the concept of novelty can be statistically related to the probability of observing abnormal data. The latter are more data-driven approaches, in which models of normality are typically learned from data in such a way as to maximise the chance of making a correct classification decision given previously-unseen examples. Typically, no statistical assumptions about the data are made, relying instead upon only the observed data to construct a model of normality.

Statistical Methods of Novelty Detection

Fundamental to statistical methods is the assumption that normal data are generated from an underlying data distribution, which may be estimated from example data. The classical statistical approach is the use of density estimation techniques to determine the underlying data distribution [2], which may be thresholded to delimit normal and abnormal areas of data space. These are parametric techniques, in which assumptions are made about the form of the underlying data distribution, and the parameters of the distribution estimated from observed data. Such assumptions may often prove too strong, leading to a poor fit between model and observed data.

More complex forms of data distribution may be assumed by using Gaussian Mixture Models [1,22], or other mixtures of basic distribution types [3,4]. Such techniques can grow the model complexity (increasing the number of distributions in the mixture model) until a good fit with data is deemed to have occurred. Typically, model parameters are estimated from data using the Expectation-Maximisation algorithm [5]. Such methods can suffer from the requirement of large numbers of training data from which to accurately perform parameter estimation [6], commonly termed the curse of dimensionality [31]. Furthermore, the number of components in a mixture model may need to become very large in order to adequately model observed data, which may lead to poor ability to generalise to previously-unseen data (over-fitting).

Non-parametric approaches to statistical novelty detection include so-called boundary- or distance-based methods, such as Parzen window estimators and clustering techniques, and are discussed in Sect. 2.4.

For statistical-based novelty detection of time-series data, Hidden Markov Models (HMMs) have been used, which provide a state-based model of a data set. Transitions between a number of hidden states are governed by a stochastic process [7]. Each state is associated with a set of probability distributions describing the likelihood of generating observable emission events; these distributions may be thresholded to perform novelty detection [8].

HMM parameters are typically learned using the Expectation-Maximisation algorithm. Novelty detection with HMMs may also be performed by constructing an abnormal state, a transition into which implies abnormal system behaviour [9].

A similar state-based approach to novelty detection in time-series data is taken by Factorial Switching Kalman Filters [10]. This is a dynamic extension of the Switched Kalman Filter [12], which models time-series data by assuming a continuous, hidden state is responsible for data generation, the effects of which are observed through a modelled noise process. As with the HMM novelty detection approach, an explicit abnormal mode of behaviour is included within the model, which is used to identify departures from normality.

Also utilising a dynamical model of time-series normal data, the Multi-dimensional Probability Evolution (MDPE) method [13] characterises normal data by using a non-linear state-space model. The regions of state space visited during normal behaviour are modelled, departures from which are deemed abnormal.

Machine Learning Methods of Novelty Detection

Non-statistical methods of novelty detection attempt to construct a model of normality without assuming an underlying data-generating distribution. One of the most well-known examples of this method is the neural network, in which a highly-connected, weighted sum of nodes is trained using observed data. Typically, neural networks are trained for multi-class classification problems, in which example data are classified into one of a pre-defined set of classes [22].

In order to be used in novelty detection (single-class classification), artificial data may be generated around the normal data in order to simulate an abnormal class [11]. Alternative approaches include using the network's instability when presented with previously-unseen abnormal data for the purposes of novelty detection [14].

The Self-Organising Map (SOM), initially proposed for the clustering and visualisation of high-dimensional data [15], provides an unsupervised representation of training data using a neural network. Various applications of the SOM to novelty detection have been proposed [16, 17], while they have been extended into generic density estimators for statistical novelty detection [18].

A more recent successor to the neural network is the Support Vector Machine (SVM), in which a hyperplane is found that best separates data from different classes, after their transformation by a kernel function [19]. In application to novelty detection, two main approaches have been taken. The first finds a hypersphere (in the transformed space) that best surrounds most of the normal data with minimum radius [20]. The second approach separates the normal data from the origin with maximum margin [21]. The latter has also been extended to novelty detection in jet engine vibration data, in which examples of abnormal data can be used to improve the model [37].

2.2 Pre-Processing

When constructing a multi-dimensional novelty detection systems, whose inputs are derived from several different sensors or parameters, normalisation is a primary pre-processing step applied to the data. The aim of normalisation in this case is to equalise the dynamic range of the different features so that all can be regarded as equally important *a priori.* It removes dependence upon absolute amplitudes, whilst preserving information about the relative "normality" of samples.

On each channel of data (or features extracted from them), presented in Sects. 3 and 4, we use the component-wise normalisation function $N(\mathbf{x}_i)$, defined [26] to be a transformation of the d elements within pattern $\mathbf{x}_i$:

$$N(\mathbf{x}_i) = \frac{\mathbf{x}_i^d - \boldsymbol{\mu}^d}{\boldsymbol{\sigma}^d}, \quad \forall d = 1 \dots D \tag{1}$$

where $(\boldsymbol{\mu}, \boldsymbol{\sigma})$ are vectors of D elements, computed component-wise across all $i = 1 \dots I$ patterns:

$$\boldsymbol{\mu}^d = \frac{1}{I}\sum_{i=1}^{I} \mathbf{x}_i^d \qquad \boldsymbol{\sigma}^d = \left(\frac{1}{I-1}\sum_{i=1}^{I}(\mathbf{x}_i^d - \boldsymbol{\mu}^d)^2\right)^{\frac{1}{2}} \tag{2}$$

In practical terms, the patterns are sets of sensor and/or parameter measurements that are most commonly represented as vectors. An example is given later in Equation 16.

2.3 Visualisation

Visualisation is a key method in exploring data sets, both in terms of confirming the results of normalisation, and in deciding which method to use when constructing a model of normality. Typically, patterns are of high dimensionality, which may contain multiple features derived from multiple channels, making the explicit visualisation of such data difficult. This section describes a method that, while allowing exploration of the data set during the construction of the novelty detection system, also provides a convenient method of describing the results of analysis to eventual users of the system.

The "usability" of the system is of great importance in monitoring of industrial systems, in which the eventual users are typically not familiar with pattern recognition techniques. Visualisation can provide a suitable method of showing the model of normality, novelty thresholds, and test patterns in such a way that makes the use of the novelty detection system more intuitive.

Topographic Projection

Unlike variance-preserving techniques such as Principle Component Analysis, topographic projection is a transformation that attempts to best preserve, in the projected space of lower-dimensionality (*latent space*, $\mathbb{R}^q$), distances between data in their original high-dimensional space (*data space*, $\mathbb{R}^d$). The *Sammon stress metric* [27] is based upon the distance d_{ij} between pairs of points (x_i, x_j) in $\mathbb{R}^d$, and the distance d^*_{ij} between the corresponding pair of points (y_i, y_j) in $\mathbb{R}^q$:

$$E_{\mathrm{sam}} = \sum_{i=1}^{N}\sum_{j>i}^{N}(d_{ij} - d^*_{ij})^2 \tag{3}$$

in which the distance measure is typically Euclidean. *Sammon's mapping* attempts to minimise (3) using gradient descent [28] techniques. However, this method provides a mapping which is only defined for the training set. This limitation is overcome using the *NeuroScale* method which provides a mapping that can be applied to datasets other than the training set, making it well suited for the proposed application of model-based novelty detection when compared with other projection methods which are only defined for the training set.

NeuroScale

With the NeuroScale method [29], a Radial Basis Function (*RBF*) neural network [22] is used to parameterise the mapping from $\mathbb{R}^d$ to $\mathbb{R}^q$, in which E_{sam} is minimised. This method allows the key advantage that new test patterns may be projected in $\mathbb{R}^q$ without generating a new mapping.

The network architecture (using a single hidden layer between the input and output layers) is typically selected such that the number of hidden nodes is an order of magnitude greater than the dimensionality of the input patterns [30]. Using the same guidelines, the number of available training patterns should be an order of magnitude greater than the number of hidden nodes, in order to adequately populate data space. Each node in the hidden layer corresponds to the centre of a radial basis function in data space, the initial positions of which we set to be those of patterns randomly selected from the training set.

The output weights of the RBF network (i.e., those from hidden layer to output layer) were initialised using Principal Component Analysis (PCA). The $N = 2$ eigenvectors of the data covariance matrix with highest corresponding eigenvalues are found, and these are the N *principal components* of the training set. They represent the projection of the data in N-dimensional space that retains maximum variance information from high-dimensional space. They provide an initial projection of the data, giving targets for the output layer from which initial values of the output weights are found, which is then refined through network training.

Training the RBF network is a two-stage process. In the first stage, the parameters of the radial basis functions are set so that they model the unconditional probability density of the data, $p(x)$ - that is, the distribution of probability mass describing the likelihood of observing normal data x. In the second stage, the output weights are set by optimising an error function using methods from linear algebra [31].

Using Visualisation

Figure 1 shows an example in which 20-dimensional patterns derived from jet engine vibration data (described in more detail later in Sect. 3) are projected into 2-dimensional space. Note that the axes of NeuroScale projections are

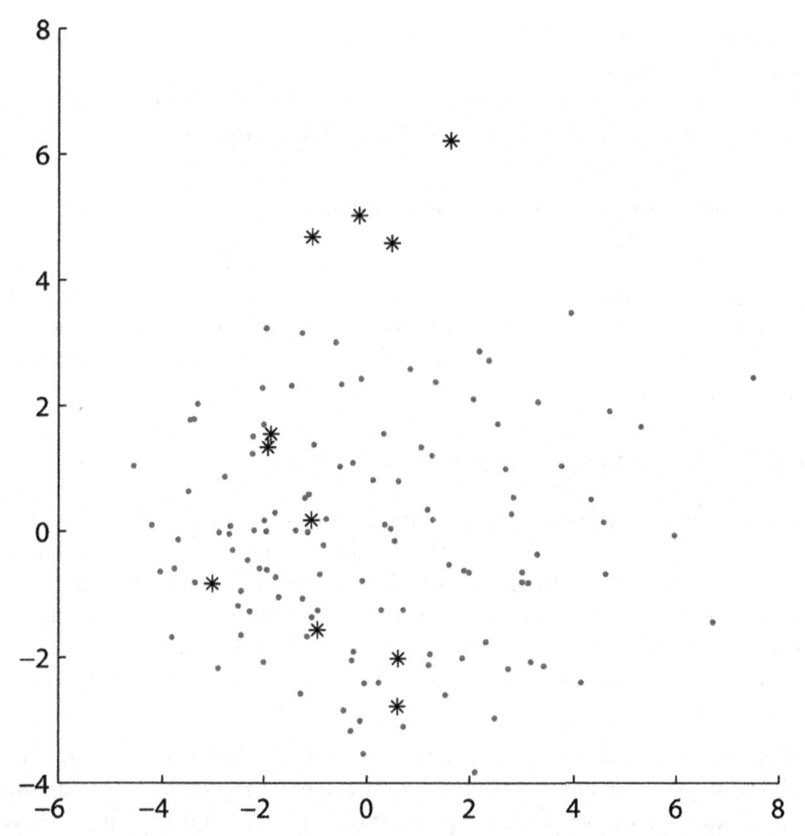

Fig. 1. NeuroScale visualisation of 20-dimensional patterns derived from jet engine vibration data. "Normal" patterns are projected as dots, "abnormal" patterns (as labelled by the data provider) are projected as asterisks. It is clear from the visualisation that seven of the "abnormal" patterns lie within the cluster of points described by the "normal" patterns, revealing that the labels supplied by the data-providers were not accurate and should be modified before selecting patterns from which to train a model of normality

unit-less. Using the component-wise normalisation scheme described above, the visualisation shows that "normal" patterns form a main cluster. 4 of the 11 patterns labelled "abnormal" by the data provider are projected significantly removed from this cluster.

This example illustrates the use of visualisation for the verification of data labels, as it is often the case that, in practice, many patterns labelled "abnormal" by domain experts may actually appear normal after feature extraction, and vice versa. On discussion with the data providers, the seven "abnormal" patterns lying within the cluster formed by normal patterns were found to contain system faults such that no vibration-based consequences were observable in the data [32]. That is, the visualisation technique (based on vibration data) correctly shows that the 7 patterns originally labelled "abnormal" should not be separate from normal data in this case (and thus may be included as "normal" data in the training set).

Using visualisation in this way allows the exploration of the data set to define which patterns should be used for training a model of normality.

2.4 Constructing a Model of Normality

As previously defined, classifying data as "abnormal" in order to raise an alert about the condition of a system requires a model of normality. This section describes the application of two main types of normal models that have proved successful in the analysis of the gas-turbine data described later in this chapter.

Distance-Based Models of Normality

Distance-based models of normality attempt to characterise the area of data space occupied by normal data, with test data being assigned a novelty score based on their (typically Euclidean or Mahalanobis) distance from that model. A novelty threshold may be set on the novelty score, above which data are deemed to be abnormal; this is further discussed in Sect. 2.5.

Here, we illustrate the process of constructing a distance-based model using two methods:

- Cluster-based models, which have previously been applied to jet engine data [33–35] and ship turbocharger data [36]. This method uses distances in the untransformed feature space, and is illustrated using vibration data in Sect. 3.
- Support Vector Machine models, which have previously been applied to jet-engine data [37] and combustion data [38,39]. This method uses distances in a transformed kernel space, and is illustrated using combustion data in Sect. 4.

Cluster-Based Models

When modelling a system whose state is represented by a large training set of patterns, it is often desirable to be able to represent the set of patterns by a smaller set of generalised "prototype" patterns, making optimisation of a model computationally-tractable. The k-means clustering algorithm [30] is an iterative method of producing a set of prototype patterns μ_j (for $j = 1 \ldots k$) that represent the distribution of a (typically much larger) set of patterns x_i.

The k-means clustering algorithm is used, as described in [40], to construct a model of normality from "normal" patterns in the data sets later described in this chapter. In this method, the distribution of "normal" patterns is defined by $\mathbf{C}_k$ cluster centres in $\mathbb{R}^d$ space, each with an associated *cluster width* σ_k. A novelty score $z(\mathbf{x})$ may be computed for shape vector $\mathbf{x}$ with respect to the K cluster centres:

$$z(\mathbf{x}) = \min_{k=1}^{K} \frac{d(\mathbf{x}, \mathbf{C}_k)}{\sigma_k} \tag{4}$$

where $d(\mathbf{x}, \mathbf{C}_k)$ is Euclidean distance. We use the definition of width σ_k from [34]

$$\sigma_k = \sqrt{\frac{1}{I_k} \sum_{i=1}^{I_k} d(\mathbf{x}_i, \mathbf{C}_k)^2}. \tag{5}$$

for the I_k points which have closest cluster centre $\mathbf{C}_k$. This allows an intuitive interpretation of the magnitude of novelty scores: $z(\mathbf{x})$ is the number of standard deviations that pattern $\mathbf{x}$ lies from its closest cluster centre, relative to the distribution of training data about $\mathbf{C}_k$.

Investigation of the placement of cluster centres is possible using the NeuroScale method. The position of cluster centres with respect to the data set may be determined by projection using the NeuroScale network previously trained using the patterns from the example data set. Selection of a candidate model of normality can be assisted through use of the projections generated by the neural network, where we ensure that the cluster centres accurately represent the distribution of training patterns.

Due to the random initialisation of cluster centres, the positions of those centres at the conclusion of the training process may vary between different models trained upon the same data set. Typically, the algorithm is run several times (each producing a different set of cluster centre locations μ_j), and one of the candidate models is selected based on some metric of fitness in characterising the data x_i.

The optimal number of centres k required to characterise the training set of normal data is often selected using a separate set of data (the validation set), to test the model's ability to generalise to previously-unseen data.

Finally, a threshold H is applied to $z(\mathbf{x})$ such that all patterns $z(\mathbf{x}) \geq H$ are classified "abnormal". This is described in Sect. 2.5.

Support Vector Machine Models

Support vector machines (SVMs) belong to a family of generalized linear classifiers, used for classification and regression. A special property of SVMs is that they simultaneously minimize the empirical classification error and maximize the geometric margin; hence they are also known as *maximum margin classifiers*.

We follow the strategy developed in a key paper for SVM novelty detection [42] that maps the data into the feature space corresponding to the kernel function, and separates them from the origin with maximum margin.

The l training data $\mathbf{x}_1, \ldots, \mathbf{x}_l \in \mathbb{R}^d$ can be mapped into another feature space $\mathbb{F}$ through a feature mapping $\mathbf{\Phi}$: $\mathbb{R}^d \rightarrow \mathbb{F}$. The kernel function operates on the dot product of the mapping function

$$k(\mathbf{x}_i, \mathbf{x}_j) = (\mathbf{\Phi}(\mathbf{x}_i) \cdot \mathbf{\Phi}(\mathbf{x}_j)). \tag{6}$$

A Gaussian kernel function is used here to suppress the growing distances for larger feature spaces [41]

$$k(\mathbf{x}_i, \mathbf{x}_j) = \exp\left(-\|\mathbf{x}_i - \mathbf{x}_j\|^2 / 2\sigma^2\right), \tag{7}$$

where σ is the width parameter associated with the kernel function.

For this investigation, we define the following function

$$g(\mathbf{x}) = \rho_0 - \sum_{i=1}^{N_s} \alpha_i k(\mathbf{s}_i, \mathbf{x}) \tag{8}$$

to assign novelty values to data, such that abnormal data (i.e., those outside the single, "normal" training class) take positive values, while normal data take zero values. In this equation, N_s is the number of support vectors, $\mathbf{s}_i$ are support vectors and α_i are non-zero Lagrangian multipliers.

A threshold is set on these novelty scores as described in Sect. 2.5.

Probabilistic Models of Normality

In contrast to distance-based modelling methods, probabilistic approaches to constructing a model of normality estimate the unconditional probability density function $p(\mathbf{x})$ of normal training data. In this investigation, we illustrate the method using Parzen window density estimation [22].

This is a non-parametric kernel-based method, in which the density function is represented by a linear superposition of kernel functions, with one kernel centred on each sample. We choose the Gaussian kernel for Parzen window methods due to its convenient analytical properties. Following [43], we set the width parameter σ for Parzen windows to be the average distance of k nearest neighbours from each sample in the normal training data. Instead of a fixed value of k used in [43], here we specify k to be a fraction of the total number of training data N (e.g., $k = N/10$), so that the value of k is adjusted depending on the number of the training data.

With the unconditional data density $p(\mathbf{x})$ estimated, we may place a probabilistic threshold to separate "normal" from "abnormal" as described in Sect. 2.5.

2.5 Novelty Scores and Thresholds

Though it is desirable for probabilities to be retained as long as possible throughout the classification process, a novelty detection scheme must ultimately decide if a value x is "normal" or "abnormal", for which a decision boundary H on x must be determined. We term this the *novelty threshold*.

Thresholds for Distance-Based Models of Normality

Typically, for distance-based models of normality, the threshold H on novelty score $z(\mathbf{x})$ is selected to best separate the "normal" and "abnormal" patterns in the data set. However, particularly in the monitoring of high-integrity systems, the number of available "abnormal" examples may be small. Furthermore, this method is heuristic, and requires manual selection of H for each new data set. This is illustrated with clustering models and the vibration data set, described in Sect. 3.

A more principled approach may be obtained by calibrating novelty scores into probabilities, and then setting the novelty threshold as with probabilistic models of normality. This approach is illustrated with the SVM model and combustion data set, described in Sect. 4.

For calibration of novelty scores, [44] proposed a non-parametric form of regression, with the restriction that the mapping from scores into probabilities is isotonic (i.e., non-decreasing); that is, the classifier ranks examples correctly. The Pair-Adjacent Violators (PAV) algorithm [44] is employed to perform the isotonic regression, which finds the stepwise-constant isotonic function $g^*(\mathbf{x})$ that fits the data according to a mean-squared error criterion.

Let $\mathbf{x}_i$ $(i = 1, \ldots, l)$ be the training examples from normal and abnormal classes $\{\mathcal{C}_0, \mathcal{C}_1\}$, $g(\mathbf{x}_i)$ be the value of the function to be learned for each training samples, and $g^*(\mathbf{x})$ be the function obtained from isotonic regression. The PAV algorithm takes the following steps:

STEP 1: Sort the examples according to their scores, in ascending order. Initialise $g(\mathbf{x}_i)$ to be 0 if $\mathbf{x}_i$ belongs normal class, and 1 if abnormal class.
STEP 2: If $g(\mathbf{x}_i)$ is isotonic, then return $g^*(\mathbf{x}_i) = g(\mathbf{x}_i)$. If not, go to STEP 3.
STEP 3: Find a subscript i such that $g(\mathbf{x}_i) > g(\mathbf{x}_{i+1})$. The examples $\mathbf{x}_i$ and $\mathbf{x}_{i+1}$ are called pair-adjacent violators. Replace $g(\mathbf{x}_i)$ and $g(\mathbf{x}_{i+1})$ with their average:

$$g^*(\mathbf{x}_i) = g^*(\mathbf{x}_{i+1}) = \{g(\mathbf{x}_i) + g(\mathbf{x}_{i+1})\}/2 \tag{9}$$

STEP 4: Set $g(\mathbf{x}_i)$ as the new $g^*(\mathbf{x}_i)$. Go back to STEP 2.

$g^*(\mathbf{x})$ is a step-wise constant function which consists of horizontal intervals, and may be interpreted as $P(\mathcal{C}_1|\mathbf{x})$, the probability that sample $\mathbf{x}$ is abnormal (i.e., belongs to $\mathcal{C}_1$, the abnormal class). For a test example $\mathbf{x}$, we first find the interval to which its score $z(\mathbf{x})$ belongs. Then we set the value of $g^*(\mathbf{x})$ in this interval to be $P(\mathcal{C}_1|\mathbf{x})$, the probability estimate of $\mathcal{C}_1$ given $\mathbf{x}$.

If the scores rank all examples correctly, then all class $\mathcal{C}_0$ examples will appear before all class $\mathcal{C}_1$ examples in the sorted data set in STEP 1. The calibrated probability estimate $g^*(\mathbf{x})$ is 0 for class $\mathcal{C}_0$ and 1 for class $\mathcal{C}_1$. Conversely, if the scores do not provide any information, $g^*(\mathbf{x})$ will be a constant function, taking the value of the average score over all examples in class $\mathcal{C}_1$.

The PAV algorithm used in isotonic regression may be viewed as a binning algorithm, in which the position and the size of the bins are chosen according to how well the classifier ranks the samples [44]. The practical use of this algorithm in calibrating novelty scores $z(\mathbf{x})$ into probabilities is illustrated in Sect. 4.

Thresholds for Probabilistic Models of Normality

Novelty threshold H can be defined using the unconditional probability distribution $p(\mathbf{x}) < H$. However, because $p(\mathbf{x})$ is a distribution function, it is necessary to integrate to give the cumulative probability $P(\mathbf{x})$. This is then used to set thresholds in relation to the actual probability of observing sensor noise (e.g., $P(\mathbf{x}) = 10^{-9}$).

A more principled method of setting thresholds for novelty detection uses Extreme Value Theory (EVT) to explicitly model the tails of the distribution of normal data.

EVT is concerned with modelling the distribution of very large or very small values with respect to a generative data distribution. Here, we consider "classical" EVT as previously used for novelty detection [45, 46], in contrast to an alternative method commonly used in financial applications, often termed the *peaks-over-threshold* approach [47]. We do not use the latter for this novelty detection application due to its requirement to adopt an arbitrary threshold above which peaks are measured.

Throughout the investigation described in this chapter, we consider the estimation of the probability of observing abnormally *large* values with respect to a set of normal data. Consideration of abnormally small values requires a simple modification of the theory, not further pursued here.

Consider a "normal" training set of m i.i.d. (independent and identically distributed) data, $\mathbf{X} = \{x_1, x_2, \ldots x_m\}$, distributed according to some function $D(x)$, with maximum $x_{\max} = \max(\mathbf{X})$. We define a distribution function for $x_{\max}$ to be $H(x_{\max} \leq x)$. I.e., our belief in the value of the maximum of the m data drawn from distribution D (over the range of x) is modelled by H.

It can be shown [48] that for H to avoid degeneration as $m \to \infty$, it must converge according to the transform

$$x_{max} \doteq \sigma_m x + \mu_m \tag{10}$$

for some location and scale parameters, $\mu_m \in \mathbb{R}$ and $\sigma_m \in \mathbb{R}^+$, respectively, and where $\doteq$ is a weak convergence of distributions. Furthermore, for any underlying data distribution D, the limit distribution must take the normalised form

$$H(x_{\max} \leq x) \doteq H(\frac{x - \mu_m}{\sigma_m}) \doteq H(y_m) \tag{11}$$

where $y_m = (x - \mu_m)/\sigma_m$ is termed the *reduced variate*. According to the Fisher–Tippett theorem [48], H must belong to one of three families of extreme value distributions (derived from the Generalised Extreme Value distribution [49]). In this investigation, we consider data distributed according to the one-sided standard Gaussian $\mathbf{X} \sim |N(0,1)|$, which converges to the Gumbel distribution for extreme values [49]. The probability of observing some $x_{\max} \leq x$ given the data is $P(x_{\max} \leq x|\mathbf{X})$, or $P(x|\mathbf{X})$ for simplicity, given by the Gumbel form:

$$\begin{aligned} P(x|\mathbf{X}) = P(x|\boldsymbol{\theta}) &\doteq H(y_m) \\ &\doteq \exp(-\exp(y_m)) \end{aligned} \tag{12}$$

where model parameters $\boldsymbol{\theta} = \{\mu_m, \sigma_m\}$ are the location and scale parameters from the reduced variate $y_m = (x_m - \mu_m)/\sigma_m$, and are derived from $\mathbf{X}$. The associated probability density function is found by differentiation:

$$\begin{aligned} p(x|\mathbf{X}) = p(x|\boldsymbol{\theta}) &= \sigma_m^{-1} \exp\{-y_m - \exp(y_m)\} \\ &= \sqrt{\lambda_m} \exp\{-y_m - \exp(y_m)\} \end{aligned} \tag{13}$$

which we term the *Extreme Value Distribution* (EVD). Note that, for later convenience, we use the precision $\lambda_m = 1/\sigma_m^2$. Classical EVT assumes that the location and scale parameters are dependent only on m [50], which has been verified via Monte Carlo simulation for $m = 2, \ldots, 1{,}000$ [45]. These take the form

$$\mu_m = \sqrt{2 \ln m} - \frac{\ln \ln m + \ln 2\pi}{2\sqrt{2 \ln m}} \qquad \lambda_m = 2 \ln m \tag{14}$$

Thus, using the EVD equation (13), we can directly set novelty thresholds in the Parzen window model, which has placed a set of Gaussian kernels in feature space. This is illustrated in Sect. 3.

3 Gas-Turbine Data Analysis

Here, we use the novelty detection techniques described previously for the analysis of jet-engine vibration data, recorded from a modern civil-aerospace engine. A cluster-based distance model and a Parzen window-based probability model are constructed as described in Sect. 2.4. Novelty thresholds are then set for these models as described in Sect. 2.5.

First, this section provides an overview of jet engine operation, and the vibration data obtained. Then, two monitoring paradigms are described:

- *Off-line Novelty Detection*, in which vibration data from an entire flight are summarised into a single pattern. Novelty detection then takes place on a flight-to-flight basis using these patterns. This scheme is suitable for ground-based monitoring of a fleet of engines.
- *On-line Novelty Detection*, in which data *within* flights are compared to a model of normality. Novelty detection takes place on a sample-by-sample basis. This scheme is suitable for "on-wing" engine monitoring.

3.1 System Description

Jet Engine Operation

Modern aerospace gas-turbine engines divide the task of air compression from atmospheric pressure to that ultimately required within the combustion chamber into several stages. Gas-turbine engines within the civil aerospace market involve up to three consecutive compression stages: the low pressure (*LP*), intermediate pressure (*IP*), and high pressure (*HP*) stages [51]. Air passes through each stage as it travels from the front of the engine to the rear, being further compressed by each, until it reaches the combustion chamber.

Each of the compressor stages is driven by its own turbine assembly, resulting in three corresponding turbine units situated within the exhaust stream at the rear of the engine. Each compressor is linked to its corresponding turbine by a separate shaft, which are mounted concentrically. In three-compressor engines, these are named the LP shaft, the IP shaft, and the HP shaft. The *operational point* of the engine is often defined in terms of the rotational speed of these shafts.

Engine Vibration Measurement

Transducers are mounted on various points of the engine assembly for the measurement of engine vibration. Vibration data used for investigations described in this report were acquired using a system [52] that computes Fast Fourier Transforms (*FFTs*) representative of engine vibration every 0.2 s, for each sensor output. Engine vibration is assumed to be pseudo-stationary over this measurement period such that the generated FFTs are assumed to be close approximations of actual engine vibration power spectra.

A *tracked order* is defined [53] to be the amplitude of engine vibration measured within a narrow frequency band centred on the fundamental or a harmonic of the rotational frequency of a shaft.

During normal engine operation, most vibration energy is present within tracked orders centred on the fundamental frequency of each rotating shaft; we define these to be *fundamental tracked orders*. Using the terms *LP*, *IP*, and *HP* to refer to engine shafts, we define fundamental tracked orders associated with those shafts to be *1LP*, *1IP*, and *1HP*, respectively.

Significant vibration energy may also be observed at harmonics of the rotational frequency of each shaft. These *harmonic tracked orders* may be expected to contain less vibration energy than corresponding fundamental tracked orders during normal engine operation. In the example of an LP shaft rotating with frequency 400 Hz, harmonic tracked orders may be observed at frequencies $400n$ Hz, for $n = 0.5, 2, 3, 4, \ldots$. We define these harmonic tracked orders of the LP shaft to be *0.5LP*, *2LP*, *3LP*, *4LP*, ...

The system used to acquire data for the investigations described within this report automatically identifies peaks in vibration spectra corresponding to fundamental and harmonic tracked orders, using measurements of the rotational frequency of each shaft. From these peaks, a time series of vibration amplitude and phase for each tracked order is generated.

3.2 Off-Line Novelty Detection

If there is no strict requirement to carry out novelty detection in real-time, then it is possible to compute summary data structures at the end of each flight, and compare the time-averaged behaviour of key features against previous engine runs. This typically suits ground-based monitoring, in which flight data from a fleet of engines is summarised and analysed by engine manufacturers.

An example of this is shown in Fig. 2 where the vibration amplitude (of a fundamental tracked order) has been averaged against speed over the known operating speed-range. This is a *vibration signature* for the flight, and is typically high-dimensional (with 400 speed sub-ranges usually considered).

Increasing dimensionality of data requires exponentially increasing numbers of patterns within the data set used to construct a general model; this is termed the *curse of dimensionality* [22]. In order to avoid this problem, each 400-dimensional vibration signature is summarised by a 20-dimensional pattern $\mathbf{x}$. This is performed by computing a weighted average of the vibration amplitude values $a(s)$ over $N = 20$ speed sub-ranges [26]. The dth dimension of shape vector $\mathbf{x}^n$, for $n = 1 \ldots N$, is defined to be:

$$\mathbf{x}^d = \int_{s_{\min}}^{s_{\max}} a(s)\omega_d(s)ds \tag{15}$$

in which the vibration amplitude $a(s)$ is integrated over the speed range $s : [s_{\min}\ s_{\max}]$, using weighting functions $\omega_d(s)$, for $d : 1 \ldots N$. Thus, each flight of a test engine results in a 20-dimensional pattern.

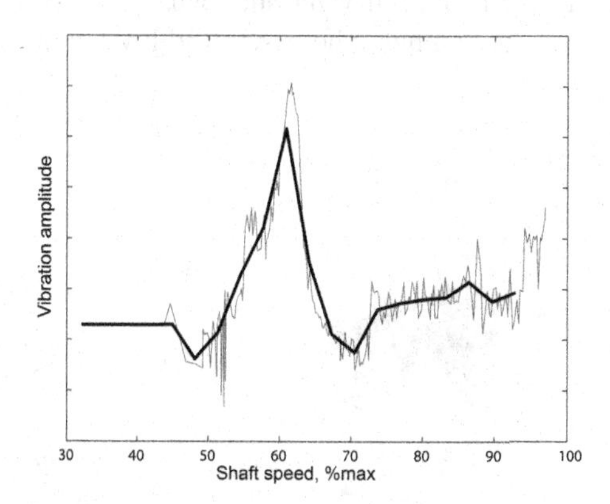

Fig. 2. Constructing a quantised 20-D pattern (*thick line*) from a high-dimensional vibration signature (*thin line*), in which average engine vibration response is plotted against a range of shaft speeds - axes have been anonymised for reasons of data confidentiality

A data set was provided for this investigation consisting of 137 flights. Flights {1...127} were "normal". A change in engine condition occurred during flights {128...135}, which was retrospectively determined to be an engine component becoming loose. An engine event was observed during flights {136...137}.

Flights {1...80} were used as the training set for both distance-based and probabilistic models. Flights {81...137} were used as a test set. From each flight, a vibration signature was constructed. Those flights which covered less than 60% of the speed range (indicating that the aircraft did not leave the ground) were deemed to be "invalid", and not considered within the analysis, to ensure that only fully-specified vibration signatures were used.

Component-wise normalisation, as described in Sect. 2.2 was applied to the signatures for each flight, and models of normality constructed using cluster-based and Parzen window methods, as described in Sect. 2.4. Novelty thresholds were then set using the methods described in Sect. 2.5.

An example of resulting models is shown in Fig. 3, in which visualisation has been performed using the NeuroScale method described in Sect. 2.3. Here, models have been trained in 2-dimensional visualisation space for the purposes of explanation. In the actual novelty detection exercise, models were constructed in feature space (here, the 20-dimensional space of the patterns derived from the vibration signatures).

In the example figures, projected normal patterns are plotted as dots (one per flight). The last flights of the engine are plotted as crosses, and are notably separated from the cluster formed by the normal patterns. These flights were known to have been abnormal in the final two flights (in which an engine event occurred), but this retrospective analysis shows that the flights immediately prior to the event are also classified as abnormal.

This is shown in Fig. 4, in which novelty scores for each flight are shown for both distance-based and probability models, with novelty thresholds (determined as described in Sect. 2.5 using heuristic and EVT methods, respectively)

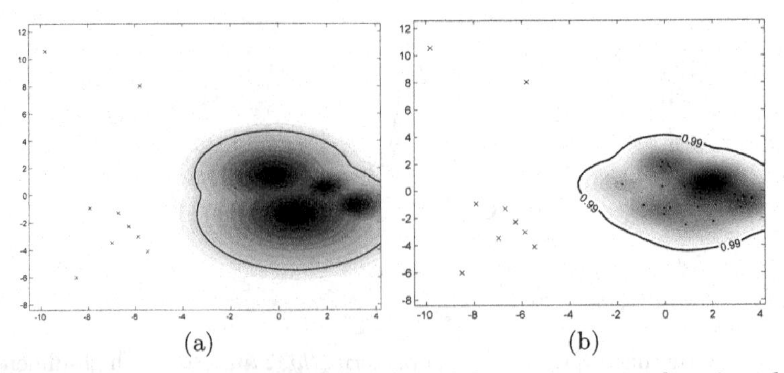

(a) (b)

Fig. 3. (**a**) Distance-based model using $k = 4$ cluster centres and projected data, constructed using cluster-based methods, (**b**) Probabilistic model and projected data, constructed using Parzen-windows method

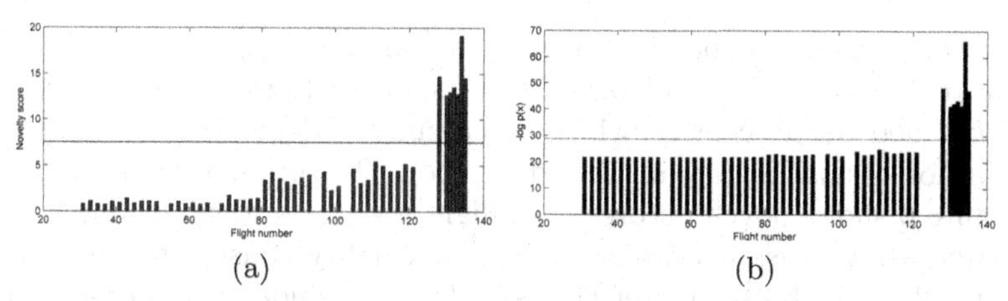

Fig. 4. (**a**) Novelty scores for the distance-based model, with novelty threshold (*dashed line*), (**b**) Novelty scores for the probabilistic model, with novelty threshold (*dashed line*)

shown as a horizontal line. It can be seen that both the event flights {136, 137} exceed the novelty threshold, but also that the preceding five flights {131...135} exceed the threshold.

An advantage of the probabilistic-based model is in the setting of the novelty threshold (via EVT), shown as a thick line for the 2-D example model, which occurs at $P(x) = 0.99$. This can be set automatically at run-time, in contrast to the heuristic setting required for the distance-based model.

Both models (distance-based and probability-based) provide early indication of eventual engine events, while the latter does so with automated training and a principled probabilistic approach.

3.3 On-Line Novelty Detection

On-line novelty detection typically takes place "on-wing". In this investigation, we form a shaft-specific pattern [54] at the sampling rate of the acquisition system, consisting of fundamental and harmonic tracked orders relating to the shaft being modelled:

$$[F,\ 1H,\ 1.5H,\ 2H,\ 3H,\ 4H,\ 5H,\ RE] \tag{16}$$

where "F" is the amplitude of the fundamental tracked order, "nH" is the amplitude of the nth harmonic tracked order, and "RE" is "residual energy", defined to be broadband energy unassigned to the fundamental tracked order, or its harmonics (and is thus "residual").

By considering the specific design of the engine, non-integer multiples of the shaft frequencies that reflect internal gearing configurations may provide specific information about the state of internal bearings in the engine.

As before, each element in the pattern is component-wise normalised to ensure that each varies over a similar dynamic range. The same training and test sets used in off-line monitoring were used to train both distance-based and probabilistic models, as described in Sect. 2.4.

Here, however, the full range of test data is available for training, rather than a summary vibration signature. In order to ensure that the training process is tractable in real-time, the number of data points in the training set is reduced using the batch k-means algorithm. The training set, consisting of patterns derived from all data in tests {1...80}, were thus summarised using $k = 500$ cluster centres, which was found to adequately characterise the distribution of normal data in the 8-dimensional feature space defined by (16).

Figure 5 shows novelty scores of in-flight data from three example flights, computed using a distance-based model of normality, constructed as described in Sect. 2.4, with thresholds set heuristically, to include all patterns from the training set as described in Sect. 2.5.

"Normal" flight 75 (Fig. 5a) has low novelty scores throughout the flight. Flight 131 (Fig. 5b) shows that this flight exceeds the novelty threshold for several transient periods. Event flight 137 (Fig. 5c) shows that this flight exceeds the novelty threshold for longer periods.

A probabilistic model was trained using the same training set, and thresholds set using EVT as described in Sects. 2.4 and 2.5.

Figure 6 shows probabilistic model output for the same three example flights, against visualisations using NeuroScale as described in Sect. 2.3. "Normal" flight 75 (Fig. 6a) has low unconditional probabilities $p(x)$ throughout the flight; the visualisation shows that flight data (grey) lies close to the model

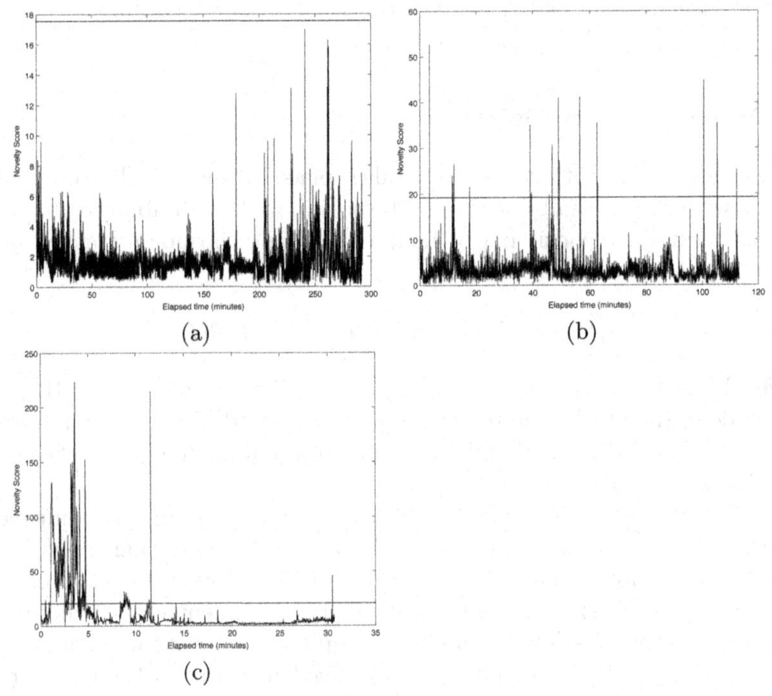

Fig. 5. In-flight novelty scores determined using distance-based model, against novelty threshold (horizontal line - using the same threshold in each case, with varying y-axis). (**a**) Flight 75, from the training set. (**b**) Flight 131, during a change in engine condition. (**c**) Flight 137, an event flight

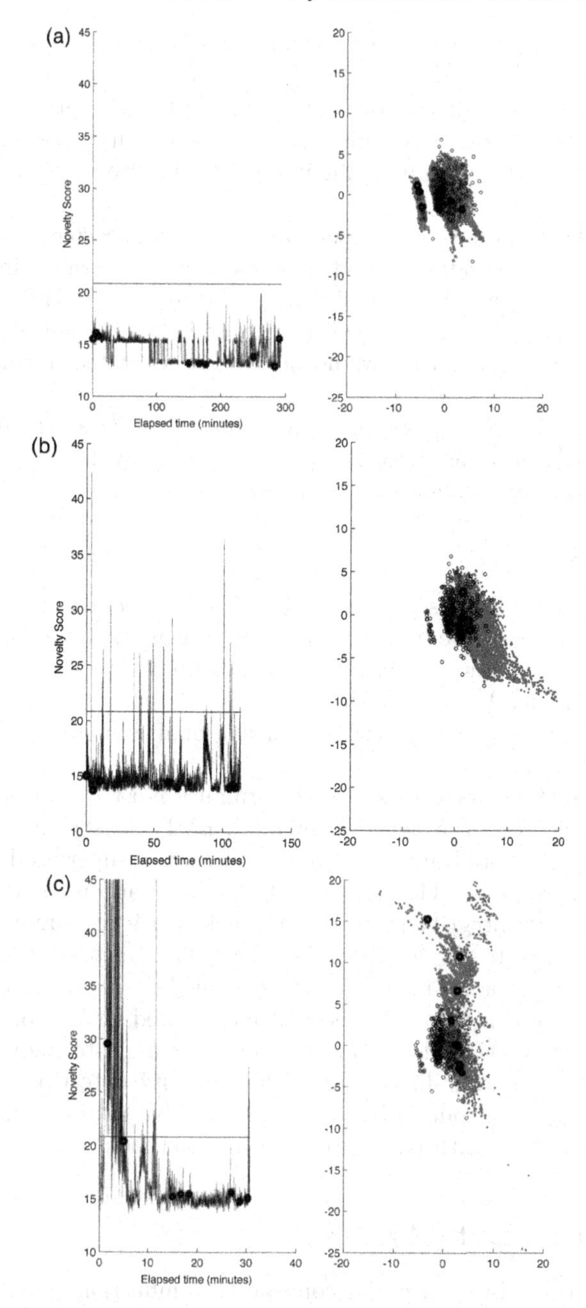

Fig. 6. *Left sub-plots*: in-flight data probabilities $p(x)$ determined using probabilistic model, against novelty threshold. *Right sub-plots*: visualisation of in-flight (*grey*) against model kernel centres (*black*). From top to bottom: (**a**) Flight 75, from the training set. (**b**) Flight 131, during a change in engine condition. (**c**) Flight 137, an event flight

kernel centres (black). Flight 131 (Fig. 6b) shows that this flight exceeds the novelty threshold for several transient periods, and that the visualised flight data begin to drift outside the locus of the model kernel centres. Event flight 137 (Fig. 6c) shows that this flight exceeds the novelty threshold for long periods, and that the visualised flight data lie far from the model kernel centres.

Thus, for both models, the event data have been correctly classified as "abnormal" with respect to the models, as has the change in engine condition during flight 131, potentially providing five flights of early warning of eventual failure. Flights $\{1...130\}$ (not shown here) were below the novelty threshold, indicating that no false alarms would have been generated during "normal" operation.

The visualisation correspondingly shows that flight data drift further from the model kernel centres as abnormality is observed, providing a useful method of communicating the system state to a user.

3.4 Discussion

Off-line and on-line analysis techniques for novelty detection in vibration data of modern jet engines have been presented, indicating that similar techniques can be used in both contexts. In both cases, distance-based and probabilistic models of normality have been shown to provide early warning up to five flights prior to an engine event (where conventional techniques detected only the event itself).

While the distance-based models of normality must be constructed (and novelty threshold set) in a heuristic manner, EVT has shown to be able to set the novelty threshold automatically, such that unsupervised model construction can take place. This is particularly advantageous in the condition monitoring of high-integrity systems, in which heuristic, manual modelling is unattractive (due to the number of systems that must be monitored) and often impractical (because models must be formed during operational use).

Visualisation techniques have been demonstrated to be able to provide meaningful interpretation of system condition from flight-to-flight (for off-line novelty detection) and in-flight (for on-line novelty detection). Such techniques allow system output to be made interpretable by users that might be potentially unfamiliar with pattern recognition methods.

4 Combustion Data Analysis

In this section, we introduce the concept of combustion instabilities, and briefly describe the combustor used for data collection. We describe a method of extracting features from time-series data, and present results of using both SVM distance-based and probabilistic methods of novelty detection. We show that early warning of combustor instability is possible using these techniques, and show the advantage of novelty score calibration.

4.1 System Description

Instabilities in combustion often originate from the resonant coupling between the heat released by the flame and the combustor acoustics, and hence are referred to as "thermo-acoustic" instabilities. The thermo-acoustic instabilities generate increased noise level (high acoustic pressure fluctuations), which in turn lead to excessive mechanical vibrations of the walls of the combustion chamber, and even cause catastrophic failure of the combustor. These instabilities also feature sub-optimal burning, increased emissions, and decreased equipment life. Thus, there is a need to predict the occurrence of combustion instabilities, in order to minimise their effect by imposing appropriate control systems. Readers are directed to [55] and [56] for detailed descriptions of the physical processes involved in combustion instabilities.

The data set used by the investigation described in this section was generated by a Typhoon G30 combustor (a component of gas-turbines), manufactured by Siemens Industrial Turbomachinery Ltd. The system was operated in stable and unstable modes of combustion. Unstable combustion is achieved by increasing fuel flow rates above some threshold, with constant air flow rate. A detailed description of the combustor can be found in [57].

The data set consists of three channels (with sampling frequency of 1 KHz), which are:

1. Gas pressure of the fuel methane (CH_4) in the main burner
2. Luminosity of C_2 radicals in the combustion chamber
3. Global intensity of unfiltered light in the combustion chamber

4.2 Pre-Processing and Feature Extraction

Each channel was normalised using the component-wise method described in Sect. 2.2.

Wavelet analysis [58] was used to extract features from each channel. Wavelet analysis represents a function in terms of basis functions, localised in both location and scale. It is capable of revealing behavioural trends or transient periods within the data. Wavelet decomposition can be regarded as a multi-level or multi-resolution representation of a function $f(t)$, where each level of resolution j (except the initial level) consists of wavelets $\Psi_{j,k}$, with the same scale but differing locations k. Wavelet decomposition of a function $f(t)$ at level J can be written as

$$f(t) = \sum_{k} \lambda_{J,k} \Phi_{J,k}(t) + \sum_{j=J}^{+\infty} \sum_{k} \gamma_{j,k} \Psi_{j,k}(t), \tag{17}$$

where $\Phi_{J,k}(t)$ are scaling functions at level J, and $\Psi_{j,k}(t)$ are wavelets functions at different levels j. $\lambda_{J,k}$ are scaling coefficients or approximation coefficients at level J. The set of $\gamma_{j,k}$ are wavelet coefficients or detail coefficients at different levels j.

Mallat [59] developed a filtering algorithm for the *Discrete Wavelet Transform*. Given a signal s, wavelet decomposition of the first level produces two sets of coefficients: approximation coefficients λ_1 and detail γ_1, by convolving s with a low-pass filter $h(k)$ for approximation, and with a high-pass filter $g(k)$ for detail, followed by dyadic decimation (down-sampling). Wavelet decomposition of the next level splits the approximation coefficients λ_1 in two parts using the same scheme, replacing s by λ_1, and producing λ_2 and γ_2, and so on.

The energy e_j of the wavelet detail coefficients $\gamma_{j,k}$ within a window of data at level j reflects the average noise level of the signal within that window [60]. Coefficient energy within the window is defined to be

$$e_j = \frac{\sum_k \gamma_{j,k}^2}{L} \tag{18}$$

for window length L, and wavelet detail coefficients $\gamma_{j,k}$ at level j.

Following [61], we divide the data set into non-overlapping windows of length $L = 64$, decomposed using the Daubauchies-3 wavelet. The mean values of approximation coefficients λ_1 and the energy of the detail coefficients γ_1 (extracted at level $j = 1$) are used as two-dimensional features in the input space, for each channel.

The combustor was operated in stable mode for the duration of 40 windows, followed by a transition into unstable operation for the duration of 48 windows, the first five of which were "transient", being neither stable nor unstable.

4.3 On-Line Novelty Detection

Both distance-based (using SVM) and probabilistic models of normality were constructed for each channel independently (as described in Sect. 2.4). 80% of the data from stable operation (32 windows) were used to construct models of normality, with all remaining data (56 windows) used as a test set.

The SVM method results in novelty scores computed using distance-based methods in the transformed feature space, which are calibrated into probabilities [0 1] using the method described in Sect. 2.5. An example of calibration for output of the classifier trained using data from channel 2 is shown in Fig. 7. Novelty scores of "normal" data (i.e., windows during which combustion was stable) are calibrated to probabilities $p(x) \approx 0$, while those of "abnormal" data (i.e., windows in which combustion was unstable) have $p(x) \approx 1$. Windows from the period of transient operation between stable and unstable modes are shown as taking calibrated probabilities $p(x) \approx 0.8$ in this case.

Figure 8a shows channel 1 SVM model uncalibrated output as a contour plot. Corresponding calibrated probabilities are shown in Fig. 8b.

It can be seen that data from the training set, windows $\{1...36\}$, are deemed "normal", with $p(x) \approx 0$, noting here that the probability of abnormality $p(x') = 1 - p(x)$ is plotted (and thus $p(x') = 1$ corresponds to an abnormal

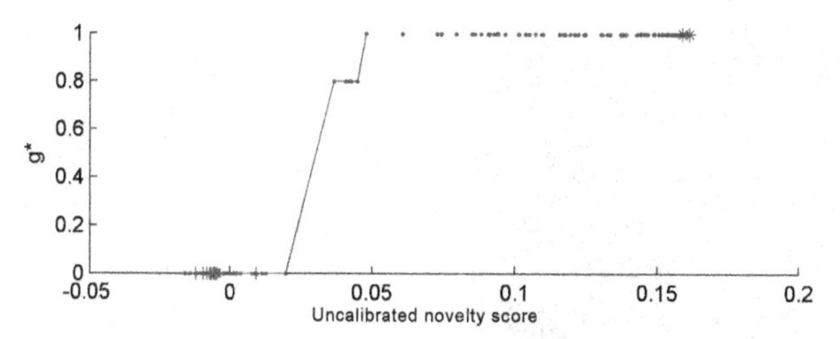

Fig. 7. Calibrating SVM novelty scores into probabilities, using SVM model trained using data from channel 2

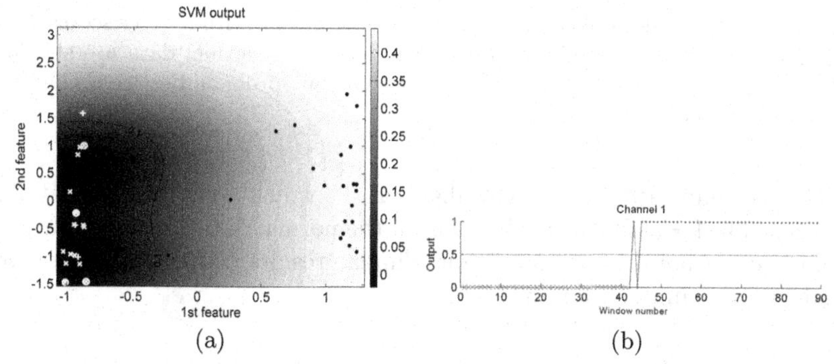

Fig. 8. (**a**) Visualisation of channel 1 SVM distance-based model, with novelty scores shown as greyscale contours. Training data are white $\times$, with support vectors shown as white $\otimes$. Test normal data are shown as white $+$, test abnormal data as black $\bullet$. (**b**) Calibrated output probabilities from channel 1 SVM model $p(x')$

window). Windows {41...45} are transient, between normal and abnormal, and show corresponding oscillation in the calibrated output probabilities. Window 42 is the first classified as abnormal, which provides three windows early warning prior to the first unstable window (46), after which all windows are abnormal. The visualisation shows this clear separation between normal and abnormal data, again providing a meaningful graphical interpretation of system state. Similar results were obtained for channels 2 and 3 (not shown here).

Figure 9a shows channel 1 probabilistic model output as a contour plot. Corresponding probabilities are shown in Fig. 9b.

Here, the true $p(x)$ of the unconditional data density is shown, in which $p(x) > 0$ for normal data, and $p(x) \approx 0$ for abnormal data. The figure shows that most normal data take non-zero probabilities, but there is a false-positive for window 37 (the first window after the 36 windows in the training set), which is incorrectly classified "abnormal". The transient data, windows {41...45} show decreasing probabilities, but only windows 43 and 45 have $p(x) \approx 0$. All unstable data, windows {46...88}, are correctly assigned outputs $p(x) \approx 0$.

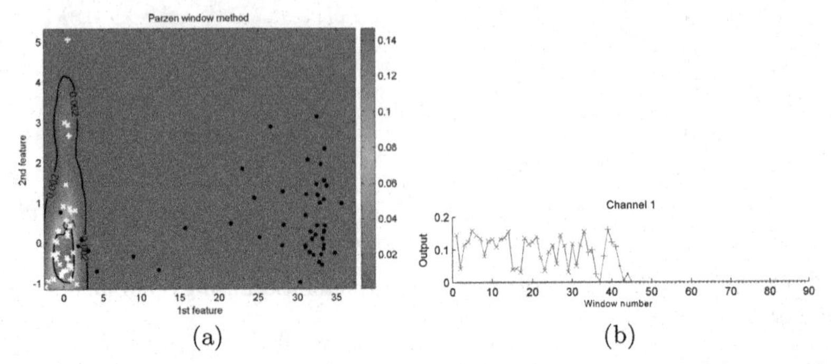

(a) (b)

Fig. 9. (**a**) Visualisation of channel 1 probabilistic model, with novelty scores shown as greyscale contours. Training data are white ×. Test normal data are shown as white +, test abnormal data as black •. (**b**) Output probabilities from channel 1 probabilistic model $p(x)$

This is confirmed by the visualisation, in which some of the test data (shown as black •) fall within the locus of the normal data, though separation to most abnormal data is significant. Similar results (not shown here) were obtained for channels 2 and 3.

4.4 Discussion

On-line analysis of combustion data has shown that early warning of system instability can be provided, by correctly classifying transient periods of operation as "abnormal" with respect to normal data. SVM distance-based models were found to outperform the probabilistic models investigated, by providing early warning of unstable operation without false-positive activations during normal combustor operation. The disadvantages of distance-based methods, in which thresholds are set heuristically, is overcome by calibrating outputs into probabilities, such that a meaningful probabilistic threshold may be set.

5 Conclusion

This chapter has compared the perform of distance-based and probabilistic methods of constructing models of normality using both case-mounted vibration sensors, and single-component monitoring (of the combustor).

Distance-based models have been shown to provide early warning of engine events, but require heuristic setting of thresholds, and manual construction of models. Typically, this can become impractical for condition monitoring of high-integrity systems, in which the number of units being monitored, and the inaccessibility of those units, require unsupervised learning of normal models.

Methods of overcoming this have been investigated, in conjunction with an SVM model, in which classifier output is calibrated into probabilities, resulting in a probabilistic novelty threshold being set automatically.

Probabilistic models have been shown to provide similar early warning, with novelty thresholds set automatically using EVT, which explicitly models the tails of the "normal" data density.

The use of visualisation techniques has been shown to provide guidance during the data evaluation and model construction phases, particularly in the verification of data labels provided by domain experts, which can be unreliable. The same techniques are shown to be able to provide meaningful representations of system state, allowing non-expert users to interpret the output of the novelty detection system.

Further analysis of fusing classifications from multiple channel-specific classifiers is on-going, as is the exploitation of automatic probabilistic threshold-setting techniques for estimates of a multi-variate data density.

Acknowledgements

The authors gratefully acknowledge discussions with Nicholas McGrogan of Oxford BioSignals Ltd.; Dennis King, Steve King, and Paul Anuzis of Rolls-Royce PLC; Hujun Yin and Yang Zhang of the University of Manchester. DAC and PRB acknowledge the support of the HECToR research project (a UK Department of Trade and Industry project), Rolls-Royce PLC., and Oxford BioSignals Ltd.

References

1. Roberts S, Tarassenko L (1994) Neural Comput 6:270–284
2. Silverman BW (1986) Density estimation for statistics and data analysis. Chapman and Hall, London
3. Mayrose I, Friedman N, Pupko T (2005) A Gamma Mixture Model Better Accounts for Among Site Heterogeneity. Bioinformatics 21(2):151–158
4. Agusta Y, Dowe DL (2003) Unsupervised Learning of Gamma Mixture Models Using Minimum Message Length. Artificial intelligence and applications proceedings 403
5. Dempster AP, Laird NM, Rubin DB (1977) Maximum Likelihood from Incomplete Data via the EM Algorithm. J R Stat Soc Series B 39:1–38
6. Markou M, Singh S (2003) Novelty Detection: A Review. Signal Processing 83:2481–2497
7. Duda RO, Hart PE, Stork DG (2001) Pattern classification. Wiley, New York
8. Yeung DY, Ding Y (2002) Host-Based Intrusion Detection Using Dynamic and Static Behavioral Models. Pattern Recognit 36:229–243
9. Smyth P (1994) Markov Monitoring with Unknown States. IEEE J Sel Areas Commun 12(9):1600–1612
10. Quinn J, Williams CKI (2007) Known Unknowns: Novelty Detection in Condition Monitoring. Proceedings of 3rd Iberian conference on pattern recognition and image analysis, Lecture Notes in Computer Science, Springer
11. Markou M, Singh S (2006) A Neural Network-Based Novelty Detector for Image Sequence Analysis. IEEE Trans Pattern Anal Mach Intell 28(10):1664–1677

12. Ghahramani Z, Hinton GE (1998) Variational Learning for Switching State-Space Models. Neural Comput 12(4):963–996
13. McSharry PE, He T, Smith LA, Tarassenko L (2002) Linear and non-linear methods for automatic seizure detection in scalp electro-encephalogram recordings. Med Biol Eng Comput 40:447–461
14. Tax DMJ, Duin RPW (1998) Outlier detection using classifier instability. Advances in pattern recognition–the joint IAPR international workshops, Sydney, Australia, 593–601
15. Kohonen T (1982) Self-Organized Formation of Topologically Correct Feature Maps. Biol Cybern 43:59–69
16. Ypma A, Duin RPW (1998) Novelty Detection Using Self-Organising Maps. Prog Connect Based Inf Syst 2:1322–1325
17. Labib K, Vemuri R (2002) NSOM: A real-time network-based intrusion detection system using self-organizing maps. Networks security
18. Yin H, Allinson NM (2001) Self-organizing mixture networks for probability density estimation. IEEE Trans Neural Netw 12(2)
19. Vapnik V (2000) The nature of statistical learning theory. Second Edition. Springer, Berlin New York Heidelberg
20. Tax DMJ, Duin RPW (1999) Data Domain Description Using Support Vectors. Proceedings of ESAN99. Brussels:251–256
21. Scholkopf B, Williamson R, Smola AJ, Shawe-Taylor J, Platt J (2000) Support vector method for novelty detection. Advances in neural information processing systems 12, (NIPS99) Solla KMSA, Leen TK (eds.), MIT: 582–588
22. Bishop CM (1995) Neural networks for pattern recognition. Oxford University Press, Oxford
23. Jennions IK (2006) Cross-platform challenges in engine health management. Proceesings of International Conference on Integrated Condition Monitoring, Anaheim, CA
24. Moya M, Hush D (1996) Neural Netw 9(3):463–474
25. Ritter G, Gallegos M (1997) Pattern Recogn Lett 18:525–539
26. Clifton DA, Bannister PR, Tarassenko L (2006) Learning shape for jet engine novelty detection. In: Wang J. et al. (eds.): Advances in neural networks III. Lecture Notes in Computer Science, Springer, Berlin Heidelberg New York, 3973:828–835
27. Sammon JW (1969) IEEE Trans Comput 18(5):401–409
28. DeRidder D, Duin RPW (1997) Pattern Recogn Lett
29. Lowe D, Tipping ME (1996) Neural Comput Appl 4:83–95
30. Tarassenko L (1998) A guide to neural computing applications. Arnold, UK
31. Nabney I (2002) Netlab: algorithms for pattern recognition. Springer, Berlin Heidelberg New York
32. Clifton DA, Bannister PR, Tarassenko L (2007) Visualisation of jet engine vibration characteristics for novelty detection. Proceedings of NCAF, London, UK
33. Nairac A, Townsend N, Carr R, King S, Cowley P, Tarassenko L (1999) Integr Comput-Aided Eng 6(1):53–65
34. Clifton DA, Bannister PR, Tarassenko L (2006) Application of an intuitive novelty metric for jet engine condition monitoring. In: Ali M, Dapoigny R (eds) Advances in applied artificial intelligence. Lecture Notes in Artificial Intelligence. Springer, Berlin Heidelberg New York 4031:1149–1158

35. Clifton DA, Bannister PR, Tarassenko L (2007) A framework for novelty detection in jet engine vibration data. In: Garibaldi L, Surace S, Holford K (eds) Key engineering materials 347:305–312
36. Clifton DA, Bannister PR, Tarassenko L (2007) Novelty detection in large-vehicle turbochargers. In: Okuno HG, Ali M (eds) New trends in applied artificial intelligence. Lecture Notes in Computer Science, Springer, Berlin Heidelberg New York, 4750
37. Hayton P, Scholkopf B, Tarassenko L, Anuzis P (2000) Support vector novelty detection applied to jet engine vibration spectra. Proceedings of Neural Information Processing Systems
38. Wang L, Yin H (2004) Wavelet analysis in novelty detection for combustion image data. Proceedings of 10th CACSC, Liverpool, UK
39. Clifton LA, Yin H, Zhang Y (2006) Support vector machine in novelty detection for multi-channel combustion data. Proceedings of 3rd International Symposium on Neural Networks
40. Nairac A, Corbett-Clark T, Ripley R, Townsend N, Tarassenko L (1997) Choosing an appropriate model for novelty detection. Proceedings of IEE 5th International Conference on Artificial Neural Networks
41. Tax D, Duin R (1999) Pattern Recogn Lett 20:1191–1199
42. Schölkopf B, Platt J, Shawe-Taylor J, Smola AJ, Williamson RC (2001) Neural Comput 13(7):1443–1471
43. Bishop CM (1994) Novelty detection and neural network validation. Proceedings of IEE Conference on Vision and Image Signal Processing
44. Zadrozny B, Elkan C (2002) Transforming classifier scores into accurate multiclass probability estimates. Pro. ACM SIGKDD 694–699
45. Roberts SJ (1999) Proc IEE 146(3)
46. Roberts SJ (2000) Proc IEE Sci Meas Technol 147(6)
47. Medova EA, Kriacou MN (2001) Extremes in operational risk management. Technical report, Centre for Financial Research, Cambridge, U.K.
48. Fisher RA, Tippett LHC (1928) Proc Camb Philos Soc 24
49. Coles S (2001) An introduction to statistical modelling of extreme values. Springer, Berlin Heidelberg New York
50. Embrechts P, Kluppelberg C, Mikosch T (1997) Modelling extremal events. Springer, Berlin Heidelberg New York
51. Rolls-Royce PLC (1996) The jet engine. Renault Printing, UK
52. Hayton P, Utete S, Tarassenko L (2003) QUOTE project technical report. University of Oxford, UK
53. Clifton DA (2005) Condition monitoring of gas-turbine engines. Transfer report. Department of Engineering Science, University of Oxford, UK
54. Bannister PR, Clifton DA, Tarassenko L (2007) Visualization of multi-channel sensor data from aero jet engines for condition monitoring and novelty detection. Proceedings of NCAF, Liverpool, UK
55. Khanna VK (2001) A study of the dynamics of laminar and turbulent fully and partially premixed flames. Virginia Polytechnic Institute and State University
56. Lieuwen TC (1999) Investigation of combustion instability mechanisms in premixed gas turbines. Georgia Institute of Technology
57. Ng WB, Syed KJ, Zhang Y (2005) Flame dynamics and structures in an industrial-scale gas turbine combustor. Experimental Thermal and Fluid Science 29:715–723

58. Daubechies I (1988) Orthonormal bases of compactly supported wavelets. Communications on Pure and Applied Mathematics 41:909–996
59. Mallat SG (1989) A theory for multiresolution signal decomposition. IEEE Trans. Pattern Analysis and Machine Intelligence 11(7):674–693
60. Guo H, Crossman JA, Murphey YL, Coleman M (2000) IEEE Trans Vehicular Technol 49(5):1650–1662
61. Clifton LA, Yin H, Clifton DA, Zhang Y (2007) Combined support vector novelty detection for multi-channel combustion data. Proceedings of IEEE ICNSC

Multiway Principal Component Analysis (MPCA) for Upstream/Downstream Classification of Voltage Sags Gathered in Distribution Substations

Abbas Khosravi[1], Joaquim Melendez[1], Joan Colomer[1], and Jorge Sanchez[2]

[1] Institut d'Informatica i Aplicacions (IIiA), Universitat de Girona (UdG), Girona, Spain, khosravi@eia.udg.es, quimmel@eia.udg.es, colomer@eia.udg.es
[2] ENDESA Distribucion, Barcelona, Spain, jslosada@fecsa.es

Summary. Occurring in transmission or distribution networks, voltage sags (transient reductions of voltage magnitude) can cause serious damage to end-use equipment (domestic appliances, precision instruments, etc.) and industrial processes (PLC and controller resets, time life reduction, etc.) resulting in important economic losses. We present a statistical method to determine whether these disturbances originate upstream (transmission system) or downstream (distribution system) of the registering point located in distribution substations. The method uses only information from the recorded disturbances and exploits their statistical properties of them in terms of covariance. Multiway Principal Component Analysis (MPCA) is proposed to model classes of sags according to their origin upstream/downstream using the RMS values of voltages and currents. New, not-yet-seen sags are projected to these models and classified based on statistical criteria measuring their consistency with the models. the successful classification of real sags recorded in electric substations is compared with other methodologies.

1 Introduction

Among the different power quality phenomena, disturbances known as voltage sags are the most common. They can be of diverse origin (network operation, faults and short circuits, weather, etc.) located in the entire power system (generation, transmission, distribution or consumers). According to IEEE standards, a voltage sag is defined as a momentary decrease (10–90%) in the Root Mean Square (RMS) voltage magnitude for a duration time of 20 ms to 1 min [1]. Lightning, motor starting, electrical facility operation, and external agents are typical causes of voltage sags. Such a sudden dip in RMS values could cause serious problems in sensitive end-user equipment, ranging from instant breakdown to more gradual deterioration over time. Sags lasting for only ten milliseconds may force an industry to be idle for several hours

A. Khosravi et al.: *Multiway Principal Component Analysis (MPCA) for Upstream/Downstream Classification of Voltage Sags Gathered in Distribution Substations*, Studies in Computational Intelligence (SCI) **116**, 297–312 (2008)
www.springerlink.com

due to the automatic shutdown of electric units by protection equipment. Consequences involve heavy costs for the company from loss of time and production. The current regulation framework assigns the responsibility for the quality of power delivered to customers to distribution companies even though the origin of the disturbances can be located anywhere in the entire power system (from generation to within the transmission and distribution systems) and even with other consumers. Consequently, utilities are very interested in developing precise methods to locate the origin of sags and to determine the consequences of badly supplied power. Moreover, voltage sag source location is necessary for power quality troubleshooting, diagnosis and mitigation strategy development. In fact, the location of the sag source is the first step toward mitigation actions and liability assessment. Generally, from the utility point of view, the causes of sags can be divided into three areas of occurrence: the transmission system (typically above 65 kV); distribution systems (65–12 kV); and, consumption points (120–400 V). Exact voltages depend on the countries and power systems. Taking into account the number of sags occurring in the transmission networks, electric utilities eagerly look for reliable techniques to determine the sag source locations: in the transmission network (High Voltage lines, HV) or in distribution network (Medium Voltage lines, MV). In this work the problem has been simplified by focusing on determining the upstream or downstream origin of sags gathered in distribution substations. Upstream correspond to the transmission system (HV) and downstream area is associated with the distribution network (MV). The goal is not to find the exact locations of the sags in the network but to associate them with incidents occurring in the transmission or distribution system. Figure 1 represents voltage RMS values of a three phase HV sag and a two-phase MV sags registered in a substation.

In the literature, different methods have been previously proposed to address this problem taking advantage of the electric characteristics of signals. For example, of a couple of methods have been developed and proposed for sag direction finding. Others use information related to protection systems. The major approaches to address this problem developed so far can be found in [2] and are briefly described below:

- Disturbance power and energy [3]: A direction finder for voltage sag source location has been introduced using disturbance power and energy computed for the whole period of disturbance. A decision about the location of an occurred disturbance is made based on changes in these two indices.
- Slope of system trajectory [4]: Here the authors assume that the relationship between the product of the voltage magnitude and power factor against current at the recording point are not the same for sags in up and down areas. The sign of a line slope, obtained by the least square fitting approach, indicates the direction of the voltage sag source.

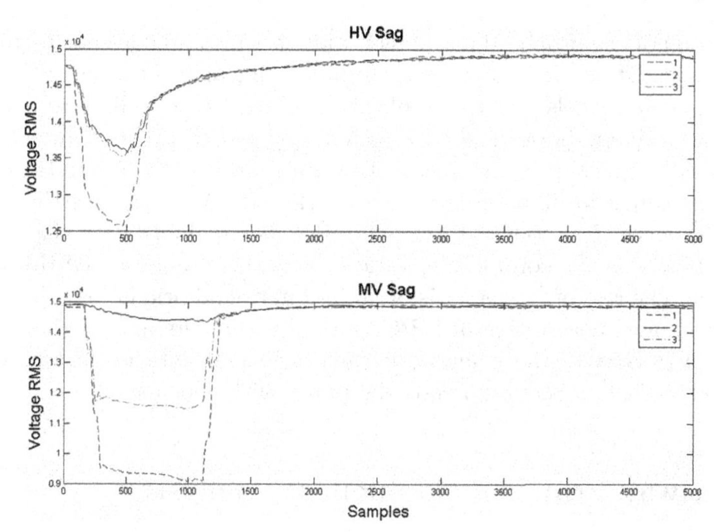

Fig. 1. Voltage RMS values of a three phase HV sag and a two-phase MV sag registered in a substation

- Resistance sign [5]: Using the positive sequence voltage and current, the authors estimate resistance and then use its sign to determine the source of voltage sags.
- Real current component [6]: The method in this paper is based on the analysis of the variation of the real part of the current. Any variation in this value is an indication of the direction of a sag with regard to the recording point.
- Distance relay [7]: The impedance seen before and during the event indicates the relative location of the fault.

As reported and claimed in [2] and [8], these methods are not totally reliable and show poor performance in detecting whether sags are coming from upstream or downstream in case of asymmetrical faults. In fact, when a sag happens in more than one phase, finding its source based on the available methods is not very straightforward and the results obtained are not reliable.

Besides the approaches mentioned, extracting some temporal descriptors and applying them to a Learning Algorithm for Multivariate Data Analysis (LAMDA)to them is an alternative method for classification of voltage sags in distribution and transmission classes [9].

Prompted by the striking applications of MPCA in other fields of engineering, especially in chemical engineering, we have developed a technique for the classification of HV and MV voltage sags based on their statistical patterns. The proposed method attempts to capture the most relevant variations of voltage and current RMS waveforms for different sags in order to set up an MPCA model that is useful for classification. In contrast to the previous methods which do some preprocessing on data to extract some features, exploitation of the proposed methods is quite simple and does not require any preprocessing stage. It is done just through a linear projection of the waveforms on built models and the application of statistical indices to check the degree of matching between the projected sags and the MPCA model.

Generally, this paper introduces a new application of MPCA in the domain of power quality to differentiate between HV and MV sags. Further details about the method as well as examination of its capabilities for classification of real voltage sags recorded in electric substations are presented throughout the paper. The rest of the paper is organized in five additional sections: Sect. 2 provides a short description of MPCA useful for understanding the methods developed in Sect. 3; the results obtained with real data are presented and discussed in Sect. 4. Section 5 ends the paper with conclusions.

2 Multiway Principal Component Analysis

Multivariate Statistical Process Control (MSPC) charts are classical multivariate quality and process monitoring tools that include a set of mathematical tools to analyze the recorded data and extract the process information or predict the values of the product quality variables. The favorite tool of MSPC for data compression and information extraction is the Principal Component Analysis (PCA) [10]. Owing to the fact that MPCA is a promising extension of PCA, explaining PCA is a prerequisite for explaining MPCA. Based on Singular Value Decomposition (SVD) of the covariance matrix of a data set, PCA is a linear unsupervised method that produces a lower dimensional representation of the data in a way that preserves the correlation structure between the process variables. Optimal in terms of capturing the variability in the data, PCA is concerned with explaining the variance-covariance structure through a few linear combinations of the original variables. Showing great performance, PCA has been widely used to explore and model multivariate data in different fields such as image processing, biology, meteorology and process control.

The concepts and mathematics used in PCA are common to many other multivariate techniques. The multivariate data can be organized in matrix form, X, representing each variable in a column and the successive experiences in rows. With n variables and m observations per variable the matrix of original data is $X \in R^{m \times n}$. Once the variables have been standardized, the covariance matrix is calculated

$$S = \frac{1}{m-1} X^T X \tag{1}$$

The matrix V which its columns are the eigenvectors of S and the diagonal matrix Λ with eigenvalues on the main diagonal are calculated as follows

$$S = V \Lambda V^T \tag{2}$$

Each eigenvalue in Λ is associated to an eigenvector in V. PCA assumes that the eigenvector with the highest eigenvalue represents the most important direction in the new space, i.e., contains the largest quantity of information; therefore, this vector is called the principal component of the data set. Ordering the eigenvectors by eigenvalue, highest to lowest, gives the components in order of significance. To reduce the dimensionality, the less important components can be eliminated, so only the r first eigenvectors are chosen ($P \in R^{n \times r}$). PCA decomposes the data matrix as the sum of the outer product of vectors t_i (scores) and p_i (the i-th loading vector) plus a residual matrix, E:

$$X = \sum_{i=1}^{r} t_i p_i^T + E \tag{3}$$

In (3), the residual matrix, E, represents the variations in the observation space not captured in the new space and represented by the $n - r$ smallest eigenvalues. Whenever a new observation ($x \in R^n$) emerges, its projection to the new space can be easily computed through a linear operation as follows:

$$t_i = X p_i \tag{4}$$

Abnormal behavior of an experiment or the lack of model fit is generally identified by means of the Q-statistic or the D-statistic, which are compared with control limits determining whether the process is in control or not. These methods are based on the assumption (generally motivated by the central limit theorem) that the underlying process follows approximately a multivariate normal distribution with a null first moment vector [11]. At the occurrence of a new type of special event that was not present in the in-control data used to build the MPCA model, the new observations will move off the plane. This type of event can be detected by computing the Q-statistic or Squared Prediction Error (SPE) of the residual for new observations. In fact, Q is calculated from the error term, E, in the reduced PCA model, (3). For data in the original data matrix, it can be simply computed as the sum of squares of each row (observation) of E, for example for the $i-th$ observation in X, x_i:

$$Q_i = e_i e_i^T = x_i (I - P_r P_r^T) x_i^T, \tag{5}$$

where e_i is the $i - th$ row of E, P_r is the matrix of the first r loading vectors retained in the PCA model, and I is the identity matrix of appropriate size $(n \times n)$. The control limit for Q that guarantees a confidence level of $100(1 - \alpha)\%$ can be obtained using the following expression:

$$Q_\alpha = \theta_i \left(\frac{c_\alpha \sqrt{2\theta_2 h_0^2}}{\theta_1} + 1 + \frac{\theta_2 h_0 (h_0 - 1)}{\theta_1^2} \right)^{\frac{1}{h_0}} \tag{6}$$

where

$$\theta_i = \sum_{i=r+1}^{n} \lambda_i^j \quad \text{for} \quad j = 1, 2, 3 \tag{7}$$

and

$$h_0 = 1 - \frac{2\theta_1\theta_3}{3\theta_2^2} \tag{8}$$

In contrast to Q, which is a measure of non-captured information in terms of variation, T^2 provides a measure of variation within a PCA model. This variation can be thought of as being in the same subspace as the model. It is computed as the sum of normalized squared scores defined as:

$$T^2 = \sum_{i=1}^{r} T_i^2 = \sum_{i=1}^{r} \frac{t_i^2}{\lambda_i} \tag{9}$$

The control limit for T^2 for a confidence level of $100(1-\alpha)\%$ can be calculated assuming an F-distribution for it and using the following expression:

$$T_\alpha^2 = \frac{r(m-1)}{m-r} F(r, m-1; \alpha), \tag{10}$$

where $F(r, m-1; \alpha)$ is an F-distribution with degrees of freedom r and $m-1$ with a level of significance α.

Despite the great capability in data dimension reduction capability, the PCA application is limited to two-dimensional databases: one dimension to describe variables and another one for their observations. Since such a two-dimensional database could be generated and recorded in different experiments or runs of a process representing the evolution of the variables during these experiments, an improved methodology has to be applied to capture correlation not only between variables and samples of this experiment but also between different experiments.

With this goal MPCA has been developed based on the PCA concept and extending the representation of data matrix to cope with the additional complexity of dealing multiple experiments. MPCA considers an experiment in which J variables are recorded at K time instants throughout the experiment. And the same data is obtained for a number (I) of experiments. Adding this new dimension to the data matrix results in a three-way data matrix $X_{K\times J\times I}$. Thus, MPCA is equivalent to performing ordinary PCA on a large two-dimensional matrix constructed by unfolding the three-way data matrix. Several methodologies to unfold the three-dimensional data matrix have been suggested in [12]. Unfolding in the direction of the experiment/batch is illustrated Fig. 2. This is the strategy followed in this work.

Applying the PCA to this unfolded two-dimensional matrix obtains a new representation consisting of score vectors, T, and loading matrices, P, plus a residual three-dimensional matrix, E. In this way and in accordance with the principles of PCA, MPCA is defined as follows

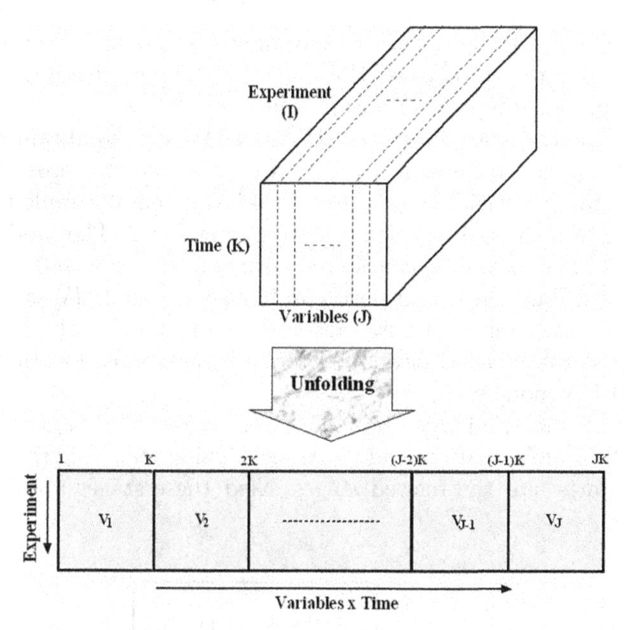

Fig. 2. Batch-wise unfolding to rearrange three-dimensional databases into two dimensional ones

$$X = \sum_{i=1}^{r} t_i \otimes P_i + E \tag{11}$$

where $\otimes$ denotes the Kronecker product or the tensor product. The systematic part, the sum of the $t_i \otimes P_i$ expresses the deterministic variation as one part (t) related only to batches and a second part (P) related to variables and their time variation.

Further information about PCA and MPCA and their application in different areas of engineering can be found in [13], [14], and [15]. Its combination with other methods can be found in [16] and [17].

3 Proposed Method for Sag Source Location

In this section we explain the main contribution of this work: in this section we explain the proposed method for classification of HV and MV sags based on MPCA models obtained following the method introduced in the previous section. First we discuss the used concepts used to develop the method and the present the steps in detail.

In process monitoring, fault detection applications of MPCA typically use T^2 and Q criteria as thresholds to define the in-control limits [13]. Exceeding these thresholds means that the expected covariance among the variables and experiments has experienced a variation reflecting a different situation. Based on this principle for the detection of abnormality, we propose the use of MPCA to classify sags into two classes, HV and MV, taking advantage of specific models to represent these two types of sags.

Two data matrices are built with HV and MV sags to obtain the MPCA model. Each matrix contains a set of sags (experiments) represented by the RMS waveforms of 6 variables (3 voltages and 3 currents) sampled at 6400 Hz and resulting in a three-dimensional matrix as in Fig. 2. The model obtained with each data matrix will be considered normal for its class and abnormal for the other class. Thus, the classification problem of HV and MV sags is reduced to testing the consistency of new sags with both models. This basic idea is the skeleton of the method developed to solve sag source location problems based on MPCA models.

The proposed method is composed of three major parts: (1) Database Construction, (2) Model Creation, and (3) Model Exploitation. Figure 3 represents these three parts and the related stages. Next these stages are explained in detail.

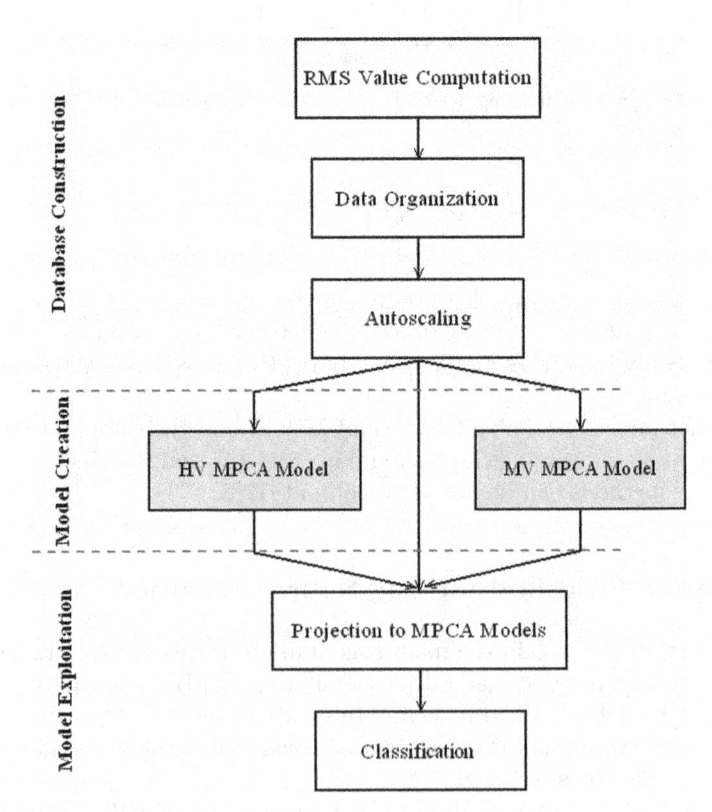

Fig. 3. Three main parts of the proposed method for sag source location in the power network based on MPCA

3.1 Database Construction

The existing data used to obtain the MV and HV models are substation registers of the three phase-voltages and three currents. Registers have a finite length of 4,993 samples with each variable corresponding to 39 periods (20 ms.) of the electric waveform (128 samples /period). This length is enough to capture the majority of sags occurring in a network. The disturbance in the registers always starts in the second period of the signal.

RMS Value Computations

The instantaneous RMS value for each variable (three voltages and three currents) is computed based on the estimation of the magnitude of the fundamental frequency (50Hz) during the sag. Short Fourier Transformation (SFT) computed in a one period (20 ms) sliding window is used to isolate the fundamental frequency during the sag.

Database Organization and Labeling

Outliers (not recovered or very long sags) and other registers not corresponding to sags (interruptions, swells, etc.) have been removed to obtain a more confident data base.

The resulting data for each class (HV and MV) have been organized into three-dimensional matrices like Fig. 2 ($X_{K\times 6\times I}$). In this way of presentation, I and K stand for the number of sags and time instances, respectively, and J=6 is the number of variables (3 voltages and 3 currents).

Unfolding

The three-dimensional matrix created in the previous stage is then unfolded, following the method depicted in Fig. 2. Thus, each block (or submatrix) in the two-dimensional matrix now contains all the information related to each variable. This organization of data will be useful to visualize the contribution of each variable to the created model (loading vectors).

Autoscaling

To avoid overweighing of voltage variables towards currents (because of their larger magnitude) in the analysis, we apply an autoscaling method over the unfolded database which makes the data in each column to be zero-mean centered with unit variance. Doing this, the effect of each of these variables may be very similar and the information content of the voltage data is not inherently greater than that of the current data.

3.2 Model Creation

HV and MV MPCA Model Creation

HV and MV MPCA models are created separately using autoscaled HV and MV databases, respectively, by following the procedure explained in the previous Sect. 2 about MPCA.

Although the model creation phase is completed in only one stage, it is important to remark that the number of principal components used in the creation of MPCA models could considerably affect the generality of the model. We will refer to this point in the next section with more explanations.

3.3 Model Exploitation

Projection

Having developed specific models representing MV and HV sags, the evaluation of new sags is performed by projecting them to the HV and MV MPCA models using (4).

Classification

Based on the use of statistical indices T^2 and Q (see (5) and (9)) we propose to analyze the matching of new sags with both models (MV and HV). Results from the testing of these indices is then evaluated in order to assign the new sag to one or the other type. No matter which criterion we use, in the ideal case, we expect those statistics to be low when types of the MPCA model and test subsets are the same. Exceeding predetermined thresholds means that the projected sag is from the opposite type of the MPCA model; e.g., if the model is HV, the projected sag is MV and vice versa. Needless to say, the classification rate is dependent on how these thresholds are adjusted.

The classification stage completes the description of the proposed method for sag direction location (upstream/downstream). As can be seen, there is no special preprocessing stage to extract features from the data. This is often computationally massive, making extraction undesirable in online applications. All we do is compute RMS values, which is quite common in power area for analysis. Additional benefits of this PCA-based approach with respect to other classification strategies reside in its generality and the fact that no additional parameters are need to be adjusted. Another significant aspect of the proposed method is its robustness. No matter how severely noise corrupts some data, it captures and considers its effects as meaningless variation.

4 Classification Results with Sags Gathered in Distribution Substations

In this section we examine the capability of the proposed method to classify HV and MV sags registered in three substations (110/25kV) in Catalonia, Spain. As these sags have occurred in the course of one year in different seasons, they well cover and reflect a wide variety of unusual behaviors after sag occurrences in power transmission and distribution networks of this area.

The data have been preprocessed and organized according to the previous section resulting in two three dimensional matrices for each substation. To evaluate the the performance of the method, four fold cross validation has been applied to report numerical results. Thus, HV and MV databases have been split into four folds. According to cross-validation, training is performed with three subsets (75% of data) and the remaining is used for testing (25%). Repeating the procedure, with all possible combinations of the four folds, assures that all data is used for training and testing purposes.

Based on the developed procedure, we project the test sets as future sags onto the created models and then classify them. With a view to maximizing the classification rates of HV and MV sags, we need to address the following questions:

- How many principal components should we choose in our models?
- Which discriminant function or criteria should we employ to get best classification rate?

To address these questions, we have varied the number of principal components used to build the model in a wide range and at the same time different confidence levels of T^2 and Q have been evaluated to define the the threshold for sags fitting the projection model. A priori T^2, Q, and T^2-Q are the potential criterion for classification of HV and MV sags after being projected onto the MPCA models but the experience tells us that the including T^2 among the the considered criteria leads to poor results. Applying this rule of thumb excludes two criteria from the candidate pool and introduces Q as the best criterion for classification. As the classification rates are considerably sensitive to the confidence limit of this criterion, looking for the best point is essential. Taking into account all these illustrations, we set up a grid of principal component number and confidence limit and then calculate the classification rates in each region of this grid seeking those regions in which the adjustment of MPCA parameters based on them leads to the highest classification rates.

Figure 4 depicts these regions for data registered in one substation. Since we applied cross validation, the rates presented in all figures are the average ones. The investigated ranges for the number of principal components and Q confidence limit, respectively, are [1,12] and [80%, 97.5%]. The top figure shows the projection of HV and MV test subsets onto HV the MPCA model and the bottom ones are the projection of those sets over the MV MPCA models. Obviously, in the right figures for the HV model, we can find some

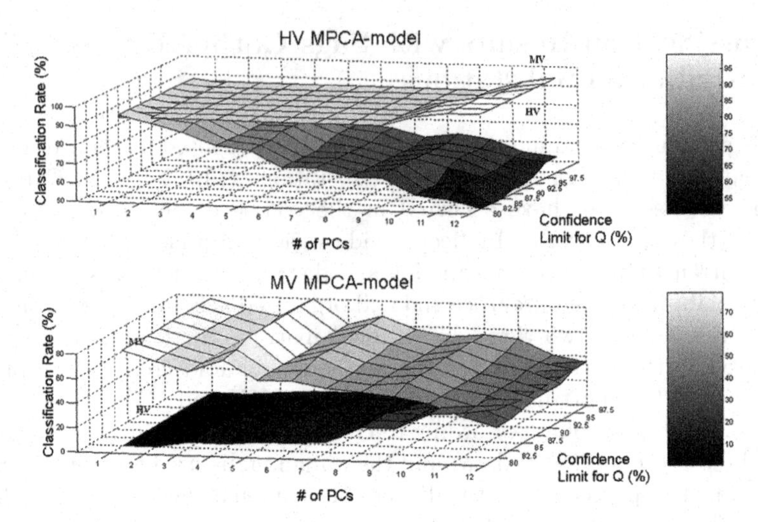

Fig. 4. Three-dimensional visualization of average classification rates for HV and MV MPCA models of substation A

regions where the classification ranges for both HV and MV data sets are satisfactorily high. On the contrary, finding such a region for the MV MPCA model is impossible. For this type of model, the satisfactory classification rates for both HV and MV sags are mutually exclusive. In other words, at the same time that we increase the number of principal components in the MV model to classify the HV sags correctly, the model becomes narrower and more sensitive to small variations which means not being able to recognize the MV. The inverse scenario also leads to general models classifying both HV and MV sags as similar.

First principal component of HV models captures a kind of variation which well separate HV and MV sags well. On the contrary, first principal component of MV models are not capable of discriminating between both types of sags. In spite of this, these components capture common information existing in both, HV and MV sags, which is unsatisfactory for classification purposes.

To have a better visualization of the classification rates, we plot the surface of those planes in Fig. 5. In this figure, the classification rates in each region are represented by the color (grey level) of that region with reference to the color map. In this figure, the left-hand-side graphs show that taking a small enough number of principal components guarantees classification rates above 92%. Choosing only one principal component and fixing Q confidence limit to a value in $[80\%, 97.5\%]$ results in the maximum classification rate. Taking into account the number of HV and MV sags for this substation, the overall average classification rate is 93.6%, which is widely acceptable for electric utilities. Also, the left-hand-side graphs clearly show the weakness of the MV MPCA model for the classification of sags. Obviously, varying either the number

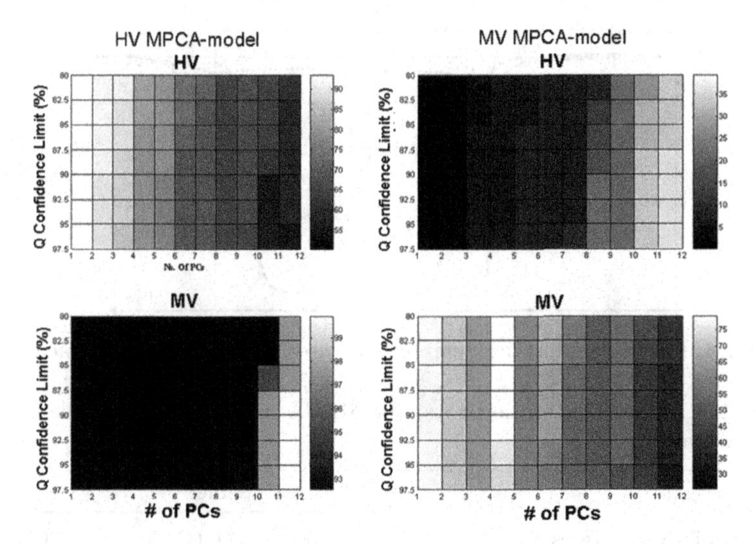

Fig. 5. Presentation of average classification rates for substation A in different regions of the grid

of principal components or the Q confidence limit leads to concurrent high classification rates.

To better demonstrate the performance and capabilities of the proposed method for classification of HV and MV sags, it would be interesting to compare the results obtained in this work with those obtained using other classification techniques based on temporal features extracted from waveforms. Typically, these features are magnitude, duration, plateau duration, and slopes of voltage fall and recovery [9]. After several tests using those attributes, sag magnitude and duration have proven to be the most significant; and an increasing number of attributes results in poorer classification rates. Consequently only duration and magnitude defined as follows have been used in the comparison:

- Sag magnitude: the minimum RMS value among three phases.
- Sag duration: the time interval between the instant when the RMS voltage of one phase crosses the voltage sag threshold (90% of normal voltage) and the instant when the last phase returns to its normal level.

These two temporal attributes have been schematically shown in Fig. 6 for a typical three-phase sag.

Three well-known classification techniques that have shown themselves to be useful in other engineering areas have been compared. These are: Multilayer Perceptron, Neural network with Radial Base Function (RBF) and Decision Trees [18]. Also the four-fold cross-validation scheme has been used to test of these techniques giving the results summarized in Table 1 for data gathered

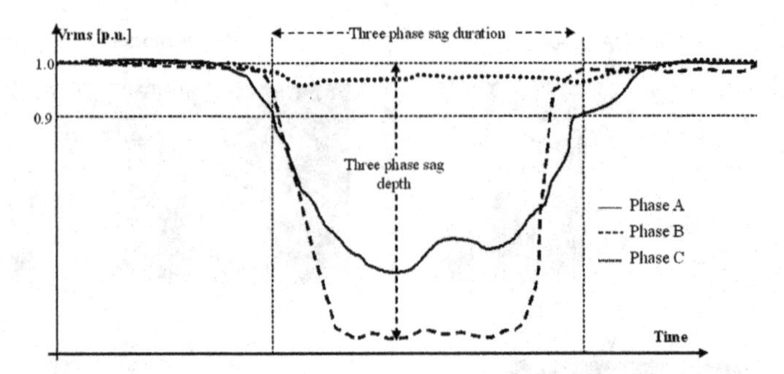

Fig. 6. Depth and duration temporal features for a three phase typical sag

Table 1. Overall classification rates for MPCA and different classifiers fed by temporal features

	Classifier			
Classification Rates (%)	MLP	RBF	DT (C4.5)	MPCA
Substation A	68.8	65.6	66.7	93.6
Substation B	80.3	73.8	78.7	92.9
Substation C	92.6	92.6	91.2	92.7
All substations	78.4	76.1	76.1	91.8

in three substations. The table includes in the last column the classification rates obtained with MPCA methodology.

The MPCA classifier has demonstrated that it performs better than others in differentiating between HV and MV sags for the three substations. An additional test has been performed to consider all the registers in one data set without differentiating the substation where they have been registered. The last row of the table shows the classification rates obtained with the different techniques. Again, MPCA demonstrates the best performance.

5 Conclusion

In this paper we have introduced a novel application of MPCA to assist power quality monitoring. With the aim of classifying HV and MV sags, we cast RMS values of three voltages and currents in a suitable format useable by MPCA. After a scaling stage, we create some statistical models which are then exploited as a classifier. To differentiate between the projection of not-yet-seen HV and MV sags over models, we use Q criterion to measure the lack of model fit. Exploring the grid, which is composed of a number of principal components and Q confidence limits, shows that the highest classification rates are obtained using a low number of principal components in the HV model. Comparing these results with those obtained using other conventional classifiers, we observe superior classification rates.

We strongly believed that we have only scratched the surface in introducing MPCA to the power quality management field. Generally speaking, we can divide further research into two domains: (1) applying the method proposed in this paper for the classification of other electrical disturbances, and (2) enhancing the obtained results. Even if we have more than two classes, we can create models and use each one to distinguishing between two classes. In this way, we build a bank of models instead of just one model.

For the second guide line, equipping the proposed method with some nonlinear transformation could lead to better results. Although the results obtained using the linear MPCA classifier in this paper (above 93%) are satisfactory enough, kernel MPCA and a combination of MPCA with other nonlinear classifiers can increase the classification rates significantly. Furthermore, as MPCA captures the most relevant information from the databases, we can employ it as a dimension reduction tool guaranteeing minimum information loss and then apply nonlinear classifiers such as neural networks.

Acknowledgement

This work has been developed in collaboration with ENDESA DISTRIBUCION on the basis of the research project "Diagnostico de Redes de Distribucin Elctrica Basado en Casos y Modelos" (DPI2006-09370) funded by the Spanish Government. Also it has been partially supported by the project DPI2005-08922-C02-02 supported by the Spanish Government. Authors from UdG, are members of the AEDS research group (http://aeds.udg.edu/) awarded as a Consolidated Research Group (SGR-002969) by the Generalitat de Catalunya.

References

1. Bollen MHJ (2000) Understanding Power Quality problems. IEEE Press, New York
2. Leborgne CR (2007) Voltage sags: single event characterisation, system performance and source location. PhD Thesis, Department of Energy and Enviroment, Chalmers University of Technology
3. Parsons AC, Grady WM, Powers EJ, Soward JC (2000) A direction finder for power quality disturbances based upon disturbance power and energy. IEEE Transactions on Power Delivery, 15(3):1081–1086
4. Li C, Tayjasanant T, Xu W, Li X (2003) Method for voltage-sag-source detection by investigatig slope of the system trajectory. In IEE Proc. Gen., Trans. & Distrib., 150(3):367–372
5. Tayjasanant T, Li C, Xu W (2005) A resistance sign-based method for voltage sag source detection. IEEE Transactions on Power Delivery, 20(4):2544–2551
6. Hamzah N, Mohamed A, Hussain A (2004) A new approach to locate the voltage sag source using real current component. Electric Power Systems Research, 72:113–123

7. Pradhan AK, Routray A (2005) Applying distance relay for voltage sag-source detection. IEEE Transactions on Power Delivery, 20(1):529–531
8. Leborgne RC, Karlsson D, Daalder J (2006) Voltage sag source location methods performance under symmetrical and asymmetrical fault conditions. In IEEE PES Transmission and Distribution Conference and Exposition Latin America, Venezuela
9. Mora J, Llanos D, Melendez J, Colomer J, Sanchez J, Corbella X (2003) Classification of Sags Measured in a Distribution Substation based on Qualitative and Temporal Descriptors. In 17th International Conference on Electricity Distribution, Barcelona, Spain
10. Wise BM, Gallagher NB, Watts S, White DD, Barna GG (1999) A comparison of pca, multiway pca, trilinear decomposition and parallel factor analysis for fault detection in a semiconductor etch process. Journal of Chemometrics, 13:379–396
11. Kourti T, MacGregor J (1996) Multivariate spc methods for process and product monitoring, Journal of Quality Technology, 28(4):409–428
12. Westerhuis JA, Kourti T, MacGregor JF (1999) Comparing alternative approaches for multivariate statistical analysis of batch process data. Journal of Chemometrics, 13:397–413
13. Russell EL, Chiang LH, Braatz RD (2000) Data-Driven Methods for Fault Detection and Diagnosis in Chemical Processes. Springer, Berlin Heidelberg New York
14. Wise BM, Gallagher NS (1996) The process chemometrics approach to process monitoring and fault detection. Journal of Process Control, 6(6):329–348
15. Nomikos P, MacGregor JF (1994) Monitoring batch processes using multi-way principal component analysis. AIChE Journal, 40(8):1361–1375
16. Lee DS,, Park JM, Vanrolleghem PA (2005)Adaptive multiscale principal component analysis for on-line monitoring of a sequencing batch reactor. Journal of Biotechnology, 116:195–210
17. Yoon S, MacGregor JF (2004) Principal-component analysis of multiscale data for process monitoring and fault diagnosis. AIChE Journal, 50(11):2991–2903
18. Witten IH, Frank E (2005) Data Mining: Practical Machine Learning Tools and Techniques, Second Edition. Morgan Kaufmann Publishers

Applications of Neural Networks to Dynamical System Identification and Adaptive Control

Xiao-Hua Yu

Department of Electrical Engineering, California Polytechnic State University, 1 Grand Avenue, San Luis Obispo, CA 93407, USA

Summary. In this chapter, we will discuss two important areas of applications of artificial neural networks, i.e., system identification and adaptive control. The neural network based approach has some significant advantages over conventional methods, such as adaptive learning ability, distributed associability, as well as nonlinear mapping ability. In addition, it does not require a priori knowledge on the model of the unknown system, so it is more flexible to implement in practice.

This chapter contains three sections. Section 1 gives a general introduction on system identification and adaptive control. Section 2 focuses on the application of artificial neural networks for rotorcraft acoustic data modeling and prediction. In Sect. 3, a neural network controller is developed for a DC voltage regulator.

1 Introduction

In order to control a particular system or analyze its properties, it is desirable to have a proper understanding of the system dynamics in advance. Therefore, system identification is a key component for adaptive filtering, prediction and control. It determines the mathematical model of a dynamical system by observing the system response to its input.

Traditional approaches for system identification include RLS (Recursive Least-Square) algorithm, MLE (Maximum Likelihood Estimation), Kalman filter, etc.; most of them are based on linear system theory. Even though the parameters of the real plant can be unknown, the system model needs to be defined in advance. The set of parameters can then be estimated based on the given structure of real plant.

In recent years, artificial neural networks have been widely used for system identification and adaptive control, especially for complicated nonlinear, time-varying dynamic systems. An artificial neural network (ANN) is a biologically inspired computational model which is similar to the real neuronal structure in a human brain. It is a large-scale, parallel-distributed information processing system which is composed of many inter-connected nonlinear computational units (i.e., neurons). The connections between the neurons

X.-H. Yu: *Applications of Neural Networks to Dynamical System Identification and Adaptive Control*, Studies in Computational Intelligence (SCI) **116**, 313–331 (2008)
www.springerlink.com

(analogy to synapses/axons) are called weights. A neural network can be trained; or in other words, it can "learn" from the given data set obtained from the input/output measurements of the plant under study. The neural network based approach does not require a priori knowledge on the model of the unknown system, and has some significant advantages over conventional methods, such as adaptive learning ability, distributed associability, as well as nonlinear mapping ability [22, 25].

In general, three issues are involved in the identification problem, they are: identification model, adaptation law and a specific performance criterion. Consider a MIMO (multi-input, multi-output) sampled dynamical system with n inputs and m outputs. Let $x(k)$ be the input signal sampled at time $t = kT$ (where T is the sampling period), $y(k)$ be the sampled output signal. The model of the system can be expressed as an operator, or a mapping (denoted by f), from the n-dimensional input space to the m-dimensional output space:

$$x \in R^n \xrightarrow{f} y \in R^m \tag{1}$$

i.e.,

$$y(k) = f(x(k)) \tag{2}$$

where f can be either linear or nonlinear.

The identification error e(k) is defined as the difference between the output of the real system $y(k)$ and the output of the identification model $\hat{y}(k)$. If we choose the norm of $e(k)$ to be the objective function to be minimized, then the identification process is completed when J is reduced to a desired level ε ($\varepsilon \geq 0$):

$$J = \|e(k)\| = \|y(k) - \hat{y}(k)\| \leq \varepsilon \tag{3}$$

This implies that the identification model follows the real plant; in other words, we can use the identification model to replicate the input/output mapping of the unknown system.

Neural networks can be organized into two topologies: recurrent and feedforward. The former allows full connections between different neurons and layers, including feedback connections; while the latter only allows the connections in forward direction. Multi-layer feedforward neural network (Fig. 1) is the class of neural networks that is used most often for system identification. This may be attributed to the fact that it has been proved to be extremely successful in representing any measurable function within any desired degree of accuracy, with the correct values of weights and sufficient number of hidden neurons.

A neural network can be trained by either supervised training, or unsupervised training. In supervised training, some selected inputs and the desired responses are provided as training examples (teachers), the network then uses these input signals to produce its own outputs which are called the actual outputs. No training is needed if there is no difference between the desired and the actual output; however, if there is a difference, the neural network

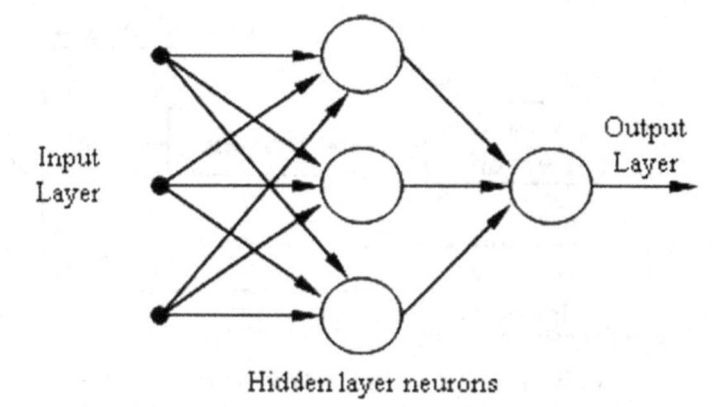

Fig. 1. Multilayer feed-forward network

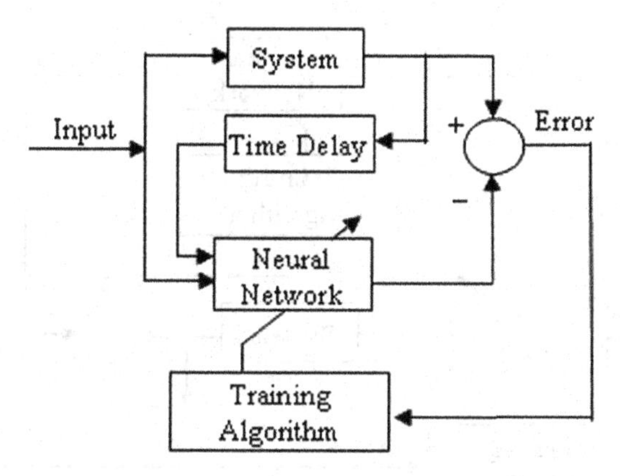

Fig. 2. Neural network for system identification

will modify its weights to reduce the difference. This procedure is repeated until the output error reaches an acceptable level. In unsupervised training, the training set consists solely of the input patterns and no output target is provided; the neural network organizes the received information within itself to response to the input signal. Figure 2 shows the block diagram of neural network based approach for system identification, where the neural network identification model is placed in parallel with the real plant and supervised learning is employed.

The objective of adaptive control is, to design a controller/regulator which can modify its behavior in response to the changes in the system dynamics and/or the environmental disturbances. Direct control and indirect control are two distinct schemes in adaptive control. In direct control, the control signals are directly adjusted to achieve certain performance criterion. However, in indirect control, the parameters of the plant have to be identified and the control signal is generated based on the assumption that the estimation represents the true value of the plant parameter. The two different approaches

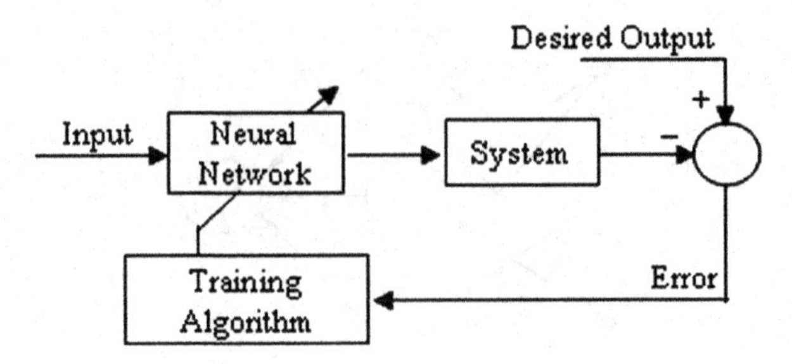

Fig. 3. Neural network control (direct control)

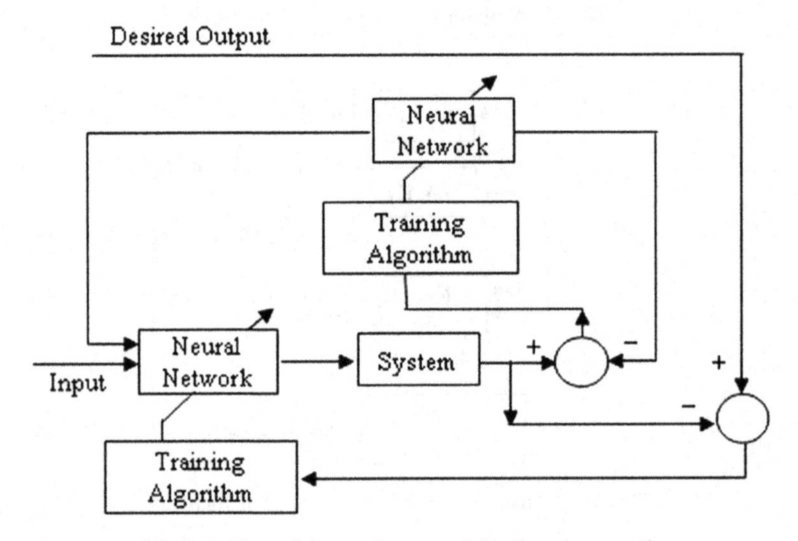

Fig. 4. Neural network control (indirect control)

are shown in Figs. 3 and 4. Note that the neural network can be chosen as the identifier, or the controller, or both. For the sake of simplicity, not all the connections are shown.

For indirect control, the identification procedure can be accomplished either on-line or off-line. In either case, the accuracy of control also depends on the accuracy of identification. That is, the identification model should be accurate enough before the control action can be taken; otherwise the whole system may become unstable. If we identify and control have to be performed simultaneously, both on-line, we should choose the control period (T_c) and the identification period (T_i) carefully such that $T_c \geq T_i$.

In the following sections, two applications on neural network based approach will be investigated: one on system identification, and the other one is on adaptive control. Also, different neural network structures and training algorithms will be discussed and compared.

2 Rotorcraft Acoustic Noise Estimation

In the studies of rotorcraft flight control and acoustics, the blade vortex interaction (BVI) noise level is an important measure of system performance. This level represents the acoustic noise generated from rotor interaction with the aerodynamic forces that are generated by previous rotor blade passages. It is the noise that is of the greatest concern for both military and civilian use [11].

BVI noise is closely related with certain flight parameters, such as the rotor thrust, the advance ratio, the angle of attack, the speed, etc. If pilots are given enough information to avoid the worst conditions under which a rotorcraft generates the greatest amount of noise, acoustic noise could be dramatically reduced in actual flight. Successful modeling and prediction of BVI noise levels may also allow for the development of future flight systems that can actively control the emission of BVI noise. This in turn may help rotorcrafts gain greater acceptance in use by general public. However, analysis on aerodynamics shows that BVI noise is a complicated nonlinear function of flight parameters, and its closed-form solution is very difficult to obtain.

To determine the highly nonlinear relationship between rotorcraft noise and flight parameters, several wind tunnel acoustic tests are conducted at NASA Ames Research Center, Moffett Field, California. A full-scale rotor model was placed in the wind tunnel and operated at a nominal revolution-per-minute (RPM) speed. Acoustic noise data are recorded in the 80×120-ft wind tunnel with microphones that are mounted on either a traverse array or fixed locations. The recorded acoustic data are characterized by the set of wind tunnel test parameters/conditions that the data was recorded at. Systematic variation of different parameters establishes the noise signature for the full range of test conditions.

The wind tunnel acoustic noise measurements come in two different forms. The first one is in time history form, which contains acoustic amplitude measurements sampled and recorded in the unit of Pascal for several revolutions of a rotorcraft under a specific set of flight test conditions. These measurements are then used to produce an averaged time history for one rotorcraft revolution. After the average is taken, the averaged time history is transformed into the frequency domain using FFT (Fast Fourier Transform) to obtain the second form of data, i.e., a sound pressure level value measured in decibels (dB). The sound pressure level is a comparative measurement that gives a magnitude comparison of the recorded sound pressure level with respect to a reference level. This reference sound level is about 2×10^{-5} (Pascal) and is considered to be the lowest sound amplitude that humans can hear. For example, a sound pressure level value of 140 dB (often cited as the sound of a jet taking off from a runway) means that the sound is 140 dB louder than the softest sound possible to be heard for human ear. The sound pressure level value gives a convenient single-value to compare acoustic noise between different operating parameters over one rotation of the test rotor. A digital bandpass filter bank can be applied to highlight the BVI frequencies.

The operating conditions of the rotor include parameters such as the rotor thrust coefficient C_T, a measure of the lifting capability of the rotorcraft; solidity ratio σ, a ratio of the total blade area of the rotor blades to the rotor disc area; the advance ratio μ, a dimension-less ratio of the true helicopter forward airspeed over the wind speed at the rotor tip; as well as the shaft angle α_s, and rotor tip-path-plane angle α_{TPP}. In a wind tunnel, a rotor is trimmed to zero flapping so we have $\alpha_{TPP} \approx \alpha_s$. It is known that rotorcraft acoustic noise is highly dependent upon these parameters [14]. In this study, these parameters are used for the neural network modeling.

There have been some developments in the past several years in the application of neural networks for wind tunnel experiments, such as [7–9,14,19,24]. Kottapalli and Kitaplioglu investigated BVI data quality analysis and prediction for the XV-15 tilt-rotorcraft using backpropagation and radial basis function neural networks [14]. Lo and Schimtz demonstrated the prediction of BVI sound pressure levels using neural networks and the Higher Harmonic Control Aeroacoustics Rotor Test (HART) database [19]. Two parameters that are related with flight conditions, including rotor tip-path-plane angle and advance ratio, are considered. Holger et al. used neural networks to model data time histories gathered from in-flight tests conducted by Eurocopter Deutschland on the BO 105 helicopter [9]. The BVI noise is modeled as a function of phase angles sent to the helicopter's individual blade controller.

2.1 The Time History Data Modeling

Figure 5 shows a set of typical measurements of the BVI of a modern four-bladed rotorcraft, plotted with respect to time. In this research, feedforward neural networks are chosen to model the BVI noise in time-domain. Let k

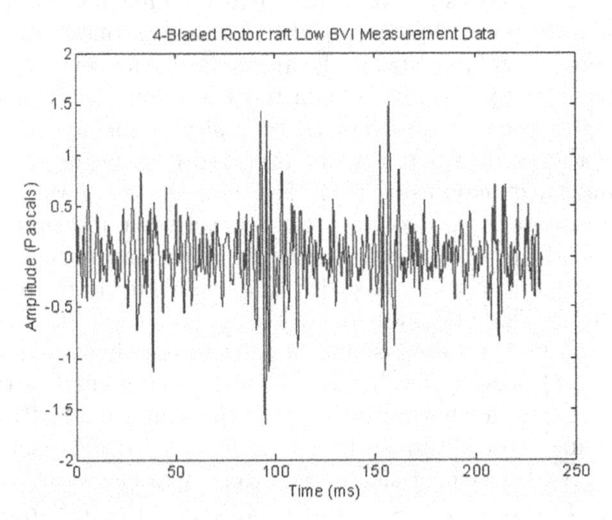

Fig. 5. Four-bladed rotorcraft time history (low BVI case)

be the discrete-time index and let $y(k)$ be the BVI acoustic level at time instant k. The inputs of the neural network include three previous values of $y(k)$, i.e., $y(k-1)$, $y(k-2)$, and $y(k-3)$. Three different neural network topologies and training algorithms are considered. The first one is a multi-layer ADALINE (adaptive linear neuron) neural network trained by LMS (least-mean-square) algorithm. All the neurons are linear and there is no hidden layer in this case. For the other two neural networks, each of them is constructed with a hidden layer of ten nonlinear neurons (sigmoid activation function) and a linear output neuron. They are trained by back-propagation and the extended Kalman filter algorithm, respectively. The weights of neural network can be adjusted using:

$$w(k) = w(k-1) - \Delta w(k) \tag{4}$$

where k is the index of training iteration.

Learning in back-propagation is performed in two steps, i.e., the forward pass and the backward pass. In the latter, the error signal (i.e., the difference between the neural network output and the target) is feedback from the output to the input, layer by layer, and all the weights are adjusted in proportion to this error, i.e.,

$$\Delta w(k) = \eta \frac{\partial J}{\partial W} \tag{5}$$

where η is called the learning/training rate; J is the objective function defined in (3) and W is the weight matrix of the neural network. For sigmoid function, we have:

$$g(u) = \frac{1}{1+e^{-au}} \tag{6}$$

where g is the output of the nonlinear neuron ($a > 0$) and

$$u = WX \tag{7}$$

with X is the input vector to the neuron. For the sake of simplicity, the dimensions of matrices are not included in the above formula.

It is well known that the extended Kalman filter algorithm (EKF) can be used to estimate the state of a nonlinear system. If the neural network is considered as the nonlinear system, then its weights can be updated through the Kalman gain K in each training iteration k:

$$w(k) = w(k-1) + K(k)[y(k) - \hat{y}(k)] \tag{8}$$

$$K(k) = P(k-1)H(k)[R(k) + H^T(k)P(k-1)H(k)]^{-1} \tag{9}$$

$$P(k) = P(k-1) - K(k)H^T(k)P(k-1) \tag{10}$$

where H is the first order derivative of the neural network output with respect to the weights; R is the covariance matrix of the identification error.

The simulation results are shown in Fig. 6 (zoomed in for better illustration). It is shown that the two nonlinear networks are able to achieve lower

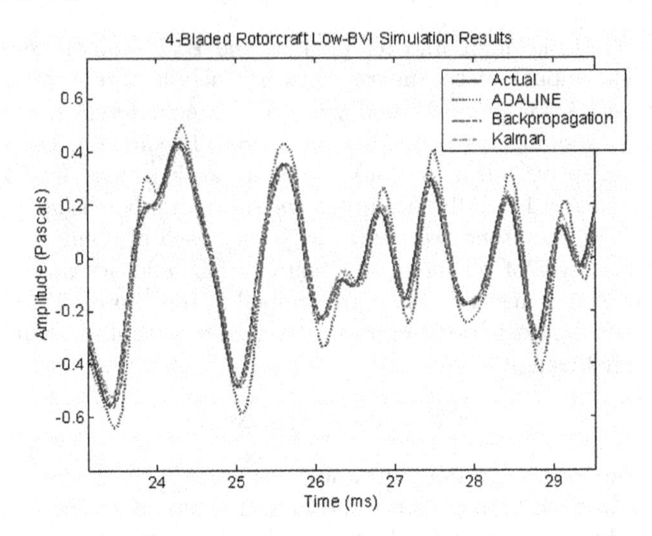

Fig. 6. Simulation results for four-bladed rotorcraft (low BVI case)

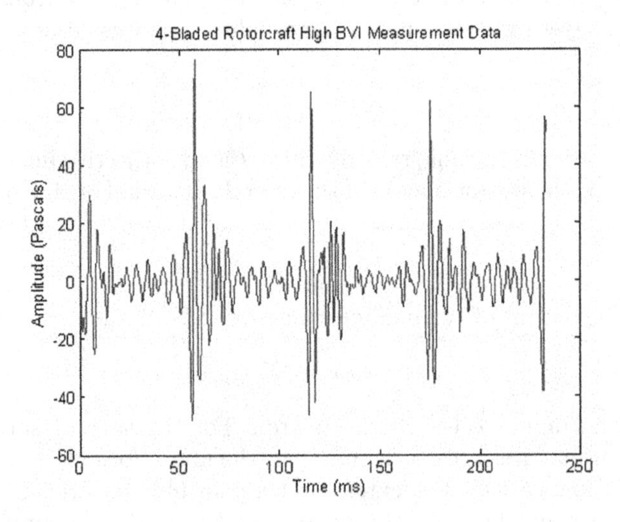

Fig. 7. Four-bladed rotorcraft time history (high BVI case)

MSE (mean-square-error) values than the linear ADALINE network, and the EKF trained neural network outperforms the backpropagation network. However, it should be noticed that comparing with the backpropagation algorithm, the computational complexity of the EKF algorithm increases dramatically for the neural network with larger size.

Another case involves the simulation of high BVI time history data. The measurement data, show in Fig. 7, contains four pulses which represent the BVI noise in high amplitude.

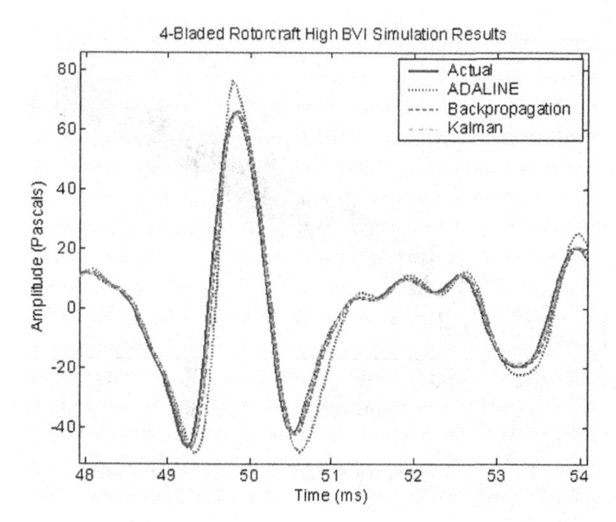

Fig. 8. Simulation results for four-bladed rotorcraft (high BVI case)

The simulation results for this case are similar to the results of the previous case. Again, the EKF trained network has the best overall performance with lowest MSE, even though the computation time is relatively longer. A comparison plot for the three algorithms and the original time history measurement data is shown in Fig. 8 (also zoomed in for better illustration).

2.2 The Sound Pressure Level Modeling

The sound pressure level metric provides a convenient way for rotorcraft researchers to study the overall BVI acoustic noise for a given set of test parameters. Instead of having to compare many sampled time history data points, the BVI acoustic noise for a set of wind tunnel test parameters may be represented by a single value that gives an overall measure of the BVI acoustic noise at those test conditions. In this study, the BVI sound pressure level metric is derived from the time history data by summing up the power spectrum for one rotor revolution of bandpass-filtered time history data and converting to dB from Pascals^2. The pass-band of the filter is between the tenth and 50th blade passage harmonic, which is the frequency range that preserves highest amounts of BVI noise [14].

The neural network employed to model the four-bladed rotorcraft BVI sound pressure level has 15 hidden neurons in the first hidden layer and five hidden neurons in the second hidden layer. The inputs of the neural network include C_T/σ, μ, α_{TPP}, the microphone number, and the current traverse distance x_m. To speed up the training, the extended-bar-delta-bar adaptive learning rate algorithm is used:

$$\Delta w(k) = \rho \Delta w(k) + (1 - \rho)\, \alpha\, \delta(k) \tag{11}$$

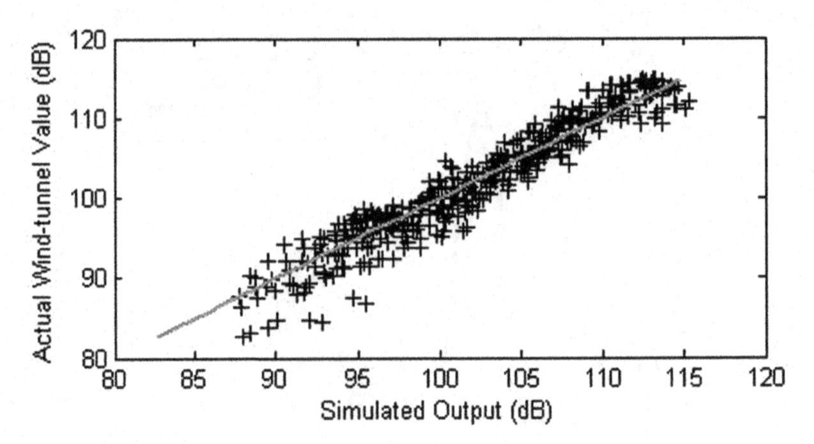

Fig. 9. Correlation plot for testing dataset

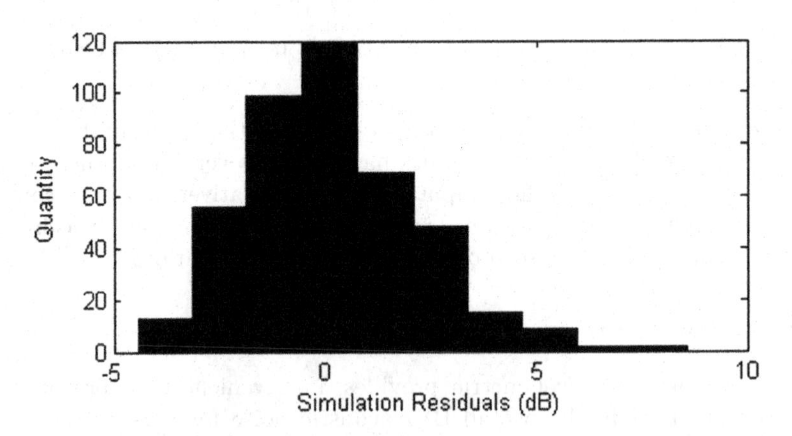

Fig. 10. Residual histogram for testing dataset

and

$$\delta(k) = \partial J / \partial W \tag{12}$$

where ρ is the momentum factor and α is called the adaptive learning rate. An average mean square error of 0.0015 is achieved in approximately 3,400 training epochs. Training to a lower mean-square-error may cause the neural network to over-fit the training data. Post-simulation scatter correlation plot for the testing dataset is shown in Fig. 9, where the straight line (with slope of 1) illustrates the exact correlation. Most points lie along the line of exact correlation, which indicates the neural network modeling is very successful.

The histogram of the residuals between the actual wind tunnel measurement and the neural network model is shown in Fig. 10. Most of the residual points (about 120) are located around 0 dB while the upper bound is about

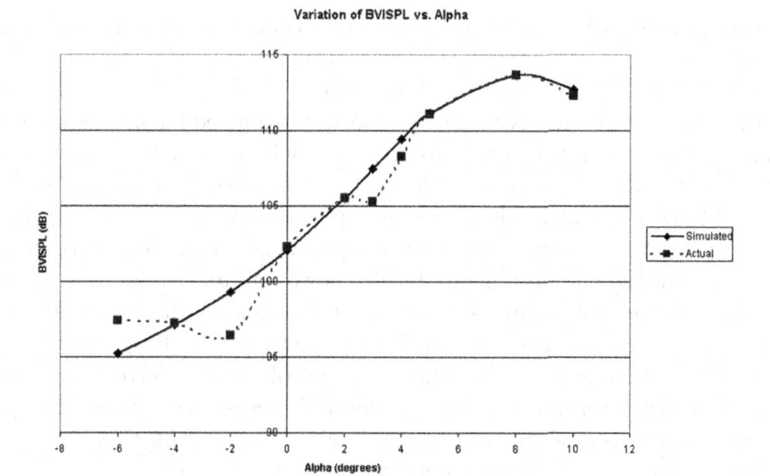

Fig. 11. Plot of BVISPL vs. α_{TPP}

9 dB (about 7.8% of the maximum BVI value) and the lower bound is less than 5 dB (about 4.3% of the maximum BVI value).

In statistics, the coefficient of determination (or goodness-of-fit, R^2 value) is often used to determine the correlation between the model prediction and the actual values [10]. Let W and $\bar{W}$ be the recorded BVI sound pressure level and its mean, respectively; S be the neural network estimated BVI sound pressure level, then

$$R^2 = 1 - \frac{\sum_{i=1}^{N} (S_i - \bar{W})^2}{\sum_{i=1}^{N} (W_i - \bar{W})^2} \tag{13}$$

The R^2 value for testing dataset is 0.961, which concludes a strong correlation between the neural network model and the actual measurements.

As a representative example, Fig. 11 shows the simulation result when a single parameter, α_{TPP}, varies. A test dataset with α_{TPP} varying from -6 to 10 degrees is used with the other input parameters fixed to test the neural network performance. It is shown that the neural network has successfully captured the relationship between α_{TPP} and BVI sound pressure level.

3 A Neural Network Controller for DC Voltage Regulator

Since all semiconductor components are powered by DC source, DC voltage regulators can be found in almost every electronic device nowadays. One of the design targets for electrical engineers is to improve the efficiency of power conversion. PWM (pulse-width modulation) technique is used in many DC-DC converters. It changes the average value (i.e., dc component) of a square waveform by changing the duty cycle.

In PWM converters, a switching network is employed for square wave modulation. Ideally, the power dissipated by the switch network is zero; however in practice, the power efficiency of a typical DC-DC converter could be as low as 70%. Many different kinds of circuit topologies [1,2,13] have been investigated in the past to reduce the switching loss. Unfortunately, those designs either need additional components for the power circuit, which may introduce some unstable factors; or operate at variable frequency, which makes the filter design at output stage very difficult. Phase-shifted zero-voltage switching full-bridge converters can overcome the above problems and have been received more and more attention recently. It employs zero-voltage-switching (ZVS) technique which allows the voltages across the transistors to swing to zero just before the start of the next conduction cycle [3,5,21].

The circuit diagram of a PSFB (Phase-Shifted Full-Bridge) DC-DC converter is shown in Fig. 12. Based on circuit analysis, the control scheme to drive the switching MOSFET (Q1, Q2, Q3 and Q4) is very complicated [21]. The circuit can be operated in one of its eight modes, depending on the states of the four switching MOSFET.

The relationship between the input/output voltage and the duty cycle can be described as:

$$V_0 = \left(\frac{2nV_{in}}{T_s}\right)\left(\Phi \cdot \frac{T_s}{2} - T_1 + \frac{T_2}{2}\right) \tag{14}$$

where V_0 is the output voltage, V_{in} is the input voltage, both in RMS value; n is the transformer ratio, T_s is the switching period constant, and Φ is the duty cycle. T_1 and T_2 can be obtained from the following equations:

$$T_1 = \frac{(nI_0 + I_c)L_r}{V_{in}} \tag{15}$$

$$T_2 = \frac{V_{in}C_r}{nI_0} \tag{16}$$

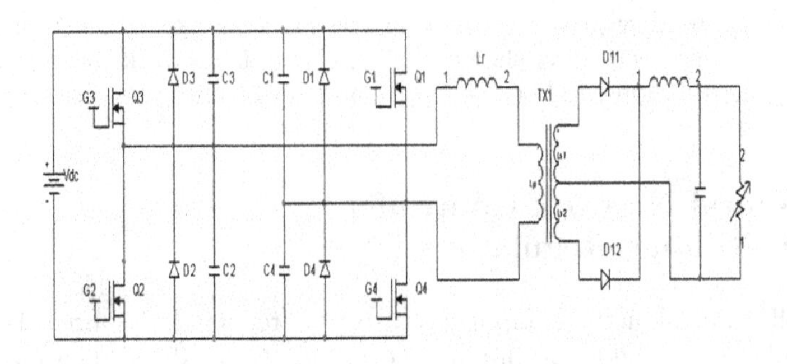

Fig. 12. PSFB DC-DC converter

where I_0 is the load current, C_r is the resonant capacitance, L_r is the resonant inductance. I_c can be calculated as follows:

$$I_c = V_{in}\sqrt{\left(\frac{nI_0}{V_{in}}\right)^2 - \frac{C_r}{L_r}} \tag{17}$$

The controller determines the duty cycle based on the input voltage and load current to achieve the desired output voltage. As shown in the above equations, the control law is highly nonlinear and as the result, it leads to a very complicated design in conventional approach.

The conventional controller design is based on the assumption that the converter is operated around its equilibrium state and thus can be modeled by a set of linear equations [4, 23]. However, in practice, both the supply voltage and load current have a wide range of variation. Therefore, an adaptive controller which can modify its behavior in response to these changes is desired. Because of its learning ability and nonlinear mapping ability, neural network is an excellent candidate for solving this problem [20]. However, no prior work has been done to design and implement a neural network controller for this type of PSFB (Phase-Shifted Full-Bridge) converter yet.

Some related research can be found in [18], in which a neural network controller is proposed for a two-quadrant DC/DC converter with switched inductor. The circuit consists of two switches, two diodes and one inductor. A three-layer feedforward neural network is trained by back-propagation algorithm and Gaussian function is employed as the activation function for the nodes in the hidden layer. In [6], the application of a neural network controller for a stand-alone fuel cell power plant (FCPP) is considered. The purpose of control is to provide the active and reactive power output from an FCPP that match with the terminal load. A three-layer feedforward neural network with a log-sigmoid activation function for hidden nodes is used. The inputs of the neural network include the load active power, load reactive power, the AC output voltage magnitude, and the hydrogen input to the FCPP. The outputs of the neural network include the inverter modulation index (which controls the voltage level and the reactive power output from the FCPP) and the phase angle of the AC output voltage (which controls the active power output from the FCPP). The neural network controller is trained off-line by the Levenberg-Marquardt algorithm. In [12], the control of a high-power-factor rectifier regulator is investigated. In this application, the neural network actually serves as a time-series predictor of the input current reference, rather than being a direct controller. The neural network is trained on-line, with a new random weight change (RWC) algorithm. The computer simulation results are very promising; however, a hardware implementation of the neural network predictor is not accomplished yet.

In this research, to design a neural network controller, a MATLAB® Simulink model is developed based on the above circuit analysis for neural network training [15–17]. Furthermore, to investigate the performance of different neural network controllers, three feed-forward networks are employed to estimate the desired duty cycle. These networks are implemented with similar structures but different activation functions. For the linear network, there are ten neurons in the hidden layer and the activation function of each neuron is defined as ($a > 0$):

$$f(x) = ax \tag{18}$$

For the non-linear tangent network, there are six neurons in the hidden layer and the activation function of each neuron is hyperbolic tangent function:

$$g(u) = \frac{1 - e^{-au}}{1 + e^{-au}} \tag{19}$$

where $a > 0$. The non-linear logistic sigmoid network has the same structure as the previous one, except the activation function of each neuron is the logistic sigmoid function, as shown in (6).

The weights of the three neural networks are initialized at random. To speed up the training, Levenberg-Marquardt algorithm is employed:

$$\Delta W = (J_a^T J_a + \mu I)^{-1} J_a^T e \tag{20}$$

where J_a is the first order derivative of the output error with respect to the weights of neural network (also called the Jacobian matrix), e is the output error (i.e., the difference between the neural network output and the desired output), and μ is a learning parameter.

The simulation results in Fig. 13 (zoomed in for better illustration) show that the two trained non-linear networks are both able to estimate the duty cycle with high accuracy, but the linear network fails. Detailed investigation shows that the linear network converges after only two iterations and cannot be trained further. This is due to the fact that the system under study is highly

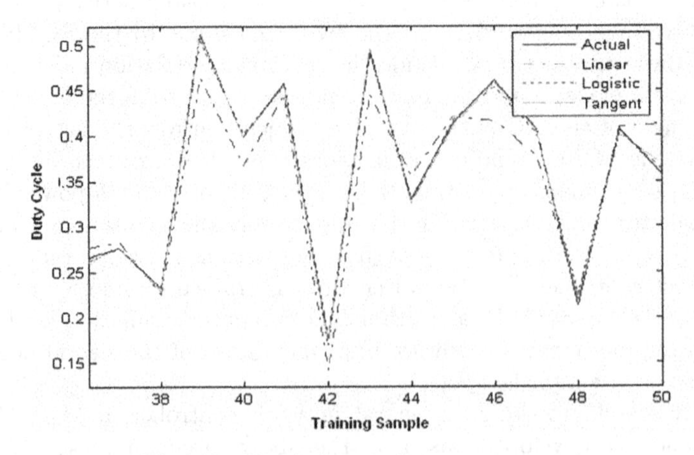

Fig. 13. Simulation performance comparison

nonlinear and cannot be approximated by a linear network. The training errors of the two nonlinear networks are both reduced to the desired accuracy after about 100 iterations.

The decision of choosing the appropriate activation function for actual hardware and software implementation now depends on the simplicity of each network. Comparing the definitions of two functions, one can see that the logistic sigmoid function is simpler and requires less computational power. Therefore, by choosing the logistic function, processing time can be reduced and faster dynamic response can be achieved for real-time implementation.

In order to implement the neural network controller in hardware and investigate its performance, a DSP (Digital Signal Processor) evaluation module eZdsp® F2812 and a phase-shifted zero-voltage-switching evaluation board UCC3895EVM are employed for the neural controller and the power circuit, respectively. The on-board digital signal processor TMS320F2812 is a 32-bit CPU with 150 MIPS (million instructions per second) operating speed. Its software development tool provides a code-efficient environment for both C/C++ and assembly language, which ensures fast processing speed for calculation intensive applications. That is, the microprocessor can be programmed in a high level language, yet still be optimized for quick response.

The phase-shifted full-bridge converter operates on 48 V nominal input and can provide an output of 3.3 V. Its input voltage ranges from 36 to 72 V so that we can test the circuit with wide range of input variation. The maximum load current is 15 A.

Some modifications are necessary before the performance of neural network controller can be tested. First, the original analog controller UCC3895 on the board needs to be removed. Next, the analog signals from the power circuit, which include the input voltage, output voltage, and load current, need to be connected to the ADC (Analog to Digital Conversion) input channels on the DSP evaluation module eZdsp® F2812. However, the power circuit is driven by 400 kHz high frequency PWM signals; thus the raw output voltage signal is quite noisy during the switch turn-on and turn-off time. To solve this problem, multiple samples are taken to obtain the average value of the signal over a certain period of time. Also, the input and output voltage signals from the power circuit have to be scaled down, due to the fact that the maximum input voltage the ADC channel can accept without overflow is only 3 V.

The inputs to the neural network include input voltage, load current, and the absolute value of the change of output voltage $|\Delta V_0|$. If $|\Delta V_0|$ is greater than a preset threshold, a new control signal (i.e., duty cycle) is then calculated by the neural network controller to reduce the voltage fluctuation; otherwise, the neural network controller is not activated. This way, the system can save unnecessary run time and resources, and respond to system dynamics more quickly.

It should be noticed that the MATLAB® SimuLink model for neural network training is developed under ideal conditions which do not consider the switching loss and power consumptions in real circuits. In fact, experiment

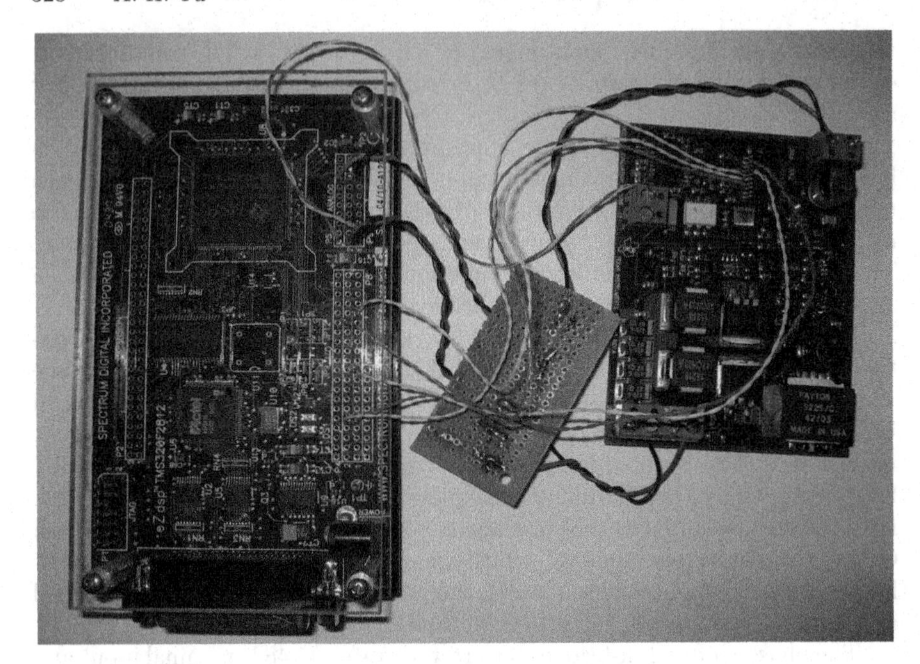

Fig. 14. The overall system

results show that the resulting output voltage is significantly lower than the desired value. Therefore, the neural network controller needs to be fine-tuned on-line to adapt to the real world situation. This is achieved with back-propagation algorithm, which is the most popular training algorithm for feedforward networks and proved to provide fast convergence for nonlinear control applications. With the correction of on-line training, the neural network controller performs as expected.

A series of tests are conducted to compare the performance of neural network controller with the on-board conventional analog controller. A picture of the overall system is shown in Fig. 14, where the power circuit board is on the right and the neural network controller (DSP board) is shown on the left.

In the design of voltage regulators, load regulation and line regulation (both in terms of percentage) are two important performance indexes. The former is an indication of how much the output voltage changes over a range of load resistance values, while the latter is an indication of how much the output voltage changes over a range of input voltages at full load. To test the controller performance, the output voltage of the DC–DC converter is measured with different load currents ranging from 0 (A) to 15 (A). Experimental results show that at the nominal input voltage, the average output by the NN controller is about 3.323 (V) with standard deviation of only 0.009 (V); while the average output by the analog controller is about 3.263 (V) with a larger standard deviation of 0.037 (V). The maximum deviation from nominal value by the NN controller is 0.08 (V) while the analog controller is 0.1 (V). The difference between the average output by the NN controller and the desired output (3.3 V) is only 0.68% while its analog counterpart yields 1.14%. Over the full range, the percentage of load regulation is about 2.1% under the control of neural network controller while load regulation is 4.5% using the conventional controller.

The performance of the neural network controller is also measured at different input voltages (from 48 to 72 V) under full load. Experimental results show that the average output by the NN controller is about 3.311 (V) with standard deviation of 0.0177 (V), which outperforms the results by analog controller (average output voltage 3.274 (V), standard deviation of 0.0458 (V)). The maximum deviation from nominal value by the NN controller is 0.04 (V) while the analog controller is 0.1 (V). The percentage difference between the average output by the NN controller is also smaller (0.346%) comparing with its analog counterpart (0.779%). In addition, the neural network controller achieves better line regulation (0.9%) over the full range than the analog controller (3.9%). From the above results, we can conclude that the neural network controller outperforms the traditional analog controller for this class of DC–DC converter.

Summary

In this chapter, two important areas of applications of artificial neural networks are discussed, i.e., system identification and adaptive control. The first case study shows how neural networks can be employed to model the acoustic data gathered from a wind tunnel test for a four-bladed rotorcraft; and the second one involves the design of a neural network controller for a DC–DC voltage converter.

Although artificial neural networks have shown feasibility and great potential in many applications, more research works are still needed to be done, such as global convergence and stability analysis, network topology study, as well as training algorithm development.

References

1. Carrasco JM, Galván E, Valderrama GE, Ortega R, Stankovic A (2000) Analysis and experimentation of nonlinear adaptive controllers for the series resonant converter. IEEE Trans Power Electron 15(3): 536–544
2. Chetty PR (1992) Resonant power supplies: Their history and status. IEEE Aerospace and Electronics System Magazine 7(4): 23–29
3. Choi HS, Cho BH (2002) Novel zero-current-switching (ZCS) PWM switch cell minimizing additional conduction loss. IEEE Trans on Industrial Electronics 49(1): 165–172

4. Choi B, Kim J, Cho BH, Choi S, Wildrick CM (2002) Designing control loop for DC-to-DC converters loaded with unknown AC dynamics. IEEE Trans on Industrial Electronics 49(4): 925–932
5. Choi HS, Kim JW, Cho BH (2002) Novel zero-voltage and zero-current-switching (ZVZCS) full-bridge PWM converter using coupled output inductor. IEEE Trans on Power Electronics 17(5): 641–648
6. El-Sharkh MY, Rahman A, Alam MS (2004) Neural networks-based control of active and reactive power of a stand-alone PEM fuel cell power plant. Journal of Power Resources 135 (1–2): 88–94
7. Fu J (2004) Modeling and analysis of rotorcraft acoustic data with neural networks. Thesis, California Polytechnic State University, San Luis Obispo
8. Fu J, Yu XH (2006) Rotorcraft Acoustic Noise Estimation and Outlier Detection. In: Proceedings of the IEEE World Congress on Computational Intelligence (International Joint Conference on Neural Networks). pp. 8834–8838
9. Holger G et al. (2003) Neural networks for BVI system identification. In: Proceedings of the 29th European Rotorcraft Forum, Germany
10. Iglewicz B, Hoaglin DC (1993) How to detect and handle outliers. ASQC Quality Press, Wisconsin
11. Johnson W (1980) Helicopter theory. Princeton University Press, New Jersey
12. Kamran F, Harley RG, Burton B, Habetler TG, Brooke MA (1998) A fast on-line neural-network training algorithm for a rectifier regulator. IEEE Trans on Power Electronics 13(2): 366–371
13. Kim MG, Youn MJ (1991) An energy feedback control of series resonant converters. IEEE Trans Power Electron 6(4): 338–345
14. Kitaplioglu C. Blade-Vortex interaction noise of a full-scale XV-15 rotor tested in the NASA Ames 80- by 120-foot wind tunnel, NASA technical memorandum 208789. National Aeronautics and Space Administration, Moffett Field
15. Li W (2006) A neural network controller for a class of phase-shifted full-bridge DC–DC converter. Thesis, California Polytechnic State University, San Luis Obispo
16. Li W, Yu XH (2007) Improving DC power supply efficiency with neural network controller. In: Proceedings of the IEEE International Conference on Control and Automation. pp. 1575–1580
17. Li W, Yu XH (2007) A self-tuning controller for real-time voltage regulation. In: Proceedings of the IEEE International Joint Conference on Neural Networks. pp. 2010–2014
18. Lin F, Ye H (1999) Switched inductor two-quadrant DC–DC converter with neural network control. In: IEEE International Conference on Power Electronics and Drive Systems. pp. 1114–1119
19. Lo C, Schmitz F (2000) Model-Based Neural Networks for Rotorcraft Ground Noise Prediction. In: Proceedings of the 38th AIAA Aerospace Sciences Meeting & Exhibit, Nevada
20. Quero JM, Carrasco JM, and Franquelo LG (2002) Implementation of a neural controller for the series resonant converter. IEEE Trans on Industrial Electronics 49(3): 628–639
21. Ruan X, Yan Y (2001) A novel zero-voltage and zero-current-switching PWM full-bridge converter using two diodes in series with the lagging leg. IEEE Trans on Industrial Electronics 48(4): 777–785
22. Schalkoff RJ (1997) Artificial neural networks. McGraw-Hill, New York

23. Texas Instrument Inc (2001) BiCMOS advanced phase shift PWM controller, Data sheet of UCC3895. Texas Instrument, TX
24. Yu XH, Fu J (2004) Aeroacoustic data analysis with artificial neural networks. In: Proceedings of the IASTED International Conference on Intelligent Systems and Control. pp. 244–247
25. Zaknich A (2003) Neural Network for Intelligent Signal Processing. World Scientific, Singapore

[illegible]

[illegible]

[illegible]

A Multi-Objective Multi-Colony Ant Algorithm for Solving the Berth Allocation Problem

Chun Yew Cheong and Kay Chen Tan

Department of Electrical and Computer Engineering, National University of Singapore, 4 Engineering Drive 3, 117576, Singapore, cheongchunyew@nus.edu.sg, eletankc@nus.edu.sg

Summary. This paper considers the allocation of a fixed number of berths to a number of ships arriving at the port within the planning horizon for container handling by determining the berthing time and location, in terms of berth index, for each ship. The solution to this berth allocation problem (BAP) involves the optimization of complete schedules with minimum service time and delay in the departure of ships, subject to a number of temporal and spatial constraints. To solve such a multi-objective and multi-modal combinatorial optimization problem, this paper presents a multi-objective multi-colony ant algorithm (MOMCAA) which uses an island model with heterogeneous colonies. Each colony may be different from the other colonies in terms of the combination of pheromone matrix and visibility heuristic used. In contrast to conventional ant colony optimization (ACO) algorithms where each ant in the colony searches for a single solution, the MOMCAA uses an ant group to search for each candidate solution. Each ant in the group is responsible for the schedule of a particular berth in the solution.

1 Introduction

Major container ports in the US and Japan, such as Los Angeles, Oakland, Kobe, and Yokohama, feature dedicated terminals, which are leased on a long term basis to shipping operators [1]. In these terminals, berths are an exclusive resource for shipping operators, who are themselves directly involved and responsible for the processing of containers. For a ship operator handling a large volume of containers and ship calls, the productivity for such terminals would be high. However, overcapitalization of the port might result if the handled volume is small as operations would be costly. In this case, it would be more economical to share the berth resource through multi-user terminals, which are defined as terminals where the incoming ships, regardless of shipping operators, may not be allocated to the same berth whenever they call. In these systems, the berth allocation process for assigning berths to calling ships

C.Y. Cheong and K.C. Tan: *A Multi-Objective Multi-Colony Ant Algorithm for Solving the Berth Allocation Problem*, Studies in Computational Intelligence (SCI) **116**, 333–350 (2008)
www.springerlink.com

for container handling plays a decisive role in minimizing the turnaround time. The multi-user terminal is a widespread system in use, especially in land-scarce ports such as Singapore, Hong Kong, and Pusan.

This paper considers the berth allocation problem (BAP) in multi-user terminals. Given a collection of ships that are to arrive at the port within a planning horizon, the BAP involves determining the berthing time and location of each of the ships, while satisfying a number of spatial and temporal constraints, to optimize operations. More details of the problem are given in Sect. 2.

In the literature, a number of objectives for optimizing port throughput have been considered. Imai et al. [1–4] considered the BAP by minimizing the total service time of ships. The service time of each ship includes the waiting time between the arrival time of the ship at the port and the time the ship berths as well as the handling time for loading or unloading of containers. Guan et al. [5] developed a heuristic for the BAP with the objective of minimizing the total weighted completion time of ship services. Kim and Moon [6] solved their version of the BAP by minimizing the penalty cost resulting from delays in the departure of ships and additional handling costs resulting from non-optimal locations of ships in the port. According to their formulation, each ship has an optimal berthing location in the port. Park and Kim [7] solved the same problem by using a sub-gradient method. In their formulation, additional cost is incurred from early or late start of ship handling against their estimated time of arrivals. Li et al. [8] solved the BAP by minimizing the makespan of the schedule. Imai et al. [3] tackled the BAP with service priority, where some ships are given priority, in terms of being serviced earlier, over others. In their work, they provided several examples and arguments for differentiating the service treatment of ships. Lim [9] took a different approach to the BAP by minimizing the maximum amount of space used for berthing ships. Lai and Shih [10] proposed some heuristic algorithms for a BAP which is motivated by the need for more efficient berth usage at the HIT terminal of Hong Kong. Their problem assumes the first-come-first-serve (FCFS) allocation strategy which in most cases does not lead to optimal schedules. Imai et al. [11] considered a BAP for commercial ports. Most service queues are traditionally processed on a FCFS basis. They concluded that for high port throughput, optimal ship-to-berth assignments should be found without considering the FCFS heuristic. However, they also noted that this may result in some dissatisfaction among ship operators regarding the order of the service sequence.

From the studies of Imai et al. [11], it appears that the BAP is a multi-objective optimization problem. As port operators try to optimize their operations to obtain a high throughput, there is also a need to account for the satisfaction levels of ship operators. Therefore, it is required to minimize concurrently the multiple conflicting objectives from the points of view of the port and ship operators, which is best solved by means of multi-objective optimization. Most of the existing literature, however, uses single-objective-based heuristic methods that incorporate penalty functions or combine the different objectives by a weighting function.

In this paper, a multi-objective multi-colony ant algorithm (MOMCAA) is applied to solve the BAP. It utilizes the concepts of Pareto optimality to minimize concurrently the objectives of total service time of ships and total delay in the departure of ships. The MOMCAA uses an island model with heterogeneous colonies. Each colony may be different from the other colonies in terms of the combination of pheromone matrix and visibility heuristic used. Different from conventional ant colony optimization (ACO) algorithms where each ant in the colony searches for a single solution, the MOMCAA uses an ant group to search for each candidate solution. Each ant in the group is responsible for the schedule of a particular berth in the solution.

This paper is organized as follows: Section 2 provides the problem formulation of the BAP, which includes a description of the constraints of the problem and some assumptions that are made. Section 3 presents an introduction to the ACO algorithm, while Sect. 4 presents the implementation details of MOMCAA. Section 5 presents the extensive simulation results and analysis of the proposed algorithm. Conclusions are drawn in Sect. 6.

2 Problem Formulation

The BAP involves allocating a fixed number of berths to a number of ships arriving at the port within the planning horizon for container handling by determining the berthing time and location, in terms of berth index, for each ship. Each of the ships will come into the port, wait for the scheduled berthing time, berth at the allocated berth, load or unload containers, and leave the port. Figure 1 shows the berth operation timeline for each ship. In Fig. 1, the service time of a ship at the port includes the waiting time between the arrival time of the ship and the time the ship berths as well as the handling time for loading or unloading containers.

The handling time for each ship is different at different berths. This is to take into account the transportation time for moving the containers to be loaded onto the ship from the original storage area to the allocated berth. The handling time is also assumed to be deterministic. There is also a deadline for servicing each ship. This deadline is usually provided by ship operators, exceeding which will incur their dissatisfaction.

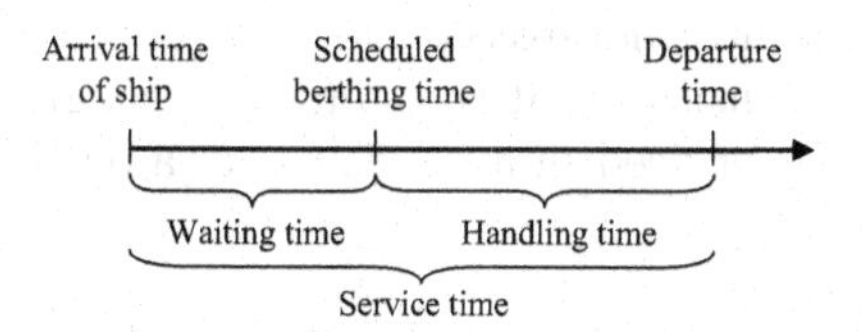

Fig. 1. Berth operation timeline

Adapting the problem studied by Nishimura et al. [12], the BAP is formulated as follows:

Minimize

$$\sum_{i \in B} \sum_{j \in V} (m_j - A_j + C_{ij}) x_{ij} \quad (1)$$

and

$$\sum_{i \in B} \sum_{j \in V} (m_j + C_{ij} - DL_j)^+ x_{ij} \quad (2)$$

where $x^+ = \max\{0, x\}$, subject to

$$\sum_{i \in B} x_{ij} = 1 \qquad \forall j \in V \quad (3)$$

$$m_j - A_j \geq 0 \qquad \forall j \in V \quad (4)$$

$$\sum_{i \in B} (WD_i - D_j) \geq 0 \qquad \forall j \in V \quad (5)$$

$$\sum_{i \in B} \left(QL_i - \sum_{j' \in V - \{j\}} L_{j'} y_{jj'} x_{ij'} - L_j \right) x_{ij} \geq 0 \qquad \forall j \in V \quad (6)$$

$$x_{ij} \in \{0, 1\} \qquad \forall i \in B, j \in V \quad (7)$$

$$y_{jj'} \in \{0, 1\} \qquad \forall j, j' \in V \quad (8)$$

$$m_j \text{ is integer} \qquad \forall j \in V \quad (9)$$

where B is the set of berths, V is the set of ships, m_j is the berthing time of ship j, A_j is the arrival time of ship j, C_{ij} is the handling time of ship j at berth i, DL_j is the deadline for servicing ship j, WD_i is the water depth of berth i, D_j is the draft of ship j including the safety vertical distance for berthing, QL_i is the length of berth i, and L_j is the length of ship j including the safety horizontal length. $x_{ij} = 1$ if ship j is serviced at berth i, $x_{ij} = 0$ otherwise. $y_{jj'} = 1$ if ship j begins its service when ship j' is being serviced at the same berth, $y_{jj'} = 0$ otherwise.

In the problem formulation above, (1) is the objective of minimizing the total service time incurred by ships, while (2) accounts for the second objective of minimizing the total delay in the departure of ships. Constraint (3) ensures that every ship can only be serviced at one berth without disruption. Constraint (4) dictates that every ship can only be berthed after it has arrived at the port. (5) imposes the physical constraint that the draft (including safety distance) of every ship must not exceed the water depth of the assigned berth, whereas constraint (6) ensures that the sum of the lengths of ships (including safety distances) being serviced simultaneously at a berth does not exceed the length of the berth.

3 Ant Colony Optimization

Ant colony optimization (ACO), which is inspired by the collective behavior of ant colonies searching for food [13, 14], is becoming an increasingly popular optimization technique. It is based on the idea that ants can collectively establish the shortest path between a food source and their nest. The search process hinges on pheromone trails, which are laid down as ants move and attract other ants. The probability that an ant chooses a path over another increases with the pheromone level on the path. As such, a path becomes more attractive as more ants use it. However, there is also a chance that an ant may not follow a well-marked path. This, coupled with the evaporation of pheromone as time passes, allows for search randomness and exploration of new regions of the search space. In addition, heuristics can be incorporated in the search process in the form of ant visibility. This section shows how the ACO can be applied to the multi-objective BAP formulated in the previous section.

3.1 Solution Encoding

A fixed-length solution representation (Fig. 2) is used in the ACO. It encodes the scheduling order of ships for each of the berths rather than the berthing times. The order of servicing the ships does not indicate that a ship can only be berthed after the completion of service of its preceding ship. It is used to specify the order in which the ships start to berth, i.e. the berthing time of a ship must be greater than or equal to the berthing time of its preceding ship. Given the scheduling order of ships, their berthing times can be easily computed by considering constraints (4)–(6) described in the problem formulation section.

3.2 Pareto Ranking

As mentioned in the introduction, the BAP is a multi-objective optimization problem where a number of objectives, such as the total service time

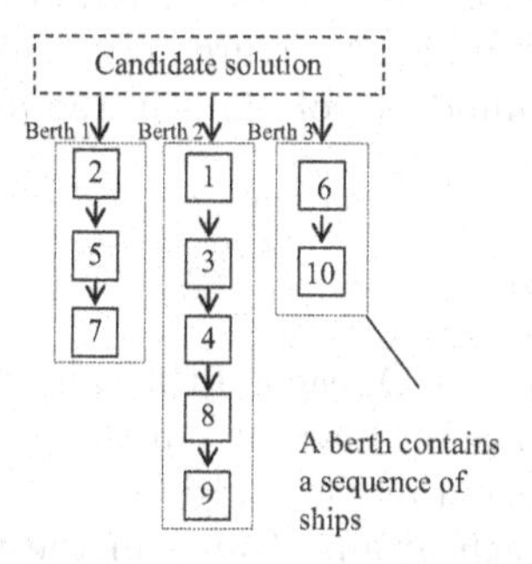

Fig. 2. Fixed-length solution representation

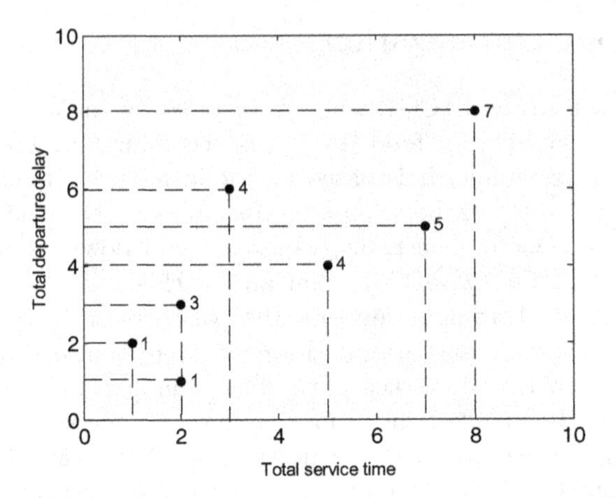

Fig. 3. Example to demonstrate the principle of dominance

of ships and the total delay in the departure of ships, need to be minimized concurrently. In contrast to single-objective optimization, the solution to a multi-objective optimization problem exists in the form of alternate trade-offs known as the Pareto optimal set. Each objective component of any non-dominated solution in the Pareto optimal set can only be improved by degrading at least one of its other objective components. Thus, the role of multi-objective optimization in the BAP is to discover such a set of Pareto optimal solutions from which the port manager can select an optimal solution based on the situation at hand. The Pareto fitness ranking scheme [15] for evolutionary multi-objective optimization is adopted to assign the relative strength of solutions. The ranking approach assigns the same smallest rank to all non-dominated solutions, while the dominated ones are inversely ranked according to the number of solutions dominating them. In Fig. 3, a population of seven hypothetical solutions is plotted in the objective domain. Each solution defines a rectangular box encompassing the origin as shown in the figure. For each solution, another solution will dominate the solution if and only if it is within or on the box defined by the first solution but not equal to the first solution in terms of the two considered objectives. The rank of each of the solutions is also shown in the figure. The rank of a solution is given by $(1 + q)$, where q is the number of solutions in the population dominating the solution.

3.3 Solution Construction

In contrast to conventional ACO, each of the candidate solutions to the BAP is searched by a group of ants. Each ant in the group is responsible for the schedule of a particular berth in the solution.

In each generation, each group of ants in the population of ant groups constructs a complete solution. For each ant group, an ant starts from a ship and moves to another ship at its turn. For example, in Fig. 2, ant 1 moves to ship 2, then to ship 5, and finally to ship 7. At each iteration, only one ant in the group gets to move. Instead of each ant in the group getting equal number of turns to make a move, which will result in each berth being allocated the same number of ships, the ant that represents the berth that is able to complete servicing the currently allocated ships fastest will get to move. This is intuitive since the berth is currently the idlest one. However, in the event that the berth is not able to accommodate any of the remaining unscheduled ships due to constraint (5), the ant representing the next idlest berth is chosen to move. During a move, the probability that an ant chooses ship j from the set S of unscheduled and feasible (based on constraint (5)) ships for the move is given as follows:

$$p_{ij} = \frac{[\tau_{ij}]^{\alpha} [\eta_{ij}]^{\beta}}{\sum_{k \in S} [\tau_{ik}]^{\alpha} [\eta_{ik}]^{\beta}} \tag{10}$$

where τ_{ij} is the amount of pheromone on the edge (i, j), η_{ij} is the visibility heuristic value, and α and β are constants that determine the relative influence of the pheromone value and the heuristic value on the decision of the ant. In this paper, two types of pheromone matrices and visibility heuristics are proposed for the BAP and explained in Tables 1 and 2, respectively. In Table 1, berth-ship allocation pheromone matrix stores the frequency that ship j is allocated to berth i, while the ship order pheromone matrix stores the frequency that ship i precedes ship j in a schedule. In Table 2, the earliest deadline first heuristic gives priority to ships with earlier deadlines, while the optimal berth heuristic gives priority to ships for which the berth represented by the ant is near their "optimal" berths.

The procedure described above is repeated until all the ships are scheduled and the algorithm moves on to the next ant group in the population. After

Table 1. Pheromone matrices for BAP

Pheromone matrix	Interpretation of τ_{ij}
Berth-ship allocation	The allocation of berth i to ship j
Ship order	Ship i precedes ship j

Table 2. Visibility heuristics for BAP

Visibility heuristic	Interpretation of η_{ij}
Earliest deadline first	$\frac{1}{DL_j}$
Optimal berth	$\frac{1}{\lvert i-OB_j \rvert+2}$ where i is the index of the berth of interest and OB_j is the index of the berth with the lowest handling time for ship j

each ant group in the population has constructed a complete BAP solution, the solutions are evaluated based on functions (1) and (2). The solutions are then ranked based on their Pareto dominance as described in the previous section. Solutions having ranking values better than $rank_{best}$ are allowed to update the pheromone matrix based on (11).

$$\tau_{ij} = \rho \cdot \tau_{ij} + \Delta\tau_{ij} \tag{11}$$

where $0 < \rho < 1$ represents the evaporation of pheromone on a trail and $\Delta\tau_{ij}$ is based on the ant-density model of Dorigo et al. [13] as follows:

$$\Delta\tau_{ij} = \begin{cases} Q & \text{if edge } (i,j) \text{ is used in the solution,} \\ 0 & \text{otherwise} \end{cases} \tag{12}$$

In this way, trails leading to more promising solutions become more attractive, directing the search towards more promising areas in the search space.

Following the pheromone updating process, an archive is updated. The archive is used to store all the best solutions found during the search. The archive updating process consists of a few steps. The current population is first appended to the archive. All repeated solutions are deleted. Pareto ranking is then performed on the remaining solutions. The larger ranked (weaker) solutions are then deleted such that the size of the archive remains the same as before the updating process.

An elitism mechanism is also employed in the ACO. At the end of each generation, solutions in the archive having ranking values better than $rank_{elite}$ are allowed to further update the pheromone matrix. However, instead of using (12), the elitism mechanism uses (13) for updating the pheromone matrix.

$$\Delta\tau_{ij} = \begin{cases} Q_{elite} & \text{if edge } (i,j) \text{ is used in the solution,} \\ 0 & \text{otherwise} \end{cases} \tag{13}$$

This is to allow the search to put more emphasis on regions in the search space around the elite solutions.

This is one complete generation of the ACO for the BAP. The algorithm iterates for a predefined number of generations.

4 Multi-Objective Multi-Colony Ant Algorithm

The previous section has described the operations for a particular ant colony for solving the BAP. The multi-objective multi-colony ant algorithm (MOMCAA) uses an island model approach based on ideas from genetic algorithms. Heterogeneous colonies are used in the model. Each colony may be different from the other colonies in terms of the combination of pheromone matrix and visibility heuristic used. These features of the MOMCAA are described in this section.

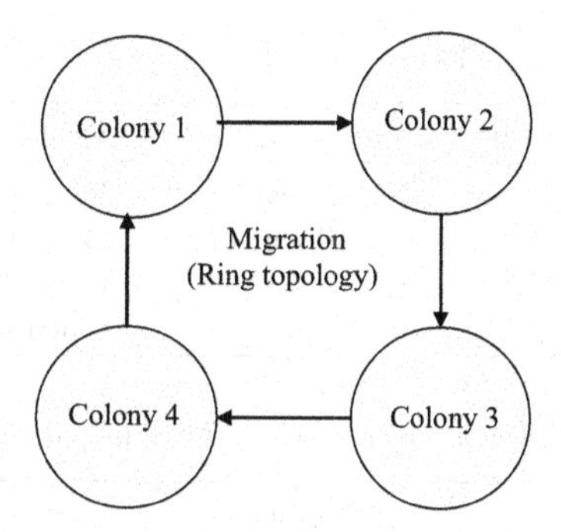

Fig. 4. Island model with ring migration topology

4.1 Island Model

The MOMCAA utilizes a number of equal-sized ant colonies. It employs an island model with a ring migration topology as shown in Fig. 4. Solution migration takes place at regular intervals of a predefined number of generations. Only solutions in the archive of each colony having ranking values better than $rank_{elite}$ are allowed to migrate. Each migration involves copying the selected solutions to the archive of the directionally adjacent colony. The pheromone matrix of each colony is then updated using the newly received solutions based on (13).

The MOMCAA maintains an archive that stores all the best solutions found by the constituting colonies during the search. The archive is updated at the end of each generation when all the colonies have completed their generation. The archive updating process uses the archive of each of the constituting colonies to update the MOMCAA archive and is identical to that for updating the archive of each individual colony.

4.2 Heterogeneous Colonies

Two types of pheromone matrices and visibility heuristics have been proposed for the BAP in Tables 1 and 2. This allows for four different types of colonies based on the different combinations of pheromone matrix and visibility heuristic used as shown in Table 3.

This leads to six settings of the MOMCAA which will be studied in this paper. ONEEACH is the setting which uses all four types of colonies. The placement of the colonies follows a fixed ordering. Referring to Fig. 4, BSAEDF is used for colony 1, SOOB for colony 2, BSAOB for colony 3, and SOEDF for colony 4. The order will be cycled if more than four colonies are used. The colony placement is chosen so that adjacent colonies do not use the same type of pheromone matrix. RANDOM is the setting that randomly chooses the colony type for each of its colonies during the initialization of the colonies. BSAEDFC, BSAOBC, SOEDFC, and SOOBC are the settings that use solely the respective types of colonies.

Table 3. Colony types

Colony type	Pheromone matrix	Visibility heuristic
BSAEDF	Berth-ship allocation	Earliest deadline first
BSAOB	Berth-ship allocation	Optimal berth
SOEDF	Ship order	Earliest deadline first
SOOB	Ship order	Optimal berth

Table 4. Parameter settings for simulation study

Parameter	Setting
Generation number	200
Colony number	4
Population size	200 ant groups over four colonies
Archive size	100
Colony archive size	25
α	1
β	5
ρ	0.7
$rank_{best}$	10
$rank_{elite}$	5
Q	1
Q_{elite}	5
Migration interval	30 generations

5 Simulation Results and Analysis

The MOMCAA was programmed in C++ and simulations were performed on an Intel Pentium 4 3.2 GHz computer. Table 4 shows the parameter settings chosen for the experiments, unless otherwise stated.

Since there is no well-established benchmark for the BAP in the literature, many researchers have generated their own test problems, with a few of them using information from their ports of study. The test problem used in this paper is generated randomly but systematically. Table 5 shows the problem parameter settings of the test problem. Ship arrivals are generated using an exponential distribution, while ship handling times are based on a 2-Erlangian distribution. Imai et al. [2] obtained these distributions from their survey on the port of Kobe.

The subsequent sections will present the extensive simulation results and analysis of the proposed MOMCAA.

5.1 Performances of Different MOMCAA Settings

Six settings of the MOMCAA have been introduced in Sect. 4.2. This section shows and compares the performances of the settings on the generated

Table 5. Problem parameter settings

Parameter	Setting
Berth number	5
Berth length	Uniformly between 350 to 700
Berth depth	Uniformly between 40 to 60
Ship number	100
Ship length	Uniformly between 100 to 350
Ship draft	Uniformly between 30 to 60
Ship arrival	Exponential interval with mean 12
Ship handling time	2-Erlangian distribution

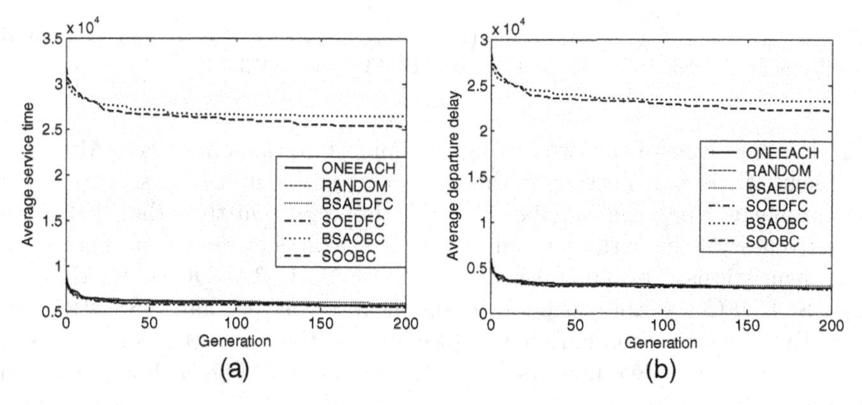

Fig. 5. (**a**) Average service time and (**b**) average departure delay of non-dominated solutions for different MOMCAA settings

test problem. To obtain statistical results, each of the settings underwent 10 independent simulation runs, each time using a different random number seed.

The convergence traces of the service time and the departure delay for the six settings are plotted in Fig. 5a, b. Figure 5a, b shows the convergence of the respective objectives, averaged over all the non-dominated solutions in the archive, over the generations. The objective values are further averaged over the 10 simulation runs performed.

From the performance comparison in Fig. 5a, b, it is obvious that the results of BSAOBC and SOOBC are not as competitive as those of the other four settings. Since the only feature that distinguishes the two settings from the rest is that they make use of the optimal berth visibility heuristic, it can be concluded that the heuristic is not effective for solving the problem. On hindsight, it is clear why the heuristic is not suitable for the particular problem. From Table 2, the heuristic assigns a value to each berth-ship pair based on the absolute difference between the index of the berth and the index of the ship's optimal berth, which is the berth that gives the lowest handling time for the ship. Since there are only five berths in the problem, the heuristic values for the berth-ship pairs of a particular ship are not very different. As such,

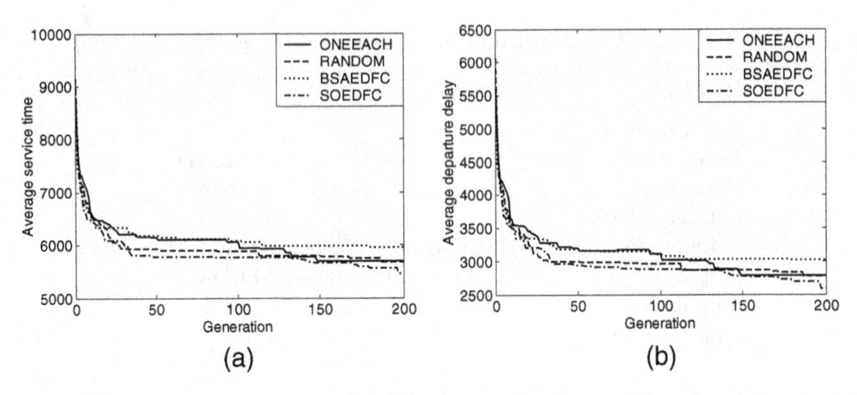

Fig. 6. (**a**) Average service time and (**b**) average departure delay of non-dominated solutions for ONEEACH, RANDOM, BSAEDFC, and SOEDFC

the search process of the two settings is similar to a random search. Although the heuristic is not effective on this particular problem, it may still be useful for problems where the number of berths is larger. On the other hand, the effectiveness of the earliest deadline first heuristic is evident from Fig. 5a, b. The heuristic is the main reason why ONEEACH, RANDOM, BSAEDFC, and SOEDFC are able to perform significantly better than BSAOBC and SOOBC. To further compare the performances of the four settings which make use of the earliest deadline first heuristic, the plots in Fig. 5a, b are enlarged in Fig. 6a, b.

From Fig. 6a, b, it can be seen that the ship order pheromone matrix is more effective than the berth-ship allocation matrix. It is able to produce solutions with lower service time and delay in departure time. The berth-ship allocation matrix suffers, to a small extent, the same problem as the optimal berth visibility heuristic. However, the larger granularity in the pheromone values lessens the extent of the problem. Overall, SOEDFC performs the best among the six settings on this particular problem since it utilizes the more effective pheromone matrix and visibility heuristic. The performances of ONEEACH and RANDOM are comparable and are slightly behind SOEDFC.

The distributions of the non-dominated solutions found by ONEEACH, RANDOM, BSAEDFC, and SOEDFC over the 10 simulation runs are shown in box plots in Fig. 7a, b. Each box plot represents the distribution of the respective objective values where the horizontal line within the box encodes the median, and the upper and lower ends of the box are the upper and lower quartiles, respectively. The two horizontal lines beyond the box give an indication of the spread of the data. A plus sign outside the box represents an outlier.

From the results in Fig. 7a, b, although RANDOM has slightly lower median objective values compared to ONEEACH, the spread in the respective objective values is significantly larger. This is to be expected given the random nature in which RANDOM picks its constituting colonies. This would undermine the reliability of the setting. In the other extreme, BSAEDFC has the

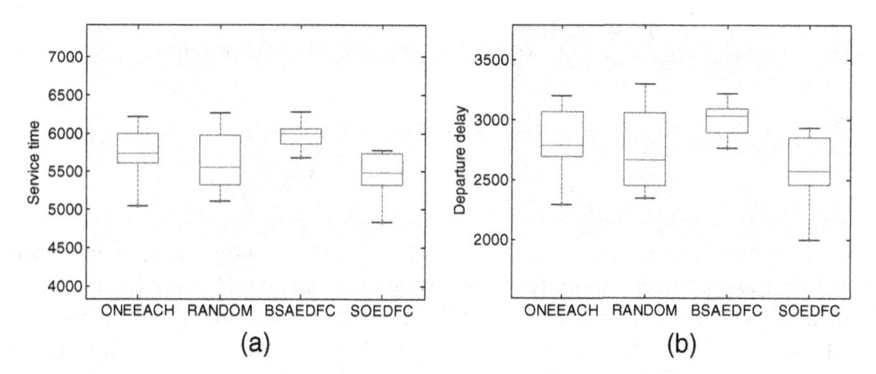

Fig. 7. Distributions of (**a**) service time and (**b**) departure delay of non-dominated solutions for ONEEACH, RANDOM, BSAEDFC, and SOEDFC

highest median objective values of the four settings but it is the most reliable setting since the spread in the respective objective values is the smallest.

Further experiments were conducted to investigate the differences between the pheromone matrices of the colonies [16] for each of the settings. Let T^k be the pheromone matrix of the kth colony of a particular setting. Then, the average pheromone matrix M is defined as follows:

$$M_{ij} = \frac{1}{N}\sum_{k=1}^{N} T_{ij}^{k} \tag{14}$$

where N is the number of ant colonies. Since there are two different types of pheromone matrices, the average pheromone matrix M only considers matrices of colonies that belong to the same type. As such, ONEEACH and RANDOM, which use two types of pheromone matrices, will have two average pheromone matrices, one for each type. To be specific, T_{BSA} and M_{BSA} respectively represent the berth-ship allocation pheromone matrix and the corresponding average pheromone matrix, while T_{SO} and M_{SO} respectively represent the ship order pheromone matrix and the corresponding average pheromone matrix.

To measure the average difference between the respective types of pheromone matrices in each of the settings, the average standard deviation σ_{avg} between single elements of the matrices is defined as follows:

$$\sigma_{avg} = \frac{1}{|B|\,|V|}\sum_{i\in B}\sum_{j\in V} S_{ij} \text{ for berth-ship allocation pheromone matrix} \tag{15}$$

where

$$S_{ij} = \sqrt{\frac{1}{N-1}\sum_{k=1}^{N}\left(T_{BSA_{ij}}^{k} - M_{BSA_{ij}}^{k}\right)} \tag{16}$$

and

$$\sigma_{avg} = \frac{1}{|V|^2} \sum_{i \in V} \sum_{j \in V} S_{ij} \text{ for ship order pheromone matrix} \tag{17}$$

where

$$S_{ij} = \sqrt{\frac{1}{N-1} \sum_{k=1}^{N} \left(T_{SO_{ij}}^{k} - M_{SO_{ij}}^{k} \right)} \tag{18}$$

Based on the definitions of σ_{avg} in (15) and (17), a large σ_{avg} value indicates that the pheromone matrices are very different, while a small σ_{avg} value indicates otherwise. The results of this experiment are shown in Fig. 8a, b.

Figure 8a, b shows that during the first few generations, the pheromone matrices in the different colonies of each of the settings evolve in different directions, resulting in the increase in the difference between the matrices. For the berth-ship allocation pheromone matrix, Fig. 8a shows that the matrices become more alike each time migration takes place. This is to be expected since only promising solutions are allowed to migrate and they update the respective pheromone matrices immediately after the migration. However, the pattern is not so obvious for the ship order pheromone matrix in Fig. 8b.

In order to show the convergence of the ACO in terms of the average number of alternatives [16] that an ant group has for the choice of the next ship to be allocated, the following experiment was conducted. In counting the number of alternatives that an ant group has, it is of interest to include only the ships that have at least some minimal chance to be chosen next. For the ant of the kth ant group that is deciding on the next ship to be allocated to the corresponding berth, where ship i is the last ship allocated to that berth, let

$$D^k(i) = |\{j | p_{ij} > \lambda, j \in V \text{ that are not yet allocated to a berth}\}| \tag{19}$$

be the number of possible successors that have a probability greater than λ to be chosen. Then, the average number of alternatives with probability greater than λ during a generation is given as follows:

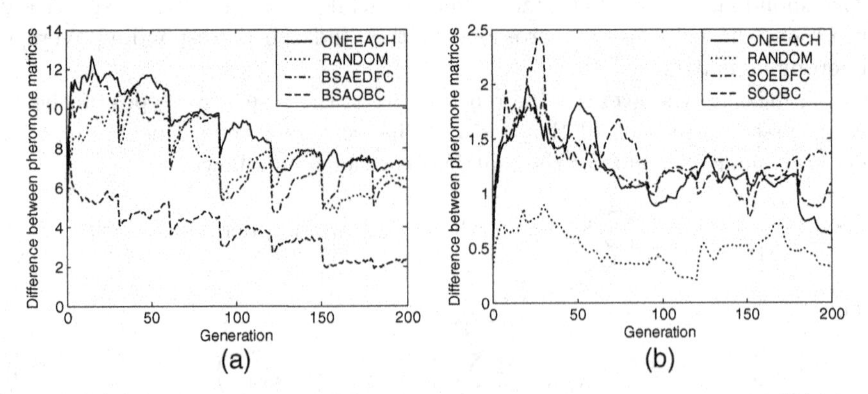

Fig. 8. Difference in (**a**) berth-ship allocation pheromone matrices and (**b**) ship order pheromone matrices

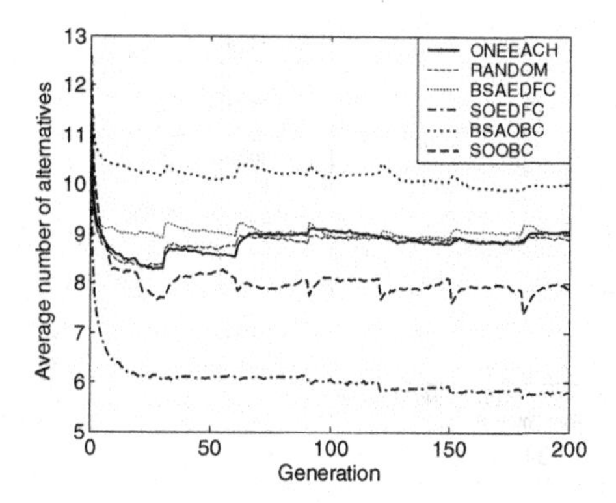

Fig. 9. Average number of alternatives with probability greater than 0.01

$$D = \frac{1}{a\,|V|} \sum_{k=1}^{a} \sum_{i=1}^{|V|} D^k(i) \tag{20}$$

where a is the number of ant groups. It is clear that $D \geq 1$ always holds for $\lambda < \frac{1}{n-1}$. A similar measure, the λ-branching factor, was used by Gambardella and Dorigo [17] to measure the dimension of a search space. In contrast to the D measure in (20), the λ-branching factor would consider all ships as possible successors, not only those that have not been allocated a berth.

The value of λ was chosen to be 0.01 and the results of this experiment are shown in Fig. 9.

From Fig. 9, it can be observed that, with the exception of SOEDFC and SOOBC, the number of alternatives increases each time migration takes place. Comparing the results in Fig. 9 with those in Fig. 8a, b, it can be concluded that migration causes pheromone matrices to be more alike and at the same time increases the number of alternatives for ants. The larger the number of alternatives, the more diverse the solutions become, allowing the algorithm to avoid being trapped in local optima. In between migrations, the pheromone matrices for different colonies evolve in different directions before the number of alternatives stagnates. During migrations, each of the colonies is perturbed such that their pheromone matrices become more alike and the number of alternatives increases.

5.2 Effects of Different Migration Intervals

While the previous section compares the performances of the six MOMCAA settings, this section focuses on the effect of different migration intervals on the performance of the ONEEACH setting. This setting was chosen for further investigation since it consists of all the four types of colonies listed in Table 3.

Ten independent simulation runs of three settings, with migration intervals of 10 and 100 generations, and without migration, respectively, were performed. The box plots for the three settings are plotted in Fig. 10a, b. The

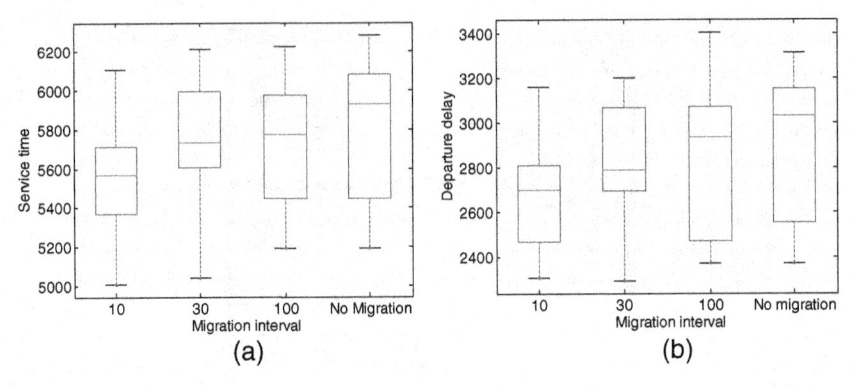

Fig. 10. Distributions of (**a**) service time and (**b**) departure delay of non-dominated solutions for ONEEACH with different migration intervals

plots for the ONEEACH setting with a migration interval of 30 in Fig. 7a, b are also included for comparison.

From the results in Fig. 10a, b, the performance of ONEEACH improves as the frequency of migration is increased, i.e. the medians of the objective values decrease with increasing frequency of migration. These results further justify the comment made in the previous section that solution migration helps the algorithm to avoid getting trapped in local optima.

6 Conclusions

A multi-objective multi-colony ant algorithm (MOMCAA) has been presented in this paper to solve the berth allocation problem (BAP). The MOMCAA adopts an island model with heterogeneous colonies, where each colony may be different from the other colonies in terms of the combination of pheromone matrix and visibility heuristic used. In contrast to conventional ant colony optimization (ACO) algorithms where each ant in the colony searches for a single solution, the MOMCAA uses an ant group to search for each candidate solution.

Without the need of aggregating the multiple objectives of the BAP into a compromise function, the MOMCAA optimizes all the objectives concurrently. The performances of the various MOMCAA settings proposed have been shown and compared. It was found that the ship order pheromone matrix and the earliest deadline first visibility heuristic are more effective for solving the BAP. The relationship between the difference in the pheromone matrices and the number of alternatives presented to ant groups during the search for candidate solutions has been shown. In addition, the effect of varying the migration interval on the performance of the MOMCAA has been demonstrated and analyzed.

The MOMCAA offers a flexible and expandable framework for the implementation of ant algorithms. On top of allowing different pheromone matrices and visibility heuristics to be incorporated for solving the problem, different parameter settings can also be used for each of the constituting colonies. The MOMCAA also offers an attractive and innovative option of incorporating other meta-heuristics, such as genetic algorithms, simulated annealing, tabu search, etc, with ant algorithms by including them as separate colonies in the algorithm. Much future work could be done in these areas to explore the potential of the MOMCAA framework.

References

1. A. Imai, X. Sun, E. Nishimura, and S. Papadimitriou, Berth allocation in a container port: using a continuous location space approach, *Transportation Research Part B: Methodological*, 39(3), 199–221, 2005.
2. A. Imai, E. Nishimura, and S. Papadimitriou, The dynamic berth allocation problem for a container port, *Transportation Research Part B: Methodological*, 35(4), 401–417, 2001.
3. A. Imai, E. Nishimura, and S. Papadimitriou, Berth allocation with service priority, *Transportation Research Part B: Methodological*, 37(5), 437–457, 2003.
4. A. Imai, E. Nishimura, M. Hattori, and S. Papadimitriou, Berth allocation at indented berths for mega-containerships, *European Journal of Operational Research*, 179(2), 579–593, 2007.
5. Y. Guan, W.-Q. Xiao, R. K. Cheung, and C.-L. Li, A multiprocessor task scheduling model for berth allocation: heuristic and worst case analysis, *Operations Research Letters*, 30, 343–350, 2002.
6. K. H. Kim and K. C. Moon, Berth scheduling by simulated annealing, *Transportation Research Part B: Methodological*, 37(6), 541–560, 2003.
7. Y.-M. Park and K. H. Kim, A scheduling method for berth and quay cranes, *OR Spectrum*, 25, 1–23, 2003.
8. C.-L. Li, X. Cai, and C.-Y. Lee, Scheduling with multiple-job-on-one-processor pattern, *IIE Transactions*, 30, 433–445, 1998.
9. A. Lim, The berth planning problem, *Operations Research Letters*, 22(2–3), 105–110, 1998.
10. K. K. Lai and K. Shih, A study of container berth allocation, *Journal of Advanced Transportation*, 26, 45–60, 1992.
11. A. Imai, K. Nagaiwa, and W. T. Chan, Efficient planning of berth allocation for container terminals in Asia, *Journal of Advanced Transportation*, 31, 75–94, 1997.
12. E. Nishimura, A. Imai, and S. Papadimitriou, Berth allocation planning in the public berth system by genetic algorithms, *European Journal of Operational Research*, 131(2), 282–292, 2001.
13. M. Dorigo, V. Maniezzo, and A. Colorni, Ant system: optimization by a colony of cooperating agents, *IEEE Transactions on Systems, Man, and Cybernetics – Part B: Cybernetics*, 26(1), 1996.
14. M. Dorigo and L. M. Gambardella, Ant colony system: a cooperative learning approach to the traveling salesman problem, *IEEE Transactions on Evolutionary Computation*, 1, 53–66, 1997.

15. C. M. Fonseca, Multiobjective genetic algorithms with application to control engineering problems, Dept. Automatic Control and Systems Eng., University of Sheffield, Sheffield, UK, Ph.D. Thesis, 1995.
16. M. Middendorf, F. Reischle, and H. Schmeck, Information exchange in multi colony ant algorithms, in *Proceedings of the Workshop on Bio-Inspired Solutions to Parallel Processing Problems*, Cancun, Mexico, Springer Lecture Notes in Computer Science, vol. 1800, pp. 645–652, 2000.
17. L. M. Gambardella and M. Dorigo, Ant-Q: a reinforcement learning approach to the traveling salesman problem, in *Proceedings of the 12th International Conference on Machine Learning*, pp. 252–260, 1995.

Query Rewriting for Semantic Multimedia Data Retrieval

Samira Hammiche, Bernardo Lopez, and Salima Benbernou, and Mohand-Saïd Hacid

LIRIS – University of Claude Bernard Lyon 1, 43, bld du 11 Novembre 1918, 69622 Villeurbanne, France, shammich@liris.univ-lyon1.fr, blopez@liris.univ-lyon1.fr, sbenbern@liris.univ-lyon1.fr, mshacid@liris.univ-lyon1.fr

1 Introduction

Recently, the increasing availability of multimedia data, especially on the Web, has led to the development of several tools to manage such data. The management of multimedia data requires to consider features extraction algorithms, content representation models and efficient search engines. Multimedia content can be described in many different ways depending on context, users, purpose of use and application domain. To address these description requirements of a wide range of applications, the MPEG committee[1] developed MPEG-7[2] standard in 2001. It aims to describe multimedia content at several levels of granularity and abstraction, to include description of features, structure, content and metadata about the content.

The development of feature-based techniques for similarity-based retrieval of multimedia information is relatively a mature area, at least for images. We should now concentrate on using semantics so that retrieval using concept based queries can fits user's requirements. Multimedia annotations should be semantically rich, and should exhibit the multiple semantics of the content.

In our work, we deal with the semantic description and efficient querying of multimedia data. To provide rich description of multimedia content, we use MPEG-7 standard. However, since MPEG-7 uses XML Schema to define data structure and data constraints, there is neither formal semantics, nor inference capabilities associated with the elements of MPEG-7 documents. Consequently, MPEG-7 cannot represent the meaning (semantics) of multimedia content and it cannot be used as an efficient multimedia description language. Moreover, it cannot integrate and exploit domain specific knowledge in the descriptions, which represent a shared and known user's vocabulary.

[1] http://www.chiariglione.org/mpeg/.

[2] http://www.chiariglione.org/mpeg/standards/mpeg-7/mpeg-7.htm/.

S. Hammiche et al.: *Query Rewriting for Semantic Multimedia Data Retrieval*, Studies in Computational Intelligence (SCI) **116**, 351–372 (2008)
www.springerlink.com

To retrieve multimedia data described using MPEG-7, we use the W3C recommendation to query XML documents: XQuery,[3] since MPEG-7 descriptions are XML documents. However, XQuery is not flexible and requires to know the details of MPEG-7 documents vocabulary to formulate precise and relevant queries. Furthermore, XQuery does not allow the extraction of implicit information since it is based on exact matching evaluation. Then, there is a need to use a flexible query language to formulate user's queries and extract implicit information from the content to fully exploit the MPEG-7 descriptions expressiveness.

In short, this chapter deals with the semantic and efficient access to multimedia content described using MPEG-7. To ensure that, we need to:

- Associate a formal semantics to the MPEG-7 vocabulary
- Integrate domain knowledge in the MPEG-7 descriptions
- Provide semantic and efficient access to the multimedia content, using a flexible query language and efficient evaluation methods

This chapter is organized as follows: in Sect. 2, we give a brief description of the MPEG-7 standard and an MPEG-7 description example. In this section, we discuss the MPEG-7 limitations also. Section 3 deals with the querying of MPEG-7 descriptions. We introduce XQuery language and discuss its ability to perform efficient retrieval in the MPEG-7 descriptions. Section 4 describes the framework architecture used to alleviate both MPEG-7 and XQuery limitations. The framework components and their inter-connection are also presented. In Sect. 5, we introduce an application example to describe how the framework processes to respond to a given user's query. Section 6 relates our work to the existing ones. Finally, we conclude in Sect. 7.

2 Preliminaries and Motivating Example

2.1 MPEG-7: Multimedia Content Description Interface

MPEG-7, formally called "*Multimedia Content Description Interface*," focuses on the description of multimedia content. It provides a generic library of description tools, to cover almost all multimedia data types, and to convey to several applications. These tools are [13]: *Descriptors* (*Ds*), *Description Schemes* (*DSs*), *Description Definition Language* (*DDL*) which is XML Schema based, and *system tools.*

To describe multimedia data using MPEG-7, it is necessary to select, then to use the required *Ds* and *DSs*. An MPEG-7 description is an XML document (cf. Fig. 1). It allows to include, in the same description, *Creation Information* (e.g., title, creator, classification), *Media Information* (e.g., media time, media location), *Structural Decomposition* (segment decomposition, segment

[3] http://www.w3.org/TR/xquery/.

attributes and relations) and *Semantic Description* (e.g., text annotation, semantic entities such as objects and events, semantic attributes and semantic relations).

2.2 Illustration Example

Figure 1 illustrates an MPEG-7 description of a soccer game video sequence. This sequence is segmented into two video scenes. The first one, identified by *ID-V1*, describes the event *Fault* and its actors (*AgentOf, PatientOf*),

```
<Mpeg7> <Description xsi:type="ContentEntityType">
  <MultimediaContent xsi:type="VideoType">
   <video id="Fault and Penalty">
    <CreationInformation> <!-- Creation Information -->
    <Creation> <Title> fault and penalty sequence </Title> </Creation>
    <Classification> <Genre Type="main"> Soccer Sport </Genre>
                     <Language type="original"> English </Language> </Classification>
   </CreationInformation>
   <!-- Media Information -->
   <MediaLocator><MediaUri>soccer.mpg</MediaUri></MediaLocator>
   <MediaTime><MediaTimePoint>T00:00:00</MediaTimePoint>
           <MediaDuration>PT2M30S</MediaDuration> </MediaTime>
   <!-- Video Segmentation -->
   <TemporalDecomposition gap="false" overlap="false">
    <VideoSegment id="ID-V1">
        <MediaTime><MediaTimePoint>T00:00:00</MediaTimePoint>
               <MediaDuration>PT0M50S</MediaDuration></MediaTime>
      <!-- Semantic Description of video segment -->
      <Semantic>
       <Label> <Name>Fault and actors of the fault </Name></Label>
       <!-- Description of the person "GoalK"  -->
       <SemanticBase xsi:type="Agent ObjectType" id="GoalK">
       <Label><Name>Goalkeeper </Name></Label>
       <Agent xsi:type= "PersonType">
         <Name> <FamilyName> Goalkeeper Name </FamilyName>
                <GivenName> ...< /GivenName> </Name> </Agent>
       </SemanticBase>
      <!-- Description of the person "Player 1" -->
       <SemanticBase xsi:type="Agent ObjectType" id="Player 1">
        <Label><Name>Player</Name></Label>
        <Agent xsi:type= "PersonType">
        <Name> <FamilyName> Player 1</FamilyName>
               <GivenName>...< /GivenName> </Name> </Agent>
       </SemanticBase>
      <!-- Description of the event  "Fault" -->
       <SemanticBase xsi:type="EventType" id="Fault">
        <Label><Name>Fault</Name></Label>
        <Relation type="cs:SemanticRelationCS:2001"  name="AgentOf"  target="#GoalK"/>
        <Relation type="cs:SemanticRelationCS:2001" name="PatientOf" target="#Player 1"/>
       </SemanticBase>
     </Semantic>
    </VideoSegment>
    <!-- Description de the second video segment -- >
   <VideoSegment id="ID-V2">
     <!-- MediaTime & MediaLocation Information -->
     <Semantic>
     <Label> <Name>Penalty event and players relationships </Name> </Label>
      <!-- Description of the event  "Penalty" -->
      <SemanticBase xsi:type="EventType" id="Penalty">
       <Label><Name> Penalty</Name></Label>
       <Relation type="cs:SemanticRelationCS:2001"  name="AgentOf"  target="#Player 1"/>
      <Relation type="cs:SemanticRelationCS:2001" name="PatientOf" target="#GoalK"/>
     </SemanticBase>
     <!-- Description of the event  "Goal" -->
      <SemanticBase xsi:type="EventType" id="Goal">
        <Label><Name>Goal</Name></Label>
     </SemanticBase
     <!-- relation between events "Penalty" and  "Goal"-->
      <Graph>
      <Relation type="EventEventRelation" name="ResultOf" source="#Penalty" target="#Goal"/>
      </Graph></Semantic> </VideoSegment> </TemporalDecomposition>
  </Video> </MultimediaContent> </Description>
</Mpeg7>
```

Fig. 1. MPEG-7 description of a video segment

whereas the second describes the *Penalty* event, its agent (*Player 1*) and its result (*Goal* event). The two scenes are instances of the MPEG-7 description scheme *VideoSegment DS*, whereas the complete sequence is represented as a multimedia content of type *VideoType*.

2.3 Querying MPEG-7 Descriptions

In its first version, MPEG-7 provides only description tools to allow rich and efficient description of multimedia content. It does not address, neither features extraction tools, nor query languages to exploit these descriptions. However, due to its spreading and increasing use, the MPEG committee constituted a working group in 2006, to define an MPEG-7 query language. This group introduced *MP7QF* (*MPEG-7 Query Format*) framework [5], to specify the requirements of an MPEG-7 query language. The framework is still in the process of specification.

In our work, since MPEG-7 descriptions are XML documents, we use an XML query language to perform the retrieval process. XPath,[4] which allows to search only path expressions, is not suitable in our case. Instead, XQuery allows to formulate complex queries, where join operations can be performed. Moreover, XQuery is the W3C recommendation to retrieve XML documents. In XQuery, Queries consist of *FLWR* (*For-Let-Where-Return*) expressions that support iteration and binding of variables to intermediate results. *For* and *Let* clauses define the research space. *Where* clause specifies query conditions and *Return* clause provides query results and form. Moreover, XQuery allows the definition of functions and their use in the clauses as personalized operators.

- *Queries examples*

Let us consider two queries Q_1 and Q_2 (the XQuery syntax of these queries is illustrated in Fig. 2) to execute on the MPEG-7 document described in Fig. 1.

```
Q1: Let $VSeg := Mpeg7//VideoSegment
    Let $Event: =$VSeg/Semantic/SemanticBase[@xsi:type="EventType" and Label/Name="Fault"]
    Let $Player: =$VSeg/Semantic/SemanticBase[@xsi:type="AgentObjectType" and Label/Name="Player"]
    Let $PlayerID := $Player[@id]
    Where  $VSeg[$Event/Relation[@name="AgentOf" @target=$PlayerID]]
    Return $VSeg

Q2: Let $VSeg:= Mpeg7//VideoSegment
    Let $Event: =$VSeg/Semantic/SemanticBase[@xsi:type="EventType" and Label/Name="Goal"]
    Let $Player: =$VSeg/Semantic/SemanticBase[@xsi:type="AgentObjectType" and Label/Name="Player"
                                              and Agent/Name/FamilyName="Player 1"]
    Let $PlayerId := $Player[@id]
    Where  $VSeg[$Event/Relation[@name="AgentOf" @target=$PlayerID]]
    Return $VSeg
```

Fig. 2. XQuery syntax of Q_1 and Q_2 queries

[4] http://www.w3.org/TR/xpath.

– Q_1: Retrieve video segments where a player made a fault?
In term of MPEG-7 vocabulary, Q_1 consists in finding *VideoSegment* elements of which semantic content is described by the event *Fault* and its agent *Player*.
– Q_2: Retrieve video segments where *Player1* scores a goal?
Using the MPEG-7 vocabulary, Q_2 consists in finding *VideoSegment* elements of which semantic content is described by the event *Goal* and its agent, called *Player1*.

- *Queries evaluation*

The expected result of Q_1 evaluation is empty because none of the players made a fault in the whole video sequence. However, the goalkeeper has made a fault in the first scene. This segment is not given as answer for two reasons: the first is that the semantic information *Goalkeeper is a Player* is not represented explicitly in the description, and the second concerns XQuery, which is a syntactic language. It has no inferential mechanism associated. Hence, it cannot deduce that *a Goalkeeper is a Player*.

The evaluation of Q_2 is an empty set too. This is due to the fact that the information *Player1 scores a goal* is not stated explicitly. It is stated as following: the event *Penalty* occurs, its agent is *Player1* and its result is the event *Goal*. XQuery can not extract this information. So, it is necessary to represent implicit informations in the descriptions, using inference rules for example, and exploit these informations during the retrieval process.

2.4 MPEG-7 and XQuery Limitations

From these two queries (Q_1 and Q_2) and the expected results, we can note that MPEG-7 descriptions need to integrate knowledge related to the structure and the content of multimedia data. Moreover, it is necessary to exploit these knowledge using appropriate query language and evaluation method. These needs are not satisfied due to the limitations of both MPEG-7 description and XQuery querying languages:

- MPEG-7 definitions (*Ds* and *DSs*) were expressed solely in XML Schema which lacks ability to represent implicit information. Indeed, XML Schema cannot represent the meaning (semantics) of MPEG-7 descriptors and the described content also, then one cannot reason to deduce new information. Hence, there is a need to formally define the semantics of MPEG-7 terms; and to express these definitions in a machine understandable and interoperable language.
- MPEG-7 does not allow the integration of domain knowledge, which represent a shared and known user's vocabulary.
- XQuery is not a flexible language. Its use requires to know details of MPEG-7 documents to formulate precise queries. Moreover, XQuery can not extract implicit information since it is not provided with any reasoning mechanism.

Hence, to allow semantic and efficient retrieval of multimedia data, described using MPEG-7, and retrieved using XQuery, it is necessary to: provide a formal semantics to the MPEG-7 vocabulary, integrate domain knowledge in the MPEG-7 descriptions and use a flexible query language, provided with inferential mechanism.

In this chapter, we take a new look at the problem of modeling and querying multimedia data and find that knowledge representation and reasoning techniques for concept languages developed in Artificial Intelligence provide an interesting angle to attack such problems. We exploit the possibility to use logical language to provide semantics to MPEG-7 vocabulary and to allow a semantic and efficient access to the multimedia content described using MPEG-7.

3 Multimedia Data Description

To provide formal semantics to MPEG-7 vocabulary and allow the integration of domain knowledge in MPEG-7 documents, several research works proposed to translate the MPEG-7 vocabulary (*Ds* and *DSs*) from the XML representation, to another formalism such as RDFS/OWL [7,17] or XDD [4], where semantics can be represented, domain knowledge can be integrated and reasoning mechanisms are provided to extract implicit information. Another approach is to use abilities of MPEG-7 and add extensions to alleviate its limitations [10,16]. In our work, we use the extension based approach. We exploit the abilities of MPEG-7 to represent structure and constraints datatype of multimedia features. Then, we propose to extend the MPEG-7 descriptions layer, with a conceptual framework, in order to associate semantics to the MPEG-7 vocabulary, and to integrate knowledge in the descriptions.

3.1 Multi-Layered Representation of Multimedia Content

Figure 3 illustrates the multi-layered architecture of the multimedia data description framework. *Physical layer* contains raw data that will be described like video sequences and images. *Metadata layer* contains the MPEG-7 descriptions produced after segmentation, features extraction and annotation process of multimedia data. *Conceptual layer*, that we introduce on top of *metadata layer*, contains domain knowledge relevant to a specific domain.

In the next section, we detail the components of the conceptual layer. We explain: how domain knowledge are represented and formalized, how these knowledge are integrated in the MPEG-7 descriptions and finally, how MPEG-7 descriptions are connected to the conceptual layer vocabulary, in order to take advantages of metadata and conceptual representation layers.

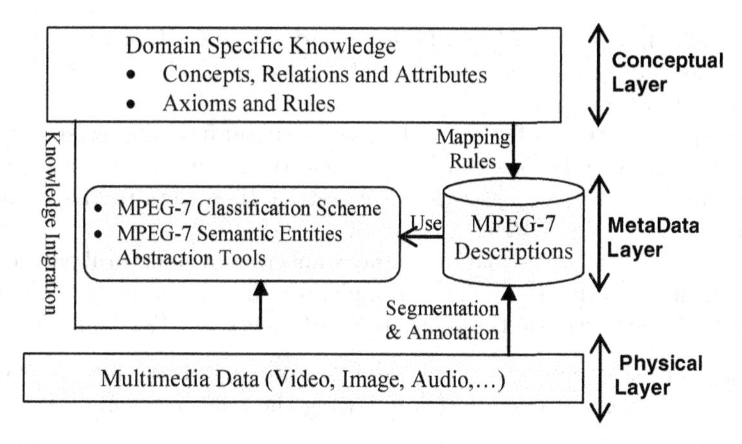

Fig. 3. Multimedia data representation framework architecture

3.2 Conceptual Layer: Domain Knowledge Representation

The most used structure to represent domain knowledge are ontologies. They facilitate knowledge sharing among users by providing a formal conceptualization of a given domain [9]. A domain ontology is composed of several components, of which the most important are concepts, relations and attributes, instances (or individuals) and axioms.

Besides the formal description and the vocabulary sharing of a specific domain, we choose to use ontologies in the framework mainly to: create semantically more accurate annotations of the multimedia content, help the users to express their queries more precisely and unambiguously and infer implicit information during the querying process since ontology languages are provided with reasoning mechanisms.

- *Domain knowledge represented in the ontology*

We consider that each domain can be represented using an ontology $O = (C_s, C_c, C_d, A_c, A_s, R_c, R_s, R_{sc})$ where:

- C_s: the set of concepts that describe the vocabulary used to characterize media structure. For example, the structure of soccer video can be defined using the concepts: *VideoSegment*, *FirstTimePeriod*, *AdditionalTime*, etc.
- C_c: the set of concepts that describe multimedia content. These concepts represent the entities perceived by human, like objects, events, persons, places, etc. We adopt the MPEG-7 semantic model [3] in the sense that semantic content is described by a set of semantic entities, their properties and their inter-relations.
- C_d: the set of concepts that describe datatypes used to represent concepts (C_s or C_c) properties. Examples of such datatypes are: *string*, *number* and *date*.

- A_s (resp. A_c): the set of attributes that describe C_s (resp. C_c) concepts. They are binary relations which relate concepts of C_s (resp. C_c) to their domain datatypes in C_d.
- R_s (resp. R_c): the set of relations between structural entities (resp. semantic entities). These relations can be symbolic (e.g., *equivalent*), spatial (e.g., *left*) or temporal (e.g., *precede*). We restrict these sets to the normative relations defined in MPEG-7.
- R_{sc}: the set of relations that allow the connection of structural entities (C_s) and semantic entities (C_c). For example, to relate an event to the segment where it occurs, we use the relation *EventOccurence (Segment, Event)*

Moreover, the concepts of C_s and C_c are organized into a concepts hierarchy, i.e., the concepts are related using the relation *is-a*.

- *Representation formalism of domain knowledge*

To formalize domain knowledge, we need to use a language able to describe the ontology vocabulary, to support query formulation and to provide decidable reasoning mechanisms. A family of languages which satisfy these requirements are CARIN (*Concept And Rules INtegrated*) languages [14]. In particular, CARIN-$\mathcal{ALN}$ combines two representation formalisms: the description logic $\mathcal{ALN}$ [1] and Datalog rules. Hence, Datalog rules can be used as a conceptual query language to formulate user queries.

The knowledge base $\triangle = (\triangle_T, \triangle_R)$ of CARIN-$\mathcal{ALN}$ language is represented using two components: terminological knowledge ($\triangle_T$) and Datalog rules ($\triangle_R$).

– *Terminological knowledge:* $\triangle_T$

In this part, domain concepts, roles and attributes are defined. More complex concepts can be defined using constructors of the $\mathcal{ALN}$ description logic. Let N_C be the set of atomic concept names, N_R the set of role names (relations or attributes). The set of $\mathcal{ALN}$ concepts is defined inductively as the smallest set for which the following holds: $\top$ (concept Top) and $\bot$ (concept Bottom) are concepts, each atomic concept name $A \in N_C$ is a concept, if C and D are concepts, R is a role and n is an integer, then $C \sqcap D$ (concept conjunction), $\neg C$ (concept negation), $\forall R.C$ (universal quantification), $\geq nR.C$ and $\leq nR.C$ (number restriction) are also concepts.

$\triangle_T$ is a set of axioms defined using two forms: $C \sqsubseteq D$ (for concept definition and concept inclusion) and $R \sqsubseteq C \times D$ (for roles and attributes definition).

– *Datalog rules:* $\triangle_R$

In this part, additional knowledge, not expressible as terminologies using $\mathcal{ALN}$ constructors, are expressed using Datalog rules. In particular, rules can be used to assert more complex relationships between concepts, roles and attributes. Moreover, Datalog rules can be used to express exclusion constraints between predicates using the following axiom $\bot \leftarrow P_1(x) \wedge P_2(y)$.

Table 1. An extract of CARIN-$\mathcal{ALN}$ knowledge base of the domain ontology $(\triangle_T, \triangle_R)$

Terminological Knowledge ($\triangle_T$): structural knowledge
$Video \sqsubseteq \top, VideoSegment \sqsubseteq Video$ $FirstTimePeriod \sqsubseteq VideoSegment, Sequence \sqsubseteq VideoSegment, Scene \sqsubseteq Sequence$ $VideoSegment \sqsubseteq \forall id\ String, VideoSegment \sqsubseteq \forall hasTitle\ String$ $VideoSegment \sqsubseteq \leq 1\ EventOccurrence \sqcap \forall EventOccurence\ Event$ $VideoSegment \sqsubseteq \leq 1\ Appear \sqcap \forall Appear\ Person$
Terminological Knowledge ($\triangle_T$): knowledge related to the described content
$Person \sqsubseteq \top, Person \sqsubseteq \forall id\ String, Person \sqsubseteq \forall hasName\ String$ $Player \sqsubseteq Person, Coach \sqsubseteq Person, Referee \sqsubseteq Person$ $PlayingPlayer \sqsubseteq Player, GoalKeeper \sqsubseteq Player, Substitute \sqsubseteq Player.$
Datalog rules ($\triangle_R$)
$score(X) \leftarrow Player(X) \wedge Goal(id) \wedge agentOf(X, id)$ $score(X) \leftarrow Player(X) \wedge Penalty(id_1) \wedge agentOf(X, id_1) \wedge Goal(id_2) \wedge resultOf(id_2, id_1)$ $\perp \leftarrow Substitute(x) \wedge Fault(y) \wedge AgentOf(y, x)$ $\perp \leftarrow Substitute(x) \wedge Goal(y) \wedge AgentOf(y, x)$

Rules defined in CARIN-$\mathcal{ALN}$ are logical sentences of the form:

$$Q(\bar{Y}) \leftarrow p_1(\bar{X_1}) \wedge ... \wedge p_n(\bar{X_n})$$

where $\bar{X_1}$,...,$\bar{X_n}$ and $\bar{Y}$ are tuples of variables or constants. Rules must be safe, i.e., a variable that appears in $\bar{Y}$ must also appear in $\bar{X_1} \cup ... \cup \bar{X_n}$. Predicates $p_1, ..., p_n$ can be either concepts or roles names (defined in the ontology), or ordinary predicates that do not appear in the ontology vocabulary. Predicate Q must be an ordinary predicate.

- *Example*

Let us consider the domain knowledge of the MPEG-7 description presented in Sect. 2.2. Knowledge concern structure and content of soccer game video sequences.

Table 1 is an extract of soccer ontology $O = (C_s, C_c, C_d, A_c, A_s, R_c, R_s, R_{sc})$, represented using CARIN-$\mathcal{ALN}$ formalism, where:

- $C_s = \{VideoSegment, FirstTimePeriod, Sequence, Scene, etc.\}$
- $C_c = \{Person, Player, Refree, GoalKeeper, Speaker, Goal, Fault, etc.\}$
- $C_d = \{String, Date, Number, etc.\}$
- $A_c = \{id, hasName, hasLocation, etc.\}$
- $A_s = \{id, hasTitle, etc.\}$
- $R_c = \{AgentOf, PatientOf, ResultOf, etc.\}$
- $R_s = \{Before, After, During, Depicts, etc.\}$
- $R_{sc} = \{Event\_Occurence, Appear, etc.\}$

A particular concept (resp. role or attribute) can be instantiated by a unary (resp. binary) predicate. To define: n-ary predicates, relations between roles or exclusion constraints, axioms of $\triangle_T$ are not enough, but Datalog rules can be used. For example, the predicate *score* is defined as disjunction of other predicates, which are defined in $\triangle_T$.

3.3 How to Integrate Domain Knowledge in MPEG-7 Descriptions

In the conceptual layer, several concept hierarchies are defined. We need to integrate these hierarchies in MPEG-7 descriptions to allow a precise description and efficient retrieval.

- *Integration of structural knowledge*

Knowledge about media structure has to be integrated in the MPEG-7 *Segment DS* since it is the description scheme used to represent multimedia content structural aspect. In the MPEG-7 *Segment DS* definition, an optional element called *StructuralUnit* is used to characterize segments decomposition. Indeed, *StructuralUnit* element allows to reference terms defined in a *Classification Scheme* (*CS*) [2]. A *CS* defines taxonomies to classify a subject area with a set of terms, organized into a hierarchy. A term represents a well defined concept in the domain and has: an identifier, a name and a definition that describes the meaning of the term.

Example: we consider structural concepts hierarchy (*VideoSegment, First Period Time, Sequence, Scene, etc.*) defined in Table 1. To integrate these structural concepts in the MPEG-7 descriptions, we first need to translate this hierarchy into a classifications schema, then to reference the adequate concept in the *StructuralUnit* element. Figure 4 illustrates how concepts are defined and related using specialization relation *NT* (*Narrower Term*), i.e., subsumption relation *is-a*, in a classification scheme.

```
<ClassificationScheme id="VideoStructure CS">
 <Term id="Video">
          <Name xml:lan= "en">Video</Name>
      <Definition> Football Game Video</Definition>
          <Term id="FirstPeriod" relation="NT" >
                <Name xml:lan= "en">First Time Period</Name>
                <Definition> the first 45 minutes of the game</Definition>
          </Term>
          <Term id="VideoSegment" relation="NT" >
                <Name xml:lan= "en">VideoSegment</Name>
                <Definition> a part of a video</Definition>
                     <Term id="Sequence" relation= "NT" >
                       <Name>Sequence</Name>
                       <Definition>a video sequence where some events take place</Definition>
                       <!-- Scene Description -->
                       ....................
                     </Term>
          </Term>
 </Term> </ClassificationScheme>
```

Fig. 4. Definition of a classification scheme

- *Integration of knowledge related to the content*

As for the structure, we need to integrate domain knowledge that describes the multimedia content. These knowledge are organized as categories of concepts hierarchies. Each category represents one of the MPEG-7 semantic entities. Therefore, these knowledge have to be integrated in the MPEG-7 semantic entities.

The MPEG-7 *SemanticBase DS* (the equivalent of *Segment DS* for the structure) allows the definition of abstraction entities using the optional element *AbstractionLevel*. This element uses an integer attribute *Dimension* set to "0" if the entity described is concrete and positif (> 0) if the entity is abstract. To link a concrete entity to its abstract entity, the relation *exampleOf* is used. To define *is-a* relation between two abstract entities, the relation *specialisationOf* is used.

Figure 5 illustrates the definition of the soccer game person hierarchy using MPEG-7 semantic abstraction tools. To use these entities in MPEG-7 descriptions, we need to reference the identifier of the corresponding entity.

3.4 How to Link the Conceptual Layer to the Metadata Layer

MPEG-7 is useful since it allows the representation of multimedia data structure and semantic content in the same description. The domain ontology of the conceptual layer provides a semantic framework to the MPEG-7 vocabulary and reasoning mechanisms. Moreover, knowledge in conceptual layer represent only abstract concepts. Instances of these concepts are represented in the MPEG-7 descriptions. Hence, it is necessary to connect conceptual layer to the metadata layer. To do so, we use a set of mapping rules of the form:

R: Conceptual Knowledge → MPEG-7 Metadata

```
<! -- Definition of Player class -->
<SemanticBase xsi:type="AgentObjectType" id= "Player">
   <AbstractionLevel dimension="1"/>
   <Label> <Name> Player </Name></Label>
   <Relation type="specializationOf" source="#Player" target="#PersonType" />
</SemanticBase>
<! -- Definition GoalKeeper class -->
<SemanticBase xsi :type="AgentObjectType" id= "GoalKeeper">
  <AbstractionLevel dimension="1"/>
  <Label> <Name> GoalKeeper </Name></Label>
  <Relation type="specializationOf" source= "#Goalkeeper" target="#Player" />
</SemanticBase>
<! -- Definition of GoalKeeper instance-->
<SemanticBase xsi:type="AgentObjectType" id= "GoalKeeper-I">
  <AbstractionLevel dimension="0"/>
  <Label> <Name> GoalKeeper-I </Name></Label>
  <Relation type="exempleOf" source= "#Goalkeeper-I" target="#GoalKeeper"/>
    <Agent xsi:type="PersonType" id="Goalkeeper-I"> ....... </Agent>
</SemanticBase>
```

Fig. 5. Definition and integration of content knowledge in MPEG-7

Table 2. The different types of mapping rules

Mapping rules for concepts
$R_{concept} : C \rightarrow E^*$ $c \rightarrow M(c) = \{e \in E \mid \exists\ extension(c) \in F \wedge e \in extension(c)\}$
Mapping rules for roles
$R_{role} : R(C \times C) \rightarrow \{0, 1\}$ $r(x, y) \rightarrow M(r(x, y)) = \{0$ if $f(r(x, y))$ is false, 1 if $f(r(x, y))$ is true. $f \in F\}$
Mapping rules for attributes
$R_{attribute} : A \rightarrow D^*$ $a(c) \rightarrow M(a(c)) = \{d \in D \mid \exists\ val\_attribut(x) \in F \wedge d \in val\_attribut(x) \wedge x = a(c)\}$

```
define function get_video_segment ($description xs:string) as element( )*
{  For $segment in $description//Mpeg7
   Let $VideoSegment := $Segment//VideoSegment
   Let $AVSegment := $Segment//AudioVisualSegment

   Return   $VideoSegment , $AVSegment}

define function person_role ($role xs:string, $role_file xs:string) as element( )*
{   Let $person := doc($role_file)// Mpeg7
    For $person_role in $person//Semantic/SemanticBase [@xsi:type="AgentObjectType"]
    Where
         $perons_role[AbstarctionLevel/@Dimension="0" ] and
         $person_role[Relation[@name="exampleOf " and  @target= $role]]

    Return    $person_role }
```

Fig. 6. XQuery functions corresponding to the concepts *VideoSegment* and *Person*

where: (1) *Conceptual Knowledge* are domain concepts, domain roles or domain attributes, (2) *MPEG-7 Metadata* are elements of the MPEG-7 descriptions obtained using a set of XQuery functions that we define.

If C is a set of domain concepts, D a set of datatype concepts, R a set of domain relations ($R \subseteq C \times C$), A a set of domain attributes ($A \subseteq C \times D$), F a set of XQuery functions (defined to retrieve a particular element, attribute or role in an MPEG-7 description) and E a set of MPEG-7 elements, then, three types of mapping rules can be defined using the mapping function M (cf. Table 2). This function returns: (1) the MPEG-7 elements instances of a given concept, (2) the attribute values of a given concept attribute and (3) a boolean value of a binary relations, which states if the relation between the entities exists or not.

Figure 6 illustrates the definition of two MPEG-7 functions using XQuery syntax. The first function corresponds to the concept *VideoSegment* of the domain ontology, and returns the set of MPEG-7 *VideoSegment* elements. Notice that we consider synonyms (*AudioVisualSegment*) of structural concepts in their functions definition. The second function defines the MPEG-7 elements of a particular person in the person hierarchy. In these two function examples, semantic abstraction tools and classification scheme of the structural unit are used to locate the most specific element in the MPEG-7 descriptions.

Using the two functions of Fig. 6, we can define the following mapping rules:

R_1: $VideoSegment \rightarrow mpeg7 : get\_video\_segment(\$file\_name)$
R_2: $Player \rightarrow mpeg7 : person\_role(Player, \$Role\_file)$

4 Querying MPEG-7 Descriptions of Multimedia Data

MPEG-7 description tools allow the integration of concept hierarchies that describe the structure and content of multimedia data. However, their XML representation does not allow their exploitation to infer implicit information since XML is not provided with reasoning mechanisms. To alleviate this limitation, we propose to exploit the domain knowledge formalism to *rewrite* user's queries and to make implicit information explicit. The set of rewritten queries are then translated into the XQuery syntax and the obtained queries are evaluated on the MPEG-7 descriptions.

4.1 Query Form and Syntax

As stated previously, we use Datalog rules of CARIN-$\mathcal{ALN}$ formalism to formulate user's queries. Datalog rules allow the formulation of conjunctive queries of the form: $Q(\bar{X}) \leftarrow P_1(\bar{X_1}) \wedge ... \wedge P_n(\bar{X_n})$ where: (1) P_i are unary predicates (concepts of $\triangle_T$), binary predicates (roles or attributes of $\triangle_T$) or ordinary predicates (predicates defined in $\triangle_R$), (2) the variables of $\bar{X}$ are called *distinguished variables* and represent the variables for which the user search the instances and (3) variables and/or constants of $\bar{X_1} \cup ... \cup \bar{X_n}$ are used to express constraints on the distinguished variables.

We assume that users are interested in retrieving a particular multimedia segment (image, video segment, audio segment, etc.) where constraints about the content are expressed. These constraints can concern person, events, objects, etc., their attributes and their inter-relations in the segment. The attributes of the structural unit are also taken into account, but we do not consider the relation between the structural segments. Given these assumptions, the syntax of multimedia queries formulated at the conceptual layer is illustrated in the Table 3.

4.2 Query Pre-Processing Algorithm

Once users formulate their queries using domain vocabulary and Datalog rules, pre-processing steps are performed to get a set of rewritten queries, where

Table 3. Syntax of multimedia conceptual queries

Q ::= "Q (" *Variables* ") ← " *Query_Expression*;
Query_Expression ::= *StructExpression*+ "∧" *ContExpression*
+ "∧" *StructContRelation*+;
StructExpression ::= *StructUnit* | *StructAttribute*;
StructUnit ::= *StructName* "(" *StructId* ")";
StructAttribute ::= *StructAttributeName* "(" *StructId* "," *AttributeValue* ")";
ContExpression ::= *ContentEntity* | *ContentAttribute* | *Content_Relation*;
ContentEntity ::= *EntityName* "(" *EntityId* ")";
ContentAttribute ::= *EntityAttributeName* "(" *EntityId* "," *AttributeValue* ")";
ContentRelation::= *ContentRelationName* "(" $EntityId_1$ "," $EntityId_2$ ")";
StructContRelation ::= *EventOccurence* "(" *StructureId* "," *EntityIid* ")" |
Appear "(" *StructureId* "," *EntityId* ")";
Variables ::= *StructureId* ;

Legend:
- *StructureName*: a terminal expression which represents the concept names of the structural units
- *StructAttributeName*: a terminal expression which represents the attribute names of a particular structural unit
- *EntityName*: a terminal expression which represents the concept names of semantic entities
- *EntityAttributeName*: a terminal expression which represents the attribute names of semantic entities
- *ContentRelationName*: a terminal expression which represents the relation name that connect two semantic entities
- *StructureId*: the structural unit identifier
- *EntityId*: the semantic entity identifier
- *AttributeValue*: the attribute value

implicit information are made explicit and more specific concepts are included.

Given a conjunctive query Q and a knowledge base $\Delta = (\Delta_T, \Delta_R)$, the rewritten algorithm produces a set of queries $Q_1, ..., Q_m$ equivalent to Q by processing the following steps:

1. For each ordinary predicate $P_i \in body(Q)$, replace P_i by its definition in Δ_R. This step is iterative. It stops when no ordinary predicate is present in $body(Q)$. Hence, the obtained queries contains only unary and/or binary predicates, defined in Δ_T.
2. Rewrite the obtained queries to integrate more specific concepts in the queries conditions.
3. Check query validity using the exclusion constraints defined in Δ_R.
4. Translate the rewritten queries into the XQuery syntax, to be evaluated on the MPEG-7 descriptions database.

4.3 Illustration Example

Let us consider again the queries Q_1 and Q_2 considered in Sect. 2.3:

- Q_1: retrieve video segments where a player made a fault?
- Q_2: retrieve video segments where player1 scores a goal?

The formulation of these queries, using Datalog rules, is illustrated in the sequel. Predicates of $\triangle_T$ and $\triangle_R$ are used in the queries.

$Q_1(X) \leftarrow VideoSegment(X) \wedge Player(y) \wedge Fault(z) \wedge AgentOf(z, y) \wedge EventOccurence(X, z)$

$Q_2(X) \leftarrow VideoSegment(X) \wedge Score(y) \wedge hasName(y, \text{“}Player1\text{”})$

Before the translation of these queries into XQuery syntax, we first process Q_1 and Q_2 at the conceptual layer. The pre-processing steps are applied.

• *Step 1*: the first step can be applied only for Q_2 since it contains the ordinary predicate *score*. This step allows to rewrite Q_2 into the two queries Q_{21} and Q_{22}, where *score* predicate is replaced by its definition in $\triangle_R$.

$Q_{21}(X) \leftarrow VideoSegment(X) \wedge Player(y) \wedge Goal(id) \wedge agentOf(id, y) \wedge hasName(y, \text{“}Player1\text{”})$

$Q_{22}(X) \leftarrow VideoSegment(X) \wedge Player(y) \wedge Penalty(id_1) \wedge agentOf(y, id_1) \wedge Goal(id_2) \wedge resultOf(id_2, id_1) \wedge hasName(y, \text{“}Player1\text{”})$

• *Step 2*: the second step consists in rewriting content concepts to include more specific concepts. Hence, in the two queries Q_{21} and Q_{22}, the concept *Player* is rewritten by its subsumers: *PlayingPlayer*, *Substitute* and *GoalKeeper*.

• *Step 3*: this step consists in checking semantic constraints validity in query conditions, using exclusion constraints defined in $\triangle_R$. In the set of rewritten queries, the exclusion constraints: $\bot \leftarrow Substitute(x) \wedge Fault(y) \wedge AgentOf(y, x)$ and $\bot \leftarrow Substitute(x) \wedge Goal(y) \wedge AgentOf(y, x)$, state that a *substitute* cannot be the agent of the events *fault* or *goal*. Queries which contain these conditions will not be evaluated since the results is always empty.

4.4 Query Translation

Datalog is a relationally complete query language that can express relational algebra in terms of the relational data model, while XQuery is a functional query language, in which a query is represented in terms of the XML's tree-like data model. In most cases, Datalog queries cannot simply be represented through simple XQuery expressions, instead they have to be translated into sets of XQuery function calls. In our framework, the set of XQuery functions has been developed during the definition of mapping rules between conceptual layer and MPEG-7 metadata layer.

The syntax used to build XQuery queries from Datalog queries is given in Table 4. The construction of XQuery queries requires the use of Datalog query and the set of mapping rules between conceptual layer (concepts, roles and attributes) and MPEG-7 metadata layer (i.e., XQuery functions defined over the MPEG-7 descriptions).

Table 4. XQuery syntax of the Datalog queries

"Let" "$" *VarStructure* ":=" *StructUnitFunction*;
"$" $VarConcept_1$ ":= $Concept_1Function$;
.....;
"$" $VarConcept_n$ ":=" $Concept_nFunction$;

"Where" (*Condition* ("and" | "or") *Condition*)*
"Return" ("$" VarStructure+)

Condition := *StructAttrCond* | *ConceptAttrCond* | *ConceptRelCond* | *StructConceptCond*
StructAttrCond := *StructAttrFunction* $Operator_1$ *AttrValue* | $Operator_2$ "(" *StructAttrFunction* "," *AttrValue* ")"
ConceptAttrCond := *ConceptAttrFunction* $Operator_1$ *AttrValue* | $Operator_2$ "(" *ConceptAttrFunction* "," *AttrValue* ")"
ConceptRelCond := *ConceptRelFunction*
ConceptRelCond := *ConceptRelFunction*
StructConceptCond := *StructConceptFunction*
$Operator_1$ $:=> \mid < \mid =$ (Arithmetic Operators)
$Operator_2$:= *contains*|*string*|*substring*|... (XQuery operator functions)

Legend:
- *VarStructure, VarConcept*$_i$: string variables expressions
- *StructUnitFunction, StructAttrFunction*: XQuery functions which deal with concepts related to structure and their attributes
- *ConceptFunction, ConceptAttrFunction*: XQuery functions related to the domain content concepts and their attributes
- *ConceptRelFunction*: XQuery functions which state if a relation exists between two concepts
- *StructConceptFunction*: XQuery function which relates the structural concepts to their content concepts

5 Implementation

To validate our approach, we developed two applications. The first allows to perform semantic annotation of multimedia data, whereas the second allows to querying multimedia described using MPEG-7.

5.1 Multimedia Data Annotation

Figure 7 illustrates a screenshot of the image annotation prototype. To ensure MPEG-7 descriptions validity, according to the description schemes, we used a JAVA API called JAXB (*Java XML Binding*).

In the annotation process, we focus on semantic annotation, supported by domain ontology concepts to allow semantic annotation of the multimedia content. In particular, the MPEG-7 semantic description tools are used.

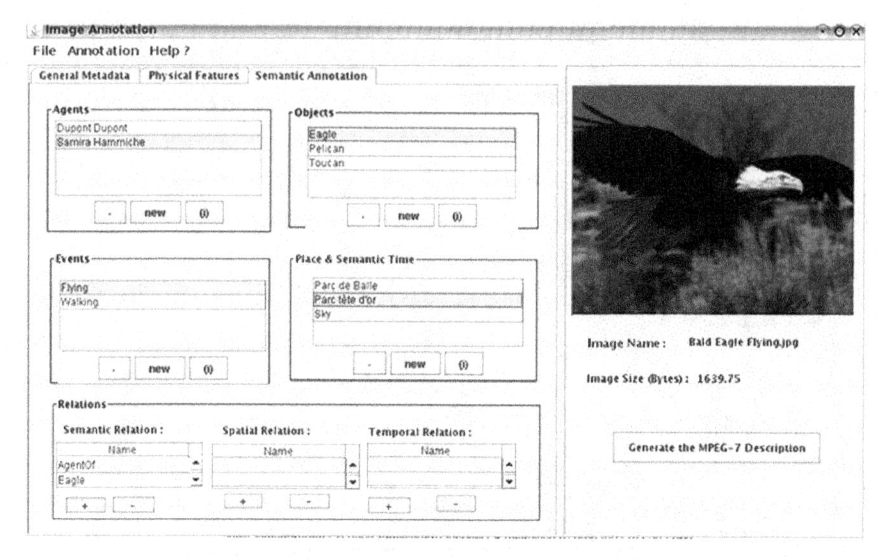

Fig. 7. Semantic annotation of images

To save the MPEG-7 descriptions generated after annotation process, we used the open source native database: *eXist*, which allows to support XQuery/ XPath queries.

5.2 Querying Multimedia Content

To retrieve the MPEG-7 descriptions, queries are formulated using Datalog rules and the conceptual vocabulary. This formulation allows to hide the MPEG-7 details to users, and to exploit semantic informations represented in the conceptual layer.

Figure 8 shows a query example, expressed using a Datalog rule. In this query, image of flying birds are retrieved. Pre-processing steps allow to exploit Bird predicate and rewrite it by different specific Bird categories. Hence, all specific birds flying are returned. The implementation of the query rewriting process makes use of domain ontology, rules and coherence rules.

6 Related Work

The presented work is related to three specific research works: adding semantics to MPEG-7, to allow semantic description of the multimedia content, semantic retrieval of MPEG-7 descriptions, and semantic query rewriting.

6.1 Adding Semantics to MPEG-7 Descriptions

The idea of adding semantics to multimedia data has been considered in several works [4, 7, 10, 16, 17]. Most of them translate MPEG-7 *Ds* and *DSs* into

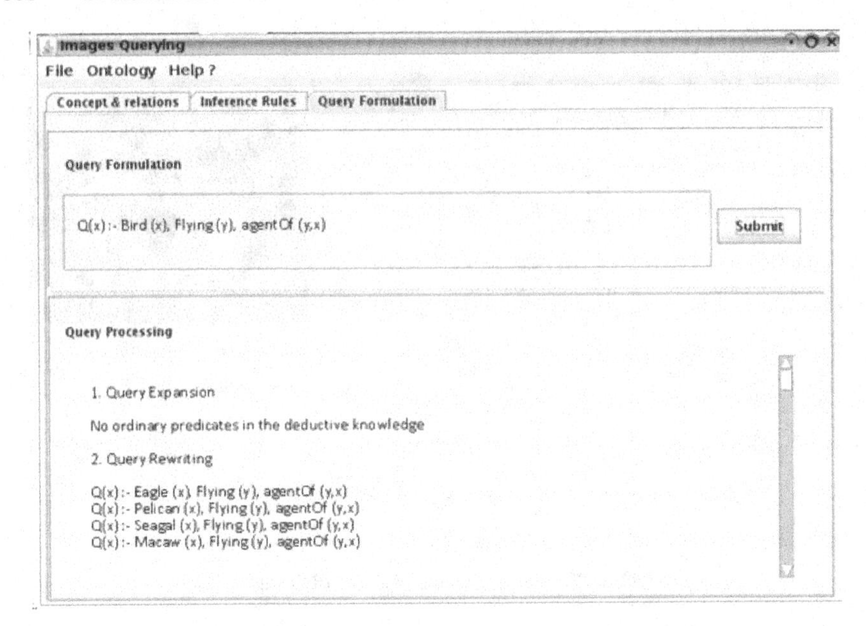

Fig. 8. Query pre-processing at the conceptual layer

knowledge representation formalism able to express semantics of the descriptions and to provide reasoning mechanisms. The formalisms used are: RDFS and DAML+OIL [10], OWL [7, 16, 17] and XDD [4]. In [10], RDF/RDF Schema and DAML+OIL are used to express semantics of MPEG-7 DSs and Ds in order to add multimedia to the semantic web. In [17], a multimedia framework is developed to support ontology-based semantic indexing and retrieval of audiovisual content following MPEG-7 and TV-Anytime standards specifications for metadata descriptions. The domain ontology is described using MPEG-7 abstraction tools which are not provided with reasoning mechanisms. Hence, it is not possible to infer implicit information.

The drawbacks of such approach are that abilities of MPEG-7 are not fully exploited especially for the low-level features, of which appropriate datatypes are defined using the *DDL*, and structural aspects expressed in XML/XML Schema. A knowledge representation formalism does not allow to deal with low-level features appropriately and the data structure is generally lost when using this formalism. Moreover, most of these works propose only the translation of the MPEG-7 *Ds* and *DSs* and do not consider domain knowledge related to the described content or the appropriate query languages to retrieve these descriptions.

In [16], OWL and MPEG-7 were combined to represent audiovisual documents. Structural aspects are expressed using MPEG-7 *DSs*, whereas content is described using RDF assertions. RQL is used to retrieve the RDF knowledge base. The two descriptions are linked by using XPath expressions. Advantage of such approach is the combination of MPEG-7 structural abilities and RDF semantic model to constrain both structure and content of multimedia data.

Our work is quite similar to [16]. However, instead of translating the MPEG-7 vocabulary to a knowledge based language, we add a conceptual layer to cope with the MPEG-7 descriptions limitations. This layer introduces domain knowledge related to structure and content of the described data. These knowledge are introduced in the MPEG-7 descriptions during the annotation process and exploited during the query formulation process, instead of the evaluation step. This approach has not been considered in previous work.

6.2 Query Languages to Retrieve the MPEG-7 Descriptions

Currently, the approaches used for querying MPEG-7 descriptions can be classified into three main groups [6].

The first group [12] uses directly XQuery as a query language. However, XQuery requires advanced knowledge of MPEG-7 details in order to express a precise query. Moreover, XQuery allows only exact matching evaluation during the execution process. Hence, using XQuery requires to define access interfaces, where MPEG-7 details are hidden to users. This is not practical since the set of MPEG-7 descriptors is very large.

The second group [8,15] defines new query languages to support retrieval of MPEG-7 descriptions. In [8], access methods based on an inference network model are proposed to search MPEG-7 video descriptions. These methods allow deduction of semantic and implicit information. In [15], a specification of crucial query issues in MPEG-7 XML queries is proposed, which takes into account the implicit information to be extracted from the MPEG-7 descriptions. A logical query language based on path calculus is proposed.

The disadvantages of such approach are that these query languages are specific to their applications and cannot be used in another context.

The third group [6] proposes to use an adaptation of XQuery called *SVQL* (*Sematic Views Query Language*) to retrieve the MPEG-7 descriptions. *SVQL* is a high level query language, which allows users to express their requirements following the semantic view model (*PhysicalView, ProductionView, ThematicView, VisualView and AudioView*) in a concise, abstract and precise way.

Our approach to retrieve multimedia data described using MPEG-7 is to use the conceptual layer to formulate queries in a flexible and precise way. Advantages of using the conceptual layer are that users can formulate concept based queries instead of keywords based queries and a query rewriting process is performed in order to make implicit information explicit. This approach will increase the retrieval effectiveness. Then, a query translation algorithm is used to translate conceptual queries to XQuery queries, which will be evaluated over the MPEG-7 descriptions.

6.3 Query Rewriting

Literatures on query rewriting fall into three general approaches [11]:

- The first approach deals with the syntactic query rewriting. User's queries can be rewritten using structure hierarchy information if syntactically exact matching is not only one available in a database schema. This approach is mostly used in the mediator based integration systems. It focuses on how to take advantage of schema semantics to generate the consistent translations from source to target by considering the constraints and structure of the target schema.
- The second approach is semantic query rewriting. In this case, user's queries can be rewritten using semantic information like view, integrity constraints, rules and semantic knowledge (synonyms, more general terms, more specific terms, etc.).
- The third approach is user interactive query rewriting. User given queries are rewritten directly by users or at least by using user inputs (preferences, profiles, etc.). This interaction stops when users are satisfied by the answer set. During the interaction, a new query is constructed. This new query involves expanding terms of the initial query and recalculating the importance of each term in the expanded query (reweighting of terms).

The first two approaches are handled internally by a system. The rewritten steps are done automatically, whereas the last needs use's interaction. In our work, we use a semantic based query rewriting approach. We exploit semantic information represented in the knowledge base (synonyms, more specific concepts), to add more precise terms or synonyms to queries.

7 Conclusion

In this paper, we presented a domain knowledge based approach to retrieve multimedia data described using MPEG-7 standard. A conceptual layer is added on top of the MPEG-7 metadata layer in order to provide MPEG-7 with a formal semantics. Furthermore, we exploit conceptual vocabulary to assist user's query formulation process.

User's queries are pre-processed before their evaluation, in order to make explicit the implicit information on one hand, and to include more specific concepts on the other hand. The rewritten queries are then checked and translated into XQuery queries using the mapping rules and a set of XQuery functions defined over the MPEG-7 descriptions.

As future work, we will investigate the optimization of query rewriting algorithm by selecting only the most similar rewritten queries to the original query. Furthermore, we can consider how to integrate semantic constraints in the MPEG-7 descriptions, since only structural constraints are considered now.

The rewriting process is based on the semantic information represented in the conceptual layer. Since MPEG-7 is provided with a preference user description scheme, we plan to exploit this *DS* to introduce user preferences during the rewriting process.

For the implementation part, we plan to test and to compare the efficiency of our approach and keywords or low-level features approaches.

References

1. F. Baader, D. Calvanese, D.L. McGuinness, D. Nardi, and P.F. Patel-Schneider (eds.). *The description logic handbook: theory, implementation, and applications.* Cambridge University Press, New York, NY, 2003.
2. A.B. Benitez, D. Zhong, S.F. Chang, and J.R. Smith. MPEG-7 MDS content description tools and applications. *Lecture Notes in Computer Science*, 2124:41–52, 2001.
3. A.B. Benitez, H. Rising, C. Jörgensen, R. Leonardi, A. Bugatti, K. Hasida, R. Mehrotra, A. Murat Tekalp, A. Ekin, and T. Walker. Semantic of multimedia in MPEG-7. In *Proceedings of IEEE 2002 Conference on Image Processing (ICIP-2002)*, Rochester, New York, 2002.
4. A. Chotmanee, V. Wuwongse, and C. Anutariya. A schema language for mpeg-7. In *Proceedings of International Conference on Asian Digital Libraries (ICADL'02)*, pp. 153–164. Springer, Berlin Heidelberg New York, 2002.
5. M. Döller, M. Gruhne, and I. Wolf. Towards an MPEG-7 Query Language. In *Proceedings of the International Conference on Signal-Image Technology and Internet Based Systems (SITIS'06)*, Hammamet, Tunisia, 2006.
6. N. Fatemi, O. Abou Khaled, and G. Coray. An XQuery adaptation for MPEG-7 documents retrieval. *XML conference and Exposition*, Decemeber 2003.
7. R. Garcia and O. Celma. Semantic Integration and Retrieval of Multimedia Metadata. In *Proceeding of the 5th International Workshop on Knowledge Markup and Semantic Annotation (SemAnnot 2005)*, Galway, Ireland, November 2005.
8. A.P. Graves and M. Lalmas. Video Retrieval using an MPEG-7 based Inference Network. In *The 25th Annual International ACM SIGIR Conference on Research and Development in Information Retrieval*, pp. 339–346. ACM, New York, 2002.
9. T.R. Gruber. A translation approach to portable ontology specifications. *Knowledge Acquisition*, 5(2):199–220, 1993.
10. J. Hunter. Adding multimedia to the semantic web – building an MPEG-7 ontology. In *Proceedings of the First Semantic Web Working Symposium (SWWS)*, pp. 261–281, 2001.
11. A. Hafez J. Yoon, and V. Raghavan. Query rewriting for multimedia xml data. In *Proceedings of the Sixth International Workshop on Multimedia Information System*, pp. 172–180, October 2000.
12. J.H. Kang, et al. An xquery engine for digital library systems. In *Proceedings of 3rd ACM/IEEE-CS Joint Conference on Digital Libraries*, Houston, TX.
13. H. Kosch. *Distributed Multimedia Database Technologies Supported by MPEG-7 and MPEG-21.* CRC, Boca Raton, 2004.

14. A.Y. Levy and M.-C. Rousset. Combining horn rules and description logics in carin. *Artificial Intelligence*, 104(1–2):165–209, 1998.
15. P. Liu and L. Sushu. Queries of digital descriptions in MPEG-7 and MPEG-21 XML documents. *XML Europe 2002, Barcelona, Spain*, May 2002.
16. R. Troncy. *Formalisation des connaissances documentaires et des connaissances conceptuelles à l'aide d'ontologies: application à la description de documents audiovisuels*. Computer science Ph.D. thesis, Joseph Fourier University, Grenoble, France, 2004.
17. C. Tsinaraki, P. Polydoros, and S. Christodoulakis. Integration of OWL Ontologies in MPEG-7 and TV-Anytime Compliant Semantic Indexing. In *Proceedings of the 16th International Conference on Advanced Information Systems Engineering (CAiSE'04)*, pp. 398–413, 2004.

Index